HETEROCYCLIC COMPOUNDS WITH INDOLE AND CARBAZOLE SYSTEMS

This is the eighth volume published in the series

THE CHEMISTRY OF HETEROCYCLIC COMPOUNDS

THE CHEMISTRY OF HETEROCYCLIC COMPOUNDS

A SERIES OF MONOGRAPHS

ARNOLD WEISSBERGER, *Consulting Editor*

HETEROCYCLIC COMPOUNDS WITH INDOLE AND CARBAZOLE SYSTEMS

WARD C. SUMPTER

Western Kentucky State College, Bowling Green, Kentucky

F. M. MILLER

University of Maryland, Baltimore, Maryland

1954

INTERSCIENCE PUBLISHERS, INC., NEW YORK

INTERSCIENCE PUBLISHERS LTD., LONDON

Library of Congress Catalog Card Number 53-10760

INTERSCIENCE PUBLISHERS, INC., 250 Fifth Avenue, New York 1, N.Y.

For Great Britain and Northern Ireland:

Interscience Publishers Ltd., 88/90 Chancery Lane, London, W.C. 2

The Chemistry of Heterocyclic Compounds

The chemistry of heterocyclic compounds is one of the most complex branches of organic chemistry. It is equally interesting for its theoretical implications, for the diversity of its synthetic procedures, and for the physiological and industrial significance of heterocyclic compounds.

A field of such importance and intrinsic difficulty should be made as readily accessible as possible, and the lack of a modern detailed and comprehensive presentation of heterocyclic chemistry is therefore keenly felt. It is the intention of the present series to fill this gap by expert presentations of the various branches of heterocyclic chemistry. The subdivisions have been designed to cover the field in its entirety by monographs which reflect the importance and the interrelations of the various compounds, and accommodate the specific interests of the authors.

Research Laboratories
Eastman Kodak Company
Rochester, New York

ARNOLD WEISSBERGER

Preface

The chemistry of indole had its beginnings in the dye industry, and thus for a number of years was intimately associated with dye chemistry. The intensely fertile research of this period produced a very diverse group of closely related compounds; today the scope of indole chemistry is indeed multiform, extending from the rather simple parent material, through the condensed systems such as carbazole, the oxygenated indole derivatives, to the highly complex materials which occur naturally.

In undertaking the preparation of a monograph on compounds containing the indole and carbazole rings our purpose has been to present a thorough and comprehensive treatment of the methods of preparation, the properties and the reactions of these compounds without attempting to duplicate the coverage of Beilstein (or of Elsevier's Encyclopedia when completed) by listing every compound. We have, however, attempted to go beyond a simple listing of basic generalities by including a sufficiently large number of compounds to indicate exceptions and trends away from the general principles. The literature coverage extends through 1952, and includes several important papers of 1953.

In the preparation of this volume, the responsibility for Chapter I (Indole), II (Carbazole), V (Isatogens), VI (Indoxyl), and VII (Indigo) was assumed by W. C. S., while Chapters III (Isatin) and IV (Oxindole) were rewritten by F. M. M. from earlier reviews on these topics by W. C. S. Chapter VIII was entirely the responsibility of F. M. M.

We desire to acknowledge our indebtedness to Miss Josephine Williams and Mr. Phil Wilken for assistance in making the literature survey and for critically reading certain chapters, and to Mrs. Joseph Stewart for typing a large portion of the manuscript.

One of us (F. M. M.) is indebted to the National Institutes of Health for a postdoctoral fellowship held at Harvard University in 1948–1949, during which time much of the work on Chapter VIII was done.

We are grateful to Drs. C. F. H. Allen and Bernhard Witkop for reading the manuscript and for making helpful suggestions for its improvement.

Bowling Green, Kentucky W. C. S.
Baltimore, Maryland F. M. M.

Contents

VII. Indigo *(continued)*

CHAPTER I

Indole

Introduction

In 1866 and 1868 Baeyer[1] published the results of his researches on the reduction of isatin. In addition to isatide,[2] Baeyer obtained dioxindole, $C_8H_7NO_2$, by the further reduction of which oxindole, C_8H_7NO, was prepared. The constitution of oxindole as the lactam of 2-aminophenylacetic acid was established[3] through its synthesis by the reduction of 2-nitrophenylacetic acid with tin and hydrochloric acid.

Indole itself was first prepared[4] by heating oxindole with zinc dust. In addition to being the first synthesis of indole this was also the first application of the zinc dust pyrolytic technique. Subsequently indole was prepared by reduction of 2,3-dichloroindole.[5] Indole (1-benzo[*b*]pyrrole, 1-benzazole, 1-azaindene, ketole) (*R.I.* 821) is a nonbasic nitrogenous compound in which a benzene ring and a pyrrole nucleus are fused together in the 2,3-positions of the pyrrole ring. Indole and skatole (β-methylindole) both possess unpleasant (fecal) odors when impure; the pure materials have a pleasant fragrance and are found in both natural and synthetic perfumes. Indole is a colorless, crystalline solid, melting at 52°, and boiling at 254°. It is volatile with steam and soluble in alcohol, benzene, ether, and ligroin and many be recrystallized from water. The formula which is generally accepted for indole was proposed by Baeyer and Emmerling.[6] This structure was suggested

[1] Baeyer, *Ber.*, **1**, 17 (1868). Baeyer and Knop, *Ann.*, **140**, 1 (1866).

[2] Laurent, *Rev. sci. ind.* (September, 1842); *J. prakt. Chem.*, [1] **47**, 166 (1849). Erdmann, *J. prakt. Chem.*, [1] **22**, 257 (1841).

[3] Baeyer, *Ber.*, **11**, 582, 1228 (1878).

[4] Baeyer, *Ann.*, **140**, 296 (1866); *Ann., Suppl. Bd.*, **7**, 56 (1870). For a recent reduction of oxindole derivatives to the corresponding indole through the agency of $LiAlH_4$ see Julian and Printy, *J. Am. Chem. Soc.*, **71**, 3206 (1949).

[5] Baeyer, *Ber.*, **12**, 459 (1879).

[6] Baeyer and Emmerling, *ibid.*, **2**, 679 (1869). Compare also Baeyer, *ibid.*, **3**, 517 (1870).

largely as a result of synthetic methods of preparation of the compound by fusion of a mixture of *o*-nitrocinnamic acid, iron filings, and sodium hydroxide, and by the action of lead peroxide upon azocinnamic acid.

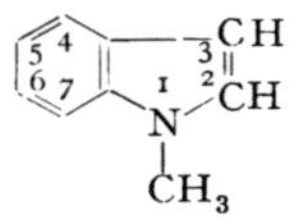

A system of nomenclature devised by Baeyer[7] and subsequently employed by Fischer[8] was cumbersome in that it employed independent numerical designations for each ring. Current practice in indole nomenclature is to number the positions as shown in the formula above. The 2- and 3-positions are also referred to as the α- and β-positions, respectively.

Indole derivatives are found in many natural products. Indole itself has been obtained from many naturally occurring materials by methods which suggest that the indole is in many cases the product of decomposition of its derivatives. Indole has been found in *Robinia pseudacacia*,[9] the jasmines[10], and certain citrus plants,[11] in the perfume of the *Hevea braziliensis*[12] and in orange blossoms.[13] Indole is also found in the wood of *Celtis reticulosa*.[14] The indole is usually obtained by repeated extraction of the blossoms with a suitable solvent with subsequent removal of the solvent by distillation.

Alkaline hydrolysis[15] and putrefaction[16] of proteins result in the formation of indole. Its formation in the putrefaction of proteins is presumed to be the result of the decomposition of tryptophan. The formation of indole from albumin may be stopped by the addition of lactose while other sugars have varying effects on its production.[17] Indole frequently accompanies pus

[7] Baeyer, *ibid.*, **17**, 960 (1884).

[8] Fischer, *Ann.*, **236**, 116 (1886).

[9] Elze, *Chem.-Ztg.*, **34**, 814 (1910).

[10] Cerighelli, *Compt. rend.*, **179**, 1193 (1924). Hesse, *Ber.*, **37**, 1457 (1904). Soden, *J. prakt. Chem.*, [2] **69**, 256 (1904).

[11] Sack, *Pharm. Weekblad*, **48**, 307 (1911).

[12] Sack, *ibid.*, **48**, 775 (1911).

[13] Hesse and Zeitschel, *J. prakt. Chem.*, [2] **66**, 481 (1902).

[14] Herter, *J. Biol. Chem.*, **5**, 489 (1909).

[15] Kühne, *Ber.*, **8**, 208 (1875). Nencki, *J. prakt. Chem.* [2] **17**, 97 (1878). Herzfeld, *Biol. Z.*, **56**, 82 (1913).

[16] Nencki, *Ber.*, **7**, 1596 (1874); **8**, 336, 725 (1875); **28**, 561 (1895). Brieger, *J. prakt. Chem.*, [2] **17**, 135 (1878). Salkowski and Salkowski, *Ber.*, **12**, 648 (1879). Brieger *Z. physiol. Chem.*, **3**, 134 (1879). Salkowski, *ibid.*, **8**, 417 (1884); **9**, 8 (1885). v. Moraczewski, *Biol. Z.*, **51**, 340 (1913).

[17] Hirschler, *Z. physiol. Chem.*, **10**, 306 (1886). Simnitzki, *ibid.*, **39**, 113 (1903).

formation[18] and is found in the liver and pancreas,[19] the brain,[20] and bile.[21] Indole, accompanied by its β-methyl homolog, skatole, is found in the feces of men and of animals[22] and in the contents of the intestines.[23]

Indole[24] and homologs of indole[25] have been found in coal tar. Indole has also been found in molasses tar.[26] Indole is present in "practical" α-methylnaphthalene.[27] Its presence was demonstrated by reaction with oxalyl chloride to give the acid chloride of 3-indoleglyoxylic acid. Indole can be prepared by the reduction of indoxyl by sodium amalgam, by zinc dust and alkali, or catalytically.[28] Indole can also be prepared by the dehydrogenation of dihydroindole.[29] In the preparation of indoxyl or of indoxylic acid in the synthesis of indigo a small amount of indole is obtained when the melt is overheated.[30] Indole has been prepared in fair yields by adding sodium amalgam or zinc dust to the alkaline melt[31] of indoxylic acid.

Synthesis of Indole

A number of more general methods for the synthesis of indole and indole derivatives involve procedures which form the pyrrole ring through ring closure. Among these the synthesis developed by Emil Fischer has proved to be the most versatile for the synthesis of indole derivatives although the reaction fails for the synthesis of indole itself. It would be expected that

[18] Porcher, *Compt. rend.*, **147,** 214 (1908).
[19] Nencki, *Ber.*, **7,** 1593 (1874).
[20] Stöckley, *J. prakt. Chem.*, [2] **24,** 17 (1881).
[21] Ernst, *Z. physiol. Chem.*, **16,** 208 (1892).
[22] Brieger, *Ber.*, **10,** 1030 (1877); *J. prakt. Chem.*, [2] **17,** 129 (1878). v. Moraczewski, *Z. physiol. Chem.*, **55,** 42 (1908).
[23] Tappeiner, *Ber.*, **14,** 2383 (1881). Ellinger, *Z. physiol. Chem.*, **39,** 44 (1903). Rosenfeld, *Beitr. zur Chem. Physiol. und Path.*, **5,** 83 (1904). Blumenthal and Jacoby, *Biol. Z.*, **29,** 472 (1910).
[24] Weissgerber, *Ber.*, **43,** 3520 (1910).
[25] Kruber, *ibid.*, **B 59,** 2752 (1926); **B 62,** 2877 (1929). Kruber, German patent 515, 543.
[26] Boes, *Pharm. Ztg.*, **47,** 131 (1902).
[27] Kharasch, Kane, and Brown, *J. Am. Chem. Soc.*, **62,** 2242 (1940).
[28] Vorländer and Apelt, *Ber.*, **37,** 1134 (1904). U.S. patent 1,891,057; *Chem. Abstr.* **27,** 1892 (1933).
[29] Sugasawa, Saloda, and Yamagisawa, *J. Pharm. Soc. Japan*, **58,** 139 (1938); *Chem. Abstr.*, **32,** 4161 (1938).
[30] German patent 260,437; *Chem. Abstracts*, **7,** 3236 (1913).
[31] Vorländer and Apelt, *Ber.*, **37,** 1134 (1904).

acetaldehyde phenylhydrazone would yield indole through the Fischer synthesis but in reality none is obtained.

In 1883 Fischer and Jourdan[32] found that, when the methylphenylhydrazone of pyruvic acid was heated with alcoholic hydrogen chloride, a small yield of a compound, $C_{10}H_9NO_2$, was obtained. This compound proved to be 1-methylindole-2-carboxylic acid.[33] In subsequent studies of the reaction, it was found that zinc chloride was a better catalyst for the reaction than hydrogen chloride. Utilizing this procedure, Fischer[34] prepared 2-methylindole (methylketole) in 60% yields from acetone phenylhydrazone, and skatole in 35% yield from propionaldehyde phenylhydrazone. The Fischer synthesis has been utilized for the synthesis of many indole derivatives from ketone phenylhydrazones.

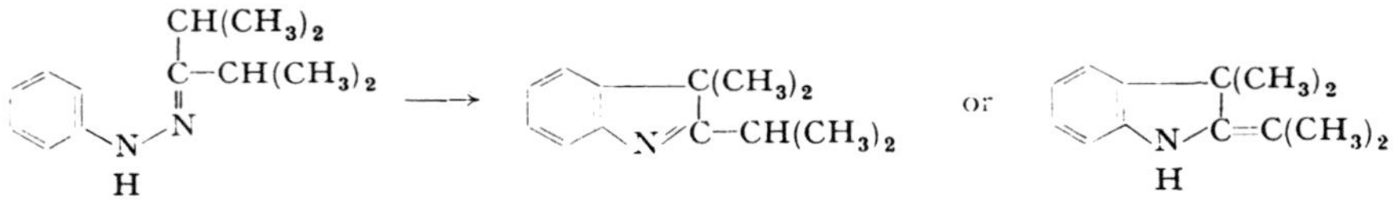

Since the early work by Fischer, a number of changes have been made in the procedure with material improvement in the yields. Thus, by employing an inert solvent such as methylnaphthalene[35] and by operating at temperatures below 150°, 2-methylindole was prepared in 75% yield from acetone phenylhydrazone, skatole in 80% yield from propionaldehyde phenylhydrazone, and indole-2-carboxylic acid in 60% yield from the phenylhydrazone of pyruvic acid. More recently it has been found[36] that the large amounts of zinc chloride used by early workers were not necessary. The reaction takes place in the presence of 1% of zinc chloride, while cuprous chloride, cuprous bromide, and platinum chloride may also be used as catalysts. Concentrated sulfuric acid has been employed as the catalyst,[37]

[32] Fischer and Jourdan, *ibid.*, **16,** 2241 (1883).

[33] Fischer and Hess, *ibid.*, **17,** 559 (1884). Compare Hegel, *Ann.*, **232,** 214 (1882).

[34] Fischer, *Ann.*, **236,** 116, 126 (1886); *Ber.*, **19,** 1563 (1886).

[35] German patent 238,138; *Chem. Abstr.*, **6,** 1659 (1912).

[36] Arbuzov and Tikhvinskii, *J. Russ. Phys.-Chem. Soc.*, **45,** 69, 649 (1915). Arbuzov and Tikhvinskii, *Ber.*, **43,** 2301 (1910).

[37] Nef, *Ann.*, **266,** 72 (1891). Walker, *Am. Chem. J.*, **14,** 576 (1892). Reissert and Heller, *Ber.*, **37,** 4378 (1904).

while alcoholic sulfuric acid[38] and alcoholic zinc chloride[39] have also been used. Nickel, cobalt, and copper powder, cobalt chloride, and many other salts also catalyze the reaction.[40] The reaction has also been catalyzed by the use of Grignard reagents.[41] Boron fluoride is also effective as the condensing agent in the Fischer synthesis.[42] Recently, polyphosphoric acid has been employed effectively as a catalyst for the reaction.[42a]

The ease of indole formation varies irregularly with the various phenylhydrazones. In some cases the reaction takes place very readily. Thus cyclohexanone phenylhydrazone undergoes indole formation when warmed gently with aqueous hydrochloric acid, yielding tetrahydrocarbazole. The methylphenylhydrazone of isopropylmethyl ketone undergoes ring closure even at room temperature in the presence of alcoholic zinc chloride.[43]

$HC(CH_3)_2$, $C{-}CH_3$, N, N, CH_3 ⟶ $C(CH_3)_2$, $N{-}C{=}CH_2$, CH_3

3,3,7-Trimethyl-2-methyleneindoline has been prepared in similar fashion by heating the *o*-tolylhydrazone of isopropylmethyl ketone with alcoholic hydrogen iodide.[44] 3,3,5-Trimethylindolenine[45] was prepared by warming isobutyraldehyde *p*-tolylhydrazone at 60° with alcoholic zinc chloride. When heated with concentrated hydrochloric acid the 3,3,5-trimethylindolenine is converted into 2,3,5-trimethylindole.[45]

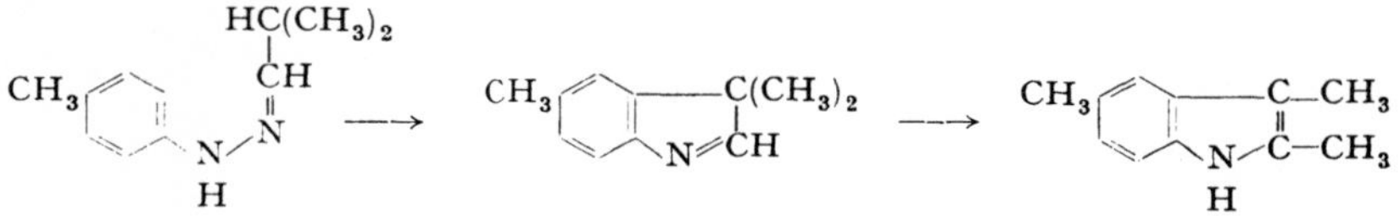

CH_3, $HC(CH_3)_2$, CH, N, N, H ⟶ CH_3, $C(CH_3)_2$, $N{=}CH$ ⟶ CH_3, $C{-}CH_3$, $N{-}C{-}CH_3$, H

Although the Fischer synthesis is the most widely applicable of the indole syntheses, there are certain limitations and exceptions. Acetaldehyde

[38] Wislicenus and Arnold, *Ann.*, **246**, 334 (1888).

[39] Plancher, *Gazz. chim. ital.*, **32**, 398 (1902); *Ber.*, **31**, 1496 (1898).

[40] Koraczynski and Kierzek, *Gazz. chim. ital.*, **55**, 361 (1925). Koraczynski, Brydowna, and Kierzek, *ibid.*, **56**, 903 (1926).

[41] Grammaticakis, *Compt. rend.*, **204**, 502 (1937).

[42] Snyder and Smith, *J. Am. Chem. Soc.*, **65**, 2452 (1943).

[42a] Kissman, Farnsworth, and Witkop, *ibid.*, **74**, 3948 (1952).

[43] Plancher, *Ber.*, **31**, 1496 (1898). Jenisch, *Monatsh.*, **27**, 1223 (1906).

[44] Plancher, *Monatsh.*, **26**, 833 (1905).

[45] Grgin, *ibid.*, **27**, 731 (1906).

phenylhydrazone should yield indole but this synthesis has not been accomplished. The phenylhydrazones of the β-ketone esters usually yield pyrazolones rather than indoles.

The catalytic decomposition of the arylhydrazones of unsymmetrical ketones can conceivably take place in two ways, yielding thereby a mixture of two products. While in many cases[46] only a single product of established structure has been obtained, in other cases the course of the reaction has not been determined.[47] In other cases two distinct products have been isolated.[48]

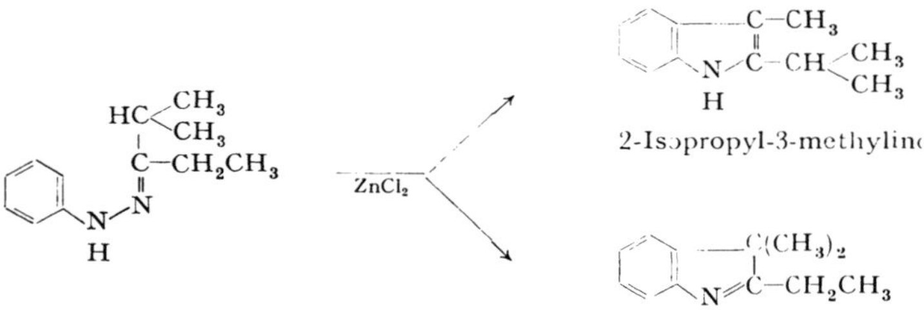

$ZnCl_2$

2-Isopropyl-3-methylindole

2-Ethyl-3,3-dimethylindolenine

The following rules governing the course of the reaction have been given by Plancher and Bonavia.[48] (*1*). Ketone phenylhydrazones containing the group $-NH-N{=}CMe-CH{<}^{R}_{R}$ yield only the corresponding indolenine. (*2*) Those with the grouping $-NH-N{=}C{<}^{CH_2R}_{CH_2R}$ give both the corresponding indole and the indolenine. (*3*) If the group $-NH-N{=}C{<}^{CH_3}_{CH_2R}$ is present the ketone phenylhydrazone is capable of yielding two indoles on condensation. Condensation by means of the $-CH_2-$ group preponderates.

The use of meta-substituted phenylhydrazones can lead to the formation of both the 4- and 6-derivatives. In a number of cases in the literature, the method of ring closure is not indicated.[49] The *m*-nitrophenylhydrazone of

[46] Arbuzov and Tikhvinskii, *J. Russ. Phys.-Chem. Soc.*, **45,** 69, 694 (1915). Fischer, *Ann.*, **236,** 116 (1886). Jenisch, *Monatsh.*, **27,** 1223 (1906). Plancher, *Ber.*, **31,** 1496 (1898). Arbuzov, Zaitzev, and Razumov, *ibid.*, **B 68,** 1792 (1935).

[47] Arbuzov, Zaitzev, and Razumov, *Ber.*, **B 68,** 1792 (1935). Arbuzov and Zaitzev, *Trans. Butlerov Inst. Chem. Technol. Kazan*, **No. 1,** 33–38 (1934); *Chem. Abstr.*, **29,** 4006 (1935).

[48] Plancher and Bonavia, *Gazz. chim. ital.*, **32,** 414 (1902).

[49] Kermack, Perkin, and Robinson, *J. Chem. Soc.*, **119,** 1622 (1921); **121,** 1880 (1922). Roder, *Ann.*, **236,** 164 (1886). Tomicek, *Chem. Listy*, **16,** 1, 35 (1922).

cyclohexanone[50] has been found to undergo ring closure, yielding both possible isomers. Drechsel[51] was the first to prepare a tetrahydrocarbazole[52] in this way from phenylhydrazine and cyclohexanone.

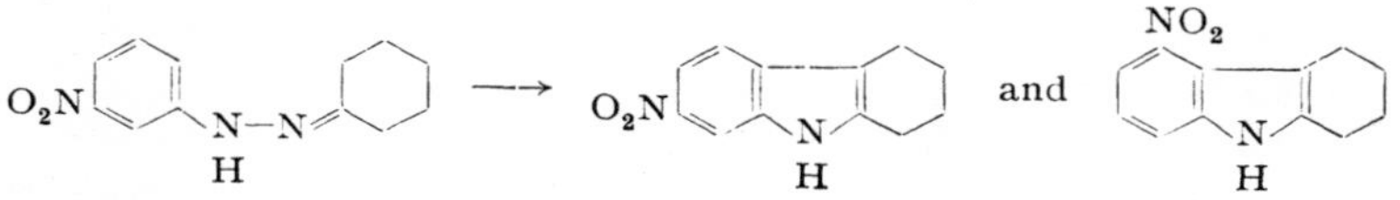

O_2N–C_6H_4–NH–N=C_6H_{10} ⟶ O_2N-tetrahydrocarbazole (N–H) and NO_2-tetrahydrocarbazole (N–H)

The study of the mechanism of the Fischer reaction has engaged the attention of a number of workers. Four distinct mechanisms have been proposed for the reaction. Reddelien[53] found that the anil of acetophenone is oxidized by phenylhydrazine or by phenylhydrazones to 2-phenylindole.

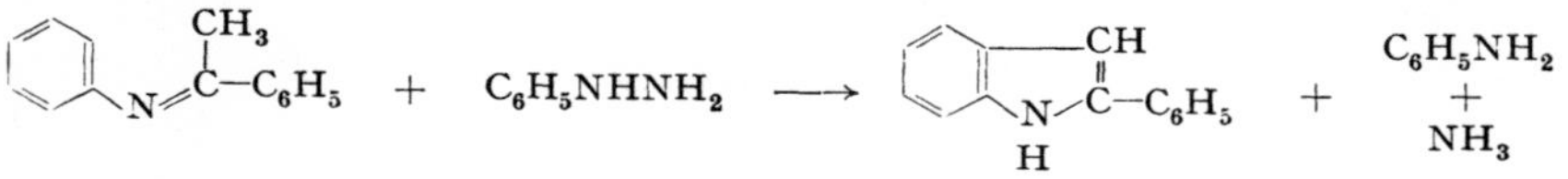

$$C_6H_5N{=}C(CH_3){-}C_6H_5 + C_6H_5NHNH_2 \longrightarrow \text{2-phenylindole (CH, N–H)} + C_6H_5NH_2 + NH_3$$

On the basis of this observation he proposed the following mechanism for the Fischer synthesis.

(I) Reduction of the phenylhydrazone during the simultaneous oxidation of stage III.

$$C_6H_5NH{-}N{=}\underset{}{\overset{R}{C}}{-}CH_2R + H_2 \longrightarrow C_6H_5NH_2 + HN{=}C\begin{cases}R\\CH_2R\end{cases}$$

(II) Condensation of the products of stage I with elimination of ammonia to yield

$$C_6H_5N{=}\underset{R}{C}{-}CH_2R$$

(III) Ring closure by oxidation of the anil (accompanied by the reduction of stage I) to yield the indolenine.

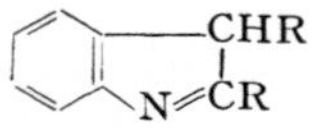

$$\text{indolenine: } C_6H_4 \text{ fused with } CHR{-}CR{=}N$$

(IV) Isomerization of the indolenine with migration of hydrogen from position 3 to position 1 yielding the indole.

This mechanism requires the assumption of an initial tautomeric

[50] Borsche, Witte, and Bothe, *Ann.*, **359**, 49 (1908). Plant, *J. Chem. Soc.*, **1936**, 899. Barclay and Campbell, *ibid.*, **1945**, 530.

[51] Drechsel, *J. prakt. Chem.*, **38**, 65 (1888).

[52] Baeyer, *Ann.*, **278**, 88 (1894). Drechsel and Baeyer, *ibid.*, **278**, 105 (1894).

[53] Reddelien, *ibid.*, **388**, 179 (1912).

hydrogen in the original phenylhydrazone (or in the ketone imide of stage I) if it is to account for the preparation of 1-alkylindoles by the Fischer method.[54] The Reddelien mechanism has met with opposition from Robinson and Robinson,[55] Bodforss,[56] and Campbell and Cooper.[57] On the other hand, Hollins[54] has favored a modification of the Reddelien mechanism.

Bamberger and Landau,[58] basing their explanation on the formation of dimethylaniline oxide by methylation of phenylhydroxylamine with methyl sulfate, assume a mechanism based upon a tautomeric form of the hydrazones. The method of elimination of the ammonia is not explained and the hypothesis fails entirely to account for the formation of *N*-methylindoles from the phenyl methyl hydrazones of ketones.[59]

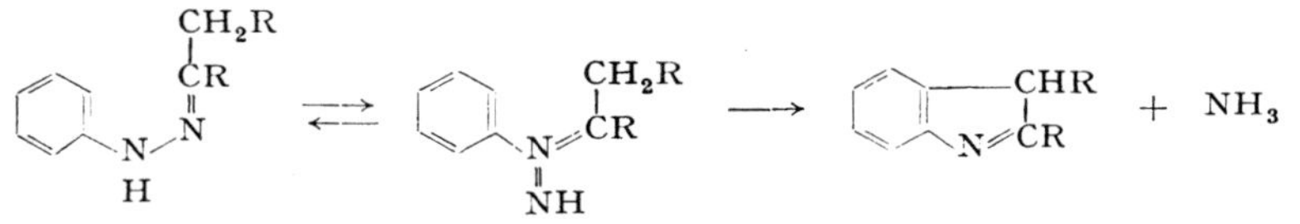

Cohn[60] suggested an ortho semidine rearrangement with subsequent loss of ammonia. This mechanism likewise fails to account for the formation of the *N*-alkylindoles and would require the formation of 6-substituted indoles from *p*-substituted phenylhydrazones, whereas 5-substituted indoles are actually obtained.

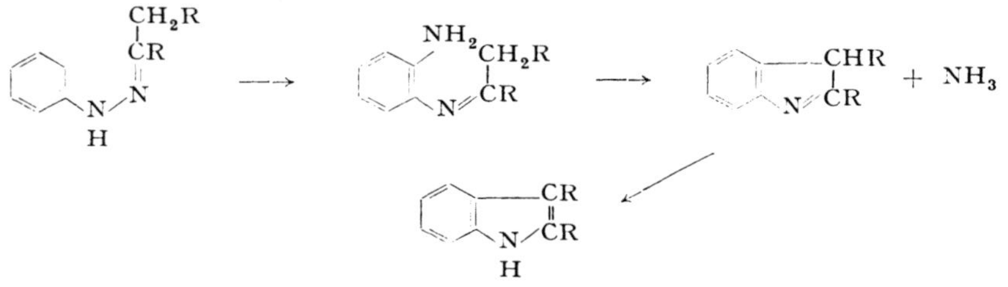

The fourth mechanism proposed for the Fischer synthesis and the one most generally accepted[61] is the one proposed by Robinson and Robinson.[55]

[54] Hollins, *J. Am. Chem. Soc.*, **44,** 1598 (1922).

[55] Robinson and Robinson, *J. Chem. Soc.*, **113,** 639 (1918; **125,** 827 (1924).

[56] Bodforss, *Ber.*, **B 58,** 775 (1925).

[57] Campbell and Cooper, *J. Chem. Soc.*, **1935,** 1208.

[58] Bamberger and Landau, *Ber.*, **52,** 1097 (footnote) (1919).

[59] Degen, *Ann.*, **236,** 153 (1886).

[60] Cohn, *Die Carbazolgruppe*. Thieme, Leipzig, 1919, p. 12.

[61] Neber, *Ann.*, **471,** 113 (1929). Campbell and Cooper, *J. Chem. Soc.*, **1935,** 1208. Plieninger, *Chem. Ber.*, **83,** 273 (1950).

This mechanism involves tautomerization followed by rearrangement before ring closure as follows:

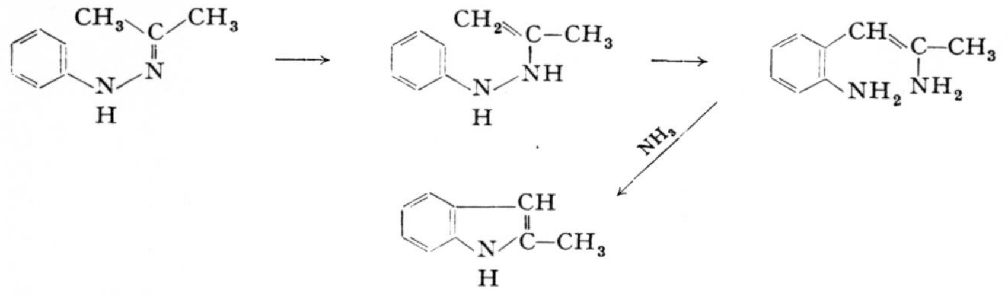

CH_3 C CH_3 N N H → CH_2 C–CH_3 NH N H → CH C–CH_3 NH_2 NH_2

NH_3

CH C–CH_3 N H

The preparation of *N*-alkylindoles from secondary hydrazines can be accounted for equally well under the Robinson mechanism. Evidence in support if the Robinson representation has been presented by Allen and

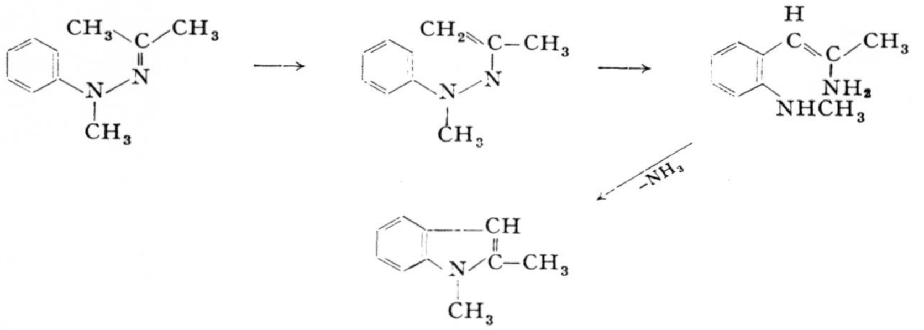

CH_3 C CH_3 N N CH_3 → CH_2 C–CH_3 N N CH_3 → H C C CH_3 NH_2 $NHCH_3$

–NH_3

CH C–CH_3 N CH_3

Wilson.[62] Through use of N^{15} isotope as a tracer element, these workers were able to show that the nitrogen farthest removed from the aromatic ring is the one eliminated as ammonia. They propose the following schemes for the elimination of ammonia:

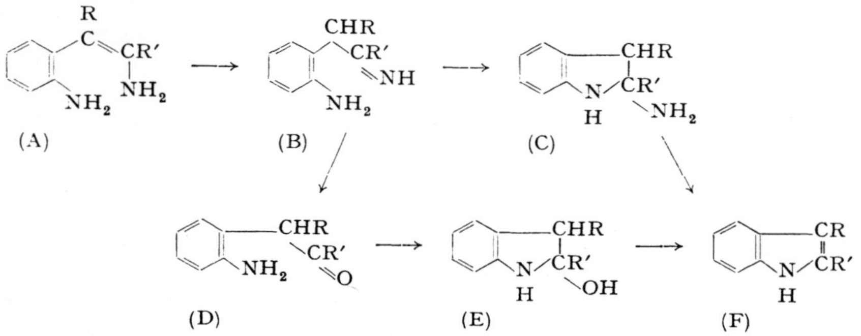

R C CR′ NH_2 NH_2 (A) → CHR CR′ NH NH_2 (B) → CHR CR′ N H NH_2 (C)

CHR CR′ NH_2 O (D) → CHR CR′ N H OH (E) → CR CR′ N H (F)

[62] Allen and Wilson, *J. Am. Chem. Soc.*, **65**, 611 (1943). Compare also Clausius and Weisser, *Helv. Chim. Acta*, **35**, 400 (1952).

This view assumes that A tautomerizes to B which may undergo ring closure to C which then eliminates ammonia to give the final indole F. The alternate hypothesis is that the ketimine B is hydrolyzed to the ketone D which through ring closure yields E which in turn loses water to give F. In support of the latter hypothesis (B → D → E → F) they cite the fact that the phenylmethylhydrazone of isopropylphenyl ketone yields 1,3,3-trimethyl-2-phenylindolin-2-ol in the Fischer synthesis.[63]

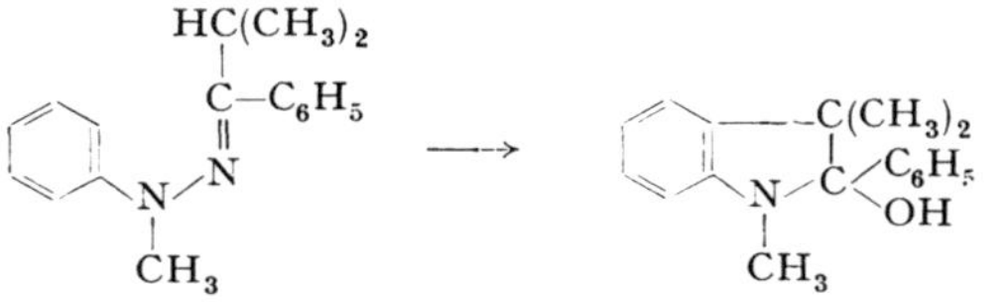

An interesting application of the Fischer synthesis in which halogen migration occurs has been reported recently.[64] The 2,6-dichlorophenylhydrazone of acetophenone on heating with zinc chloride gave 2-phenyl-5,7-dichloroindole in small yields. The same compound was also prepared (in better yields) from the 2,4-dichlorophenylhydrazone of acetophenone.

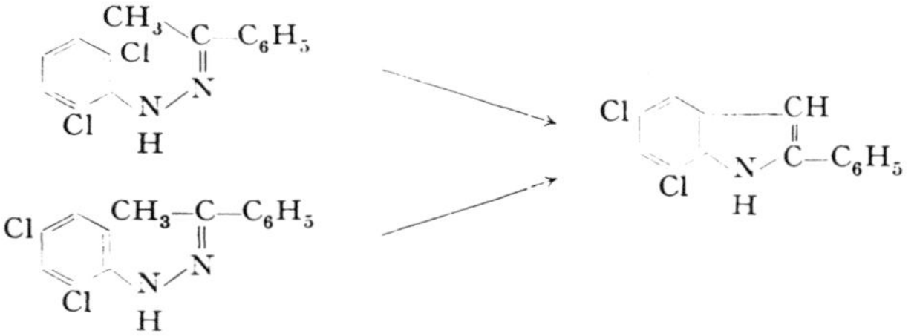

Halogen migration was not observed in the ring closure of any of the dichlorophenylhydrazones except the 2,6-dichloro derivative when displacement of a halogen atom is essential to ring closure. The 2,6-dichlorophenylhydrazones of four other ketones were converted into the corresponding 5,7-dichloroindoles by means of zinc chloride.

The hydrazones required for the synthesis of indoles by the Fischer synthesis may be prepared by means of the Japp-Klingemann reaction.[65]

[63] Jenisch, *Monatsh.*, **27,** 1223 (1906).

[64] Carlin and Fisher, *J. Am. Chem. Soc.*, **70,** 3422 (1948) Carlin, Wallace, and Fisher, *ibid.*, **74,** 990 (1952). Carlin, *ibid.*, **74,** 1077 (1952).

[65] Japp and Klingemann, *Ber.*, **21,** 549 (1888); *Ann.*, **237**, 218 (1888). Lions, *J. Proc. Soc. N.S. Wales*, **66,** 516 (1933); *Chem. Abstr.*, **27,** 2954 (1933). Hughes, Lions, *et al.*, *J. Proc. Roy. Soc. N.S. Wales*, **71,** 475 (1938); *Chem. Abstr.*, **33,** 587 (1939). Hughes and Lions, *J. Proc. Roy. Soc. N.S. Wales*, **71,** 494 (1938); *Chem. Abstr.*, **33,** 588 (1939). Hughes, Lions, and Ritchie, *J. Proc. Roy. Soc. N.S. Wales*, **72,** 209 (1939); *Chem. Abstr.*, **33,** 6837 (1939). Sempronj. *Gazz. chim. ital.*, **68,** 263 (1938).

In this reaction, benzenediazonium chloride couples with the sodium salt of a β-keto acid yielding a phenylhydrazone through elimination of the

$$C_6H_5N_2^+Cl^- + 2Na^+\left[CH_3COC^{-}(CH_2C_4H_5)-COO^-\right] \longrightarrow C_6H_5NH-N{=}C(CH_2C_6H_5)COCH_3 \longrightarrow \text{indole: } C-C_6H_5,\ C-COCH_3,\ NH$$

carboxyl group. If the carboxyl group is protected by esterification the acetyl group is eliminated rather then the carboxyl.[65,66]

$$C_6H_5N_2^+Cl^- \quad Na\left[CH_3COC(CH_2C_6H_5)-COOC_2H_5\right] \longrightarrow C_6H_5NH-N{=}C(CH_2C_6H_5)COOC_2H_5 \longrightarrow \text{indole: } C-C_6H_5,\ C-COOC_2H_5,\ NH$$

In a synthesis resembling the Fischer synthesis, Diels and Reese[67] found that acetylenedicarboxylic esters react with hydrazobenzene as well as with unsymmetrical hydrazines to give intermediates which on heating give indole

$$C_6H_5NHNH(C_6H_5) + C(COOCH_3){\equiv}C(COOCH_3) \longrightarrow C_6H_5NH-N(C_6H_5)-C(COOCH_3){=}CHCOOCH_3 \longrightarrow \text{indole: } C-COOCH_3,\ C-COOCH_3,\ NH$$

$$C_6H_5N(CH_2C_6H_5)NH_2 + C(COOCH_3){\equiv}C(COOCH_3) \longrightarrow \text{indole: } C-COOCH_3,\ C-COOCH_3,\ N-CH_2C_6H_5$$

[66] Manske, Perkin, and Robinson, *J. Chem. Soc.*, **1927**, 1.

[67] Diels and Reese, *Ann.*, **511**, 168 (1934); **519**, 147 (1935). Huntress and Hearon, *J. Am. Chem. Soc.*, **63**, 2762 (1941).

derivatives. Hydrazobenzene and acetone condense similarly in the presence of acetic acid and zinc chloride to give 1-phenyl-2-methylindole (m.p. 58–58.5°).[68]

$$\begin{matrix} C_6H_5NH \\ | \\ C_6H_5NH \end{matrix} + \begin{matrix} CH_3 \\ CH_3 \end{matrix}\!\!>C=O \xrightarrow{ZnCl_2} \text{1-phenyl-2-methylindole } (C_6H_4\text{–}CH=C(CH_3)\text{–}N(C_6H_5))$$

Another general method for the synthesis of indoles makes use of the reaction of arylamines with α-halogenated ketones or α-hydroxyketones.[69] The first product, isolable at low temperatures but not usually separated, is a phenacylaniline. Reaction with a second molecule of aniline completes the synthesis. It has been suggested by Bischler[69] that the arylamino ketone (I) formed in the first step shown above then condenses with a second

$$C_6H_5NH_2 + BrCH_2COC_6H_5 \longrightarrow C_6H_5NHCH_2COC_6H_5 \text{ (I)}$$

$$C_6H_5NH_2 + \begin{matrix} CH_2NHC_6H_5 \\ | \\ COC_6H_5 \end{matrix} \xrightarrow{\text{heat}} \text{(indole: CH=C–}C_6H_5\text{, N–H)} + C_6H_5NH_2 + H_2O$$

molecule of arylamine to yield the "aniline anil" (II). Julian and Pikl[70] were able to show that in addition to III an isomer IV was also obtained in

$$\begin{matrix} RCO \\ | \\ R'CHNHC_6H_5 \end{matrix} + C_6H_5NH_2 \xrightarrow{HX} \begin{matrix} RC=NC_6H_5 \\ | \\ R'CHNHC_6H_5 \end{matrix} \text{ (II)}$$

$$\begin{matrix} NH\text{–}C_6H_5 \\ | \\ CHR' \\ | \\ C_6H_5\text{–}N=CR \end{matrix} \xrightarrow{HX} \left[\text{(CHR'–CR=N ring) (III)}\right] \longrightarrow \text{(indole: CR'=CR, N–H)}$$

some instances. The formation of IV as well as of III in this reaction is

$$\text{(indole: CR=CR', N–H) (IV)}$$

[68] Mann and Haworth, *J. Chem. Soc.*, **1944,** 670.

[69] Mohlau, *Ber.*, **15,** 2480 (1882); **21,** 510 (1888). Bischler, *ibid.*, **25,** 2860 (1892). Nencki and Berlinerblau, German patent 40,889 (1884); *Friedl.*, **I,** 150 (1886). Bischler and Fireman, *Ber.*, **26,** 1336 (1893). Ritchie, *J. Proc. Roy. Soc. N.S. Wales*, **80,** 33–40 (1946); *Chem. Abstr.*, **41,** 3094 (1947). Cowper and Stevens, *J. Chem. Soc.*, **1947,** 1041. Mentzer, Molho, and Berguer, *Bull. soc. chim. France*, **1950,** 555.

[70] Julian and Pikl, *J. Am. Chem. Soc.*, **55,** 2105 (1933).

understandable since it has been shown[71] that two anilinoketones may be obtained in the reaction of the α-bromoketone (V) with aniline. While the

$$\begin{array}{c} R^1CHBr \\ | \\ R^2C{=}O \end{array} \quad (V)$$

$$\downarrow$$

$$\begin{array}{c} R^1CHNHC_6H_5 \\ | \\ R^2C{=}O \end{array} \rightleftharpoons \begin{array}{c} R^1C{=}O \\ | \\ R^2CHNHC_6H_5 \end{array}$$

Bischler hypothesis of the intermediate formation of the "aniline anil" (II) has been rejected by some[72] in favor of a mechanism involving direct ring closure of the anilinoketone strong evidence in favor of the intermediate formation of II has been produced.[71] When desylaniline (VIII) was heated for several hours with one molecular proportion of aniline in the presence of a few drops of hydrochloric acid 2,3-diphenylindole was obtained in good yield. Repetition of this experiment using dimethylaniline as a substitute

$$\underset{(VIII)}{\begin{array}{c} C_6H_5CHNHC_6H_5 \\ | \\ C_6H_5CO \end{array}} \xrightarrow[C_6H_5NH_2]{HCl} \underset{(IX)}{\begin{array}{c} C_6H_5CHNHC_6H_5 \\ | \\ C_6H_5C{=}NC_6H_5 \end{array}} \longrightarrow \underset{(X)}{\text{2,3-diphenylindole } (C_6H_4\text{–}NH\text{–}C(C_6H_5){=}C(C_6H_5))}$$

for aniline gave a recovery of 98% of VIII unchanged. Failure to obtain X in this experiment constitutes strong evidence against a direct ring closure and in favor of a mechanism providing for the interaction of a second molecule of aniline. In further experiments[71] in which the reaction was interrupted at definite intervals IX was isolated from the reaction mixture. By dividing the product into two portions it was possible to convert part to X as shown and to oxidize the other part to the dianil XI.[73] Support for the

$$\underset{(XI)}{\begin{array}{c} C_6H_5C{=}NC_6H_5 \\ | \\ C_6H_5C{=}NC_6H_5 \end{array}}$$

last step of the Bischler mechanism, that the indole formation takes place

[71] Julian, Meyer, Magnani, and Cole, *ibid.*, **67,** 1203 (1945). Brown and Mann, *J. Chem. Soc.*, **1948,** 858. Catch, Elliott, Hey, and Jones, *ibid.*, **1948,** 272. Catch, Hey, Jones, and Wilson, *ibid.*, **1948,** 276.

[72] Crowther, Mann, and Purdie, *ibid.*, **1943,** 58. Brown and Mann, *ibid.*, **1948,** 847 Verkade and Janetsky, *Rec. trav. chim.*, **62,** 763, 775 (1943); **64,** 129 (1945); **65,** 193 (1946).

[73] For further conversion of IX to X see also Strain, *J. Am. Chem. Soc.*, **51,** 269 (1929).

through the indolenine, has come from the observation that "aniline anils" of the type of XII do in fact yield indolenines of the type of XIII when heated in the presence of HCl.[74] The work of Julian and his collaborators

(XII)  C_6H_5HN—$C(CH_3)_2$—C—CH_3 (ring, N=) ⟶ (XIII) indolenine with $C(CH_3)_2$ and $N{=}C{-}CH_3$

establishes definitely the intermediate formation of the "aniline anils" in the Möhlau-Bischler synthesis. These investigators are careful to state, however, that their work does not completely rule out the possibility of direct ring closure in some cases. Thus 3-phenylamino-2-butanone on heating with an equal weight of zinc chloride for thirty minutes gave a 56% yield of 2,3-dimethylindole,[75] showing that direct ring closure is possible. However, the same anilinoketone on heating with twice its weight of aniline hydrochloride gave the same indole in 65% yield. It seems probable that direct ring closure takes place when the compound is heated with zinc chloride but that the Julian mechanism is the correct one under the conditions of the Bischler reaction.

Benzoin has been used also in this synthesis in place of desyl bromide for the preparation of 2,3-diphenylindole.[76] The Möhlau-Bischler synthesis has been employed by a number of workers[77] for the synthesis of indole derivatives. A variation of the synthesis[78] gives indole from aniline and

$$ClCH_2CHClOC_2H_5 \longrightarrow ClCH_2CHO$$

$$2\,C_6H_5NH_2 + ClCH_2CHO \longrightarrow C_6H_5NH{-}CH{=}CH{-}NHC_6H_5 \longrightarrow \text{Indole}$$

[74] Garry, *Compt. rend.*, **211,** 399 (1940); **212,** 401 (1941). Garry, *Ann. Chim.*, **17,** 5 (1942).

[75] Janetzky and Verkade, *Rec. trav. chim.*, **65,** 691 (1945). Brown and Mann, *J. Chem. Soc.*, **1948,** 847.

[76] Japp and Murray, *J. Chem. Soc.*, **65,** 889 (1894); *Ber.*, **26,** 2638 (1893). Lachowicz, *Monatsh.*, **15,** 402 (1894).

[77] Wolff, *Ber.*, **20,** 428 (1887); **21,** 133, 3360 (1888). German patent 533,471; *Chem. Abstr.*, **26,** 480 (1932). Richards, *J. Chem. Soc.*, **97,** 977 (1910); *Proc. Chem. Soc.*, **26,** 92 (1910). Pictet and Duparc, *Ber.*, **20,** 3415 (1887). British patent 354,392; *Chem. Anstr.*, **26,** 5431 (1932). Sircar and Guha, *J. Indian Chem. Soc.*, **13.** 704 (1936); *Chem. Abstr.*, **31,** 3911 (1937). Hell and Cohen, *Ber.*, **37,** 866 (1904). Hell and Bauer, *ibid.*, **37,** 872 (1904). Mentzer, *Compt. rend.*, **222,** 1176 (1946), Meisenheimer, Angermann, Finnand, and Vieweg, *Ber.*, **B 57,** 1774 (1924). Bauer and Bühler, *Arch. Pharm.*, **262,** 128 (1924). Emerson, Heimach, and Patrick, *J. Am. Chem. Soc.*, **75,** 2256 (1953).

[78] Berlinerblau, *Monatsh.*, **8,** 180 (1887). Berlinerblau and Poliker, *ibid.*, **8,** 187 (1887). Nencki and Berlinerblau, German patent 40,889 (1884).

α,β-dichloroethyl ether according to the preceding scheme. A modification employs the diethylacetal of chloroacetaldehyde to obtain *N*-alkylindoles.[79]

$$C_6H_5NHC_2H_5 + ClCH_2CH(OC_2H_5)_2 \longrightarrow \text{1-ethyl-3-ethoxyindoline } (CHOC_2H_5,\ CH_2,\ N\text{-}C_2H_5) \text{ and } \text{1-ethylindole } (CH{=}CH,\ N\text{-}C_2H_5)$$

The procedure does not yield indole with aniline but has been reported to give satisfactory results with secondary amines.[80] In the hands of other workers the procedure has been less satisfactory.[81]

Indole has also been prepared through a scheme in which aniline and ethylene bromide are the initial reactants.[82]

$$C_6H_5NHCH_2CH_2NHC_6H_5 \xrightarrow{CrO_3} C_6H_5NHCH{=}CHNHC_6H_5 \xrightarrow{Zn} \text{Indole}$$

An indole synthesis[83] which has found considerable use consists of an intramolecular Claisen condensation of an acyl derivative of an *o*-toluidine.

$$o\text{-}CH_3C_6H_4NHCOCH_3 \xrightarrow{NaOC_2H_5} \text{2-methylindole } (CH{=}CCH_3,\ NH)$$

Sodium amide[84] and a variety of other condensing agents[85] have also been employed in this synthesis. Indole itself has been prepared by treating *N*-formyl-*o*-toluidine with sodium amide.

Indole has been prepared[86] by heating *o*-amino-ω-chlorostyrene with

[79] Rath, *Ber.*, **B 57,** 715 (1924).

[80] Kiematsu and Inoue, *J. Pharm. Soc. Japan*, **No. 518,** 351 (1925); *Chem. Abstr.*, **19,** 2493 (1925).

[81] Koenig and Bucheim, *Ber.*, **B 58,** 2868 (1925). Janetzky, Verkade, and Meerburg, *Rec. trav. chim.*, **66,** 317 (1947); *Chem. Abstr.*, **42,** 558 (1948).

[82] Prud'homme, *Bull. soc. chim.*, [2] **28,** 558 (1877).

[83] Madelung, *Ber.*, **45,** 1128, 3521 (1912). Madelung, German patent 262,237; *Chem. Abstr.*, **7,** 3642 (1913). Verley, *Bull. soc. chim.*, [4] **35,** 1039 (1924). Verley and Beduive, *ibid.*, **37,** 189 (1925). Salway, *J. Chem. Soc.*, **103,** 351, 1988 (1913).

[84] Kiematsu and Sugasawa, *J. Pharm. Soc. Japan*, **48,** 755 (1928); *Chem. Abstr.* **23,** 834 (1929). British patent 303,478; *Chem. Abstr.*, **23,** 4484 (1929)

[85] British patent 330,332; *Chem. Abstr.*, **24,** 5770 (1930). Tyson, *J. Am. Chem. Soc.*, **63,** 2024 (1941); *Org. Synthesis*, **23,** 42 (1943). Marion and Ashford, *Can. J. Research*, **B 23,** 26 (1945). Galat and Friedman, *J. Am. Chem. Soc.*, **70,** 1280 (1948). U.S. patent 2,442,952; *Chem. Abstr.*, **42,** 6857 (1948). Tyson, *J. Am. Chem. Soc.*, **72,** 2801 (1950). Tyson and Shaw, *ibid.*, **74,** 2273 (1952).

[86] Lipp, *Ber.*, **17,** 1067, 2507 (1884).

sodium ethoxide. Indole has been prepared from *o*-nitrocinnamide[87] according to the scheme:

$$o\text{-}O_2NC_6H_4CH{=}CHCONH_2 \xrightarrow[CH_3OH]{KOCl} o\text{-}O_2NC_6H_4CH{=}CHNHCOOCH_3$$

$$\downarrow Fe + CH_3COOH$$

$$\text{Indole} \xleftarrow{KOH} H_2NC_6H_4CH{=}CHCOOCH_3$$

Heating the dianilide of tartaric acid with zinc chloride gives indole.[88] Small yields of indole are obtained when the calcium salt of phenylglycine is heated with excess calcium formate,[89] as well as when *o*-chloro-ω-chloroacetanilide is distilled with zinc dust.[90] Heating *o*-formylphenylglycine with acetic anhydride and sodium acetate yields indole-2-carboxylic acid[91] as an intermediate and through decarboxylation of this product indole is obtained. Indole has been prepared through the distillation of oxal-*o*-toluic acid with zinc dust or by dry distillation of its barium salt.[92] Small yields of indole were obtained by dropping *N*-methyl-*o*-toluidine on reduced nickel at 300–330°.[93] Other indole derivatives were prepared in similar fashion from alkyl toluidines.[94] Indole has been obtained through the pyrolysis of *N*-ethylaniline[95] as well as by catalytic dehydrogenation of *o*-ethylaniline.[96] Heating *o,o'*-diaminostilbene hydrochloride under reduced pressure gives a quantitative yield of indole.[97] Indole has been prepared by the action of alcoholic potash on the dibromide of *N*-acetyl-*o*-aminostyrene.[98]

$$C_6H_4(CHBrCH_2Br)(NHCOCH_3) \longrightarrow \text{indole (CH=CH, N–H)} + 2\,HBr + CH_3COOH$$

[87] Weermann, *Rec. trav. chim.*, **29,** 18 (1910); *Ann.*, **401,** 14 (1913); German patent 213,713 (1908).

[88] Poliker, *Ber.*, **24,** 2954 (1891).

[89] Mauthner and Suida, *Monatsh.*, **10,** 250 (1889).

[90] Schwalbe, Schulz, and Jockheim, *Ber.*, **41,** 3792 (1908).

[91] Gluud, *J. Chem. Soc.*, **103,** 1254 (1913); *Ber.*, **48,** 420 (1915); German patent 287,282 (1913).

[92] Mauthner and Suida, *Monatsh.*, **7,** 230 (1886).

[93] Carrasco and Padoa, *Gazz. chim. ital.*, **36,** ii, 512 (1906); **37,** ii, 49 (1907); *Atti accad. Lincei*, [5] **15,** i, 699 (1906); [5] **15,** ii, 729 (1906).

[94] Baeyer and Caro, *Ber.*, **10,** 1262 (1877).

[95] Baeyer and Caro, *ibid.*, **10,** 692 (1877).

[96] Gresham and Brunner, U.S. patent 2,409,676; *Chem. Abstr.*, **41,** 998 (1947). Hansch and Kelmkamp, *J. Am. Chem. Soc.*, **73,** 3080 (1951).

[97] Thiele and Dimroth, *Ber.*, **28,** 1411 (1895); German patent 84,578 (1895).

[98] Taylor and Hobson, *J. Chem. Soc.*, **1936,** 181.

Indole is the product when *o*-nitrophenylacetonitrile is reduced with stannous chloride and hydrogen chloride in anhydrous ether with subsequent decomposition of the reduction product with water.[99] Indole is also obtained

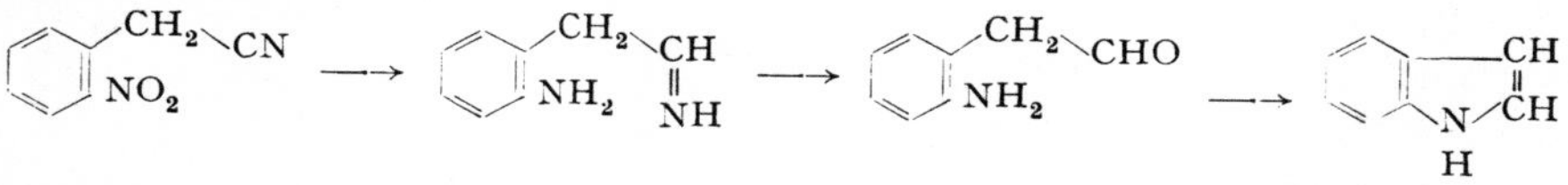

$CH_2\diagdown CN$, NO_2 ⟶ $CH_2\diagdown CH$, NH_2, $\Vert$ NH ⟶ $CH_2\diagdown CHO$, NH_2 ⟶ CH $\Vert$ $N\diagup CH$ H

through the reduction of *o*-nitrocinnamic acid[100] and of *o*-nitrophenylacetaldehyde.[101]

Indole-2-carboxylic acid (which is easily decarboxylated to indole) is prepared readily by reduction of *o*-nitrophenylpyruvic acid with zinc dust and acetic acid.[102] The *o*-nitrophenylpyruvic acid is easily prepared from *o*-nitrotoluene and ethyl oxalate. This synthesis has been utilized for the

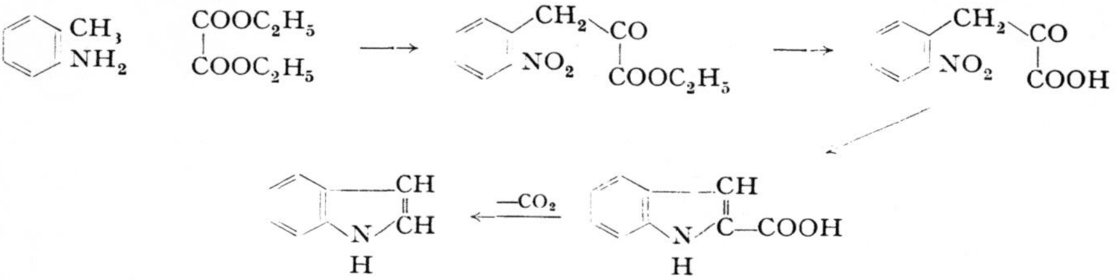

CH_3, NH_2 + $COOC_2H_5$ | $COOC_2H_5$ ⟶ $CH_2\diagdown CO$, NO_2, | $COOC_2H_5$ ⟶ $CH_2\diagdown CO$, NO_2, | $COOH$ ⟶

CH $\Vert$ $N\diagup CH$ H $\xleftarrow{-CO_2}$ CH $\Vert$ $N\diagup C—COOH$ H

preparation of a large number of indole-2-carboxylic acid derivatives which on decarboxylation give the corresponding indoles.[103] On the other hand,

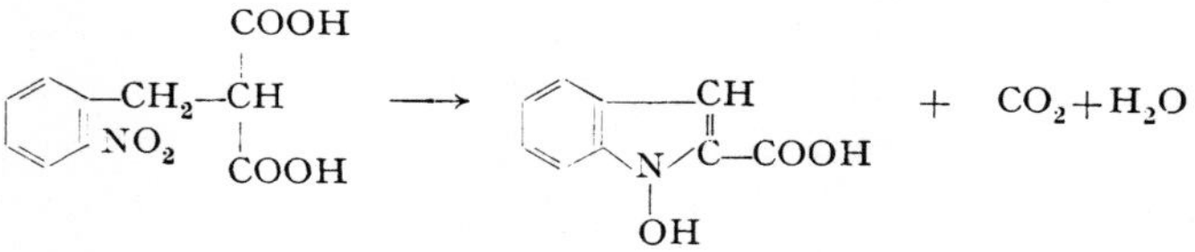

$COOH$ | $—CH_2—CH$, NO_2, | $COOH$ ⟶ CH $\Vert$ $N\diagup C—COOH$ | OH + $CO_2 + H_2O$

[99] Stephen, *ibid.*, **127,** 1874 (1925).

[100] Baeyer and Emmerling, *Ber.*, **2,** 679 (1869). Beilstein and Kuhlberg, *Ann.*, **163,** 141 (1872).

[101] Weerman, *Rec. trav. chim.*, **29,** 18 (1910).

[102] Reissert, *Ber.*, **29,** 639 (1896); **30,** 1030 (1897). Reissert and Heller, *ibid.*, **37,** 4378 (1904).

[103] Stiller, U. S. patent 2,380,479; *Chem. Abstr.*, **40,** 367 (1946). Kermack, Perkin, and Robinson, *J. Chem. Soc.*, **119,** 1625 (1921). Polyakova, *Masloboino Zhirovoe Delo*, **11,** 452 (1935); *C. A.*, **30,** 4869 (1936). Shorygin and Polyakova, *Sintezy Dushistykh Veshchesto, Sbornik Statei*, **1939,** 130–136; *Khim. Referat. Zhur.*, **1940,** No. 4, 114; *Chem. Abstr.*, **36,** 3802 (1942). Burton and Stoves, *J. Chem. Soc.*, **1937,** 1726. Blackie and Perkin, *ibid.*, **125,** 296 (1924). Kermack, *ibid.*, **125,** 2285 (1924). Bergel and Morrison, *ibid.*, **1943,** 49. Maurer and Moser, *Z. physiol. Chem.*, **161,** 131 (1936). Oxford and Raper, *J. Chem. Soc.*. **1927,** 417.

reduction of *o*-nitrophenylpyruvic acid with sodium amalgam gives 1-hydroxyindole-2-carboxylic acid (m.p. 159.5°) (dec.) which is also obtained when *o*-nitrobenzylmalonic acid is heated with NaOH, 1-Hydroxy-indole derivatives have been prepared also from ethyl *o*-nitrobenzylaceto-acetate according to the following scheme:[104]

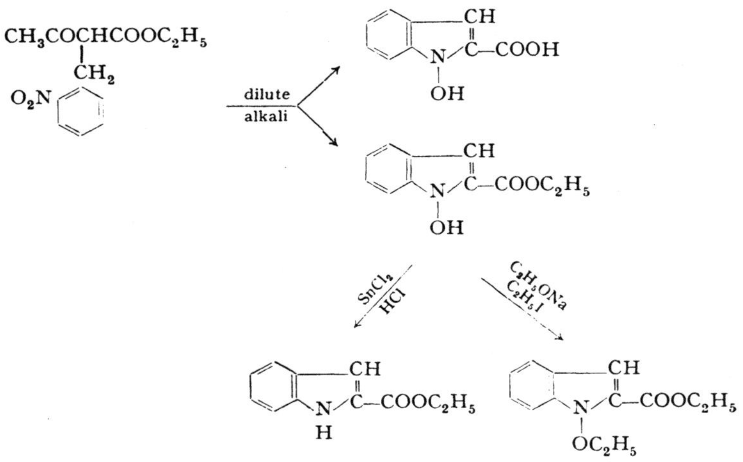

$CH_3COCHCOOC_2H_5$ / CH_2 / O_2N–C_6H_4 —(dilute alkali)→ 1-hydroxyindole-2-carboxylic acid (CH, C–COOH, N–OH) and ethyl 1-hydroxyindole-2-carboxylate (CH, $C–COOC_2H_5$, N–OH)

ethyl 1-hydroxyindole-2-carboxylate —($SnCl_2$, HCl)→ (CH, $C–COOC_2H_5$, N–H)

ethyl 1-hydroxyindole-2-carboxylate —(C_2H_5ONa, C_2H_5I)→ (CH, $C–COOC_2H_5$, $N–OC_2H_5$)

Indole has been prepared by treating acetyl-*o*-aminocinnamic acid with hydrochloric acid at high temperatures[105] as well as from *o*-aminophenyl-propiolic acid.[106] 2-Aminoindole[107] has been prepared from *o*-aminobenzyl cyanide and that product converted into indole by reduction with sodium and alcohol. Indole has been prepared also by reduction of *o*-ω-dinitro-styrene.[108] Indole has been prepared through fusion of carbostyril[109] and a number of other quinolines[110] with alkali. On the other hand, a number of methylindoles have been converted into quinolines.[111] Heating indole-2-

[104] Gabriel, Gerhard, and Walter, *Ber.*, **B 56,** 1024 (1923).

[105] Walter, *ibid.*, **25,** 1261 (1892).

[106] Richter, *ibid.*, **16,** 679 (1883).

[107] Pschorr and Hoppe, *ibid.*, **43,** 2543 (1910).

[108] Menitzescu, *ibid.*, **B 58,** 1063 (1925). Van der Lee, *Rec. trav. chim.*, **44,** 1089 (1925).

[109] Morgan, *Chem. News*, **36,** 269 (1877).

[110] Padoa and Carughi, *Atti accad. Lincei*, [5] **15,** II, 113 (1906). Padoa and Scagliarini, *ibid.*, [5] **17,** I, 728 (1908).

[111] Hofmann and Königs, *Ber.*, **16,** 727 (1883). Ciamician, *ibid.*, **37,** 4231 (1894). Pictet, *ibid.*, **38,** 1949 (1905). Plancher and Carrasco, *Atti accad. Lincei*, [5] **14,** I, 162 (1905). Ellinger, *Ber.*, **39,** 2517 (1906). Ellinger and Flamand, *ibid.*, **39,** 4388 (1906). Heller and Wunderlich, *ibid.*, **47,** 1619 (1914).

carboxylic acid[112] and quinoline-2,3-dicarboxylic acld[113] with calcium carbonate gives indole. Ketodihydrobenzene-*p*-thiazine[114] yields some indole with a red-hot mixture of zinc and copper powder. Acetylene and aniline react to form indole when passed over certain heated catalysts.[115] Indole has been obtained by passing the vapor of cumidine over hot lead oxide[116] and by distilling skatole over hot porcelain chips. Small yields of indole were obtained by distilling 3-nitro-4-isopropenylbenzoic acid over lime.[117] Fusion of albumin with potassium hydroxide yields indole, skatole, and other products.[118]

2-Phenylindole[119] has been prepared through the pyrolysis of benzal-*o*-toluidine as well as by nitrating phenyl benzyl ketone and reducing the nitro derivative with zinc dust and ammonia. Treatment of benzil dianil with methylmagnesium iodide yields an intermediate which on treatment with hydrochloric acid yields 1-methyl-2,3-diphenylindole, m.p. 139°.[120] Other diketone dianils react similarly with Grignard reagents.[121] Indole

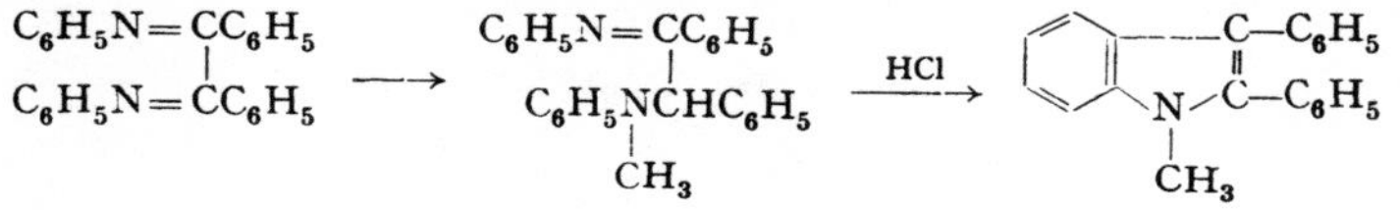

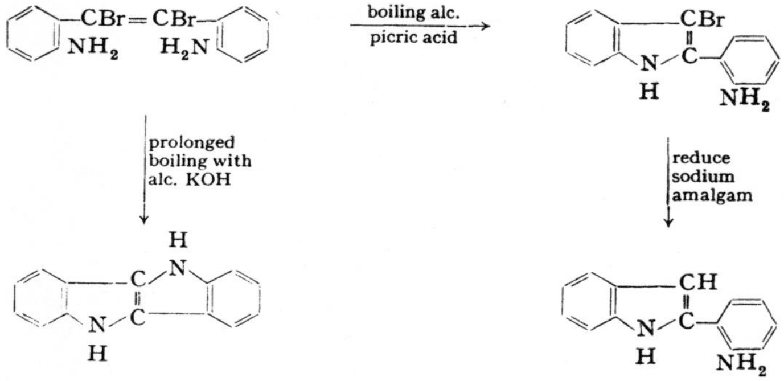

$C_6H_5N{=}CC_6H_5$ / $C_6H_5N{=}CC_6H_5$ ⟶ $C_6H_5N{=}CC_6H_5$ / $C_6H_5NCHC_6H_5$ / CH_3 —HCl⟶ C—C_6H_5 / C—C_6H_5 / N / CH_3

CBr=CBr / NH_2 H_2N —boiling alc. picric acid⟶ CBr / C / N / H / NH_2

prolonged boiling with alc. KOH ↓ H / N / C / C / N / H

reduce sodium amalgam ↓ CH / C / N / H / NH_2

[112] Ciamician and Zatti, *Ber.*, **22,** 1976 (1889).
[113] Graebe and Caro, *ibid.*, **13,** 99 (1880).
[114] Unger and Graff, *ibid.*, **30,** 2389 (1897).
[115] Majima, Unno, and Ono, *ibid.*, **B 55,** 3854 (1922).
[116] Fileti, *Gazz. chim. ital.*, **13,** 350 (1881); *Ber.*, **16,** 2928 (1884).
[117] Widmann, *ibid.*, **15,** 2547 (1872).
[118] Engler and Janecke, *ibid.*, **9,** 1411 (1876). Kühne, *ibid.*, **8,** 206 (1875). Nencki, *J. prakt. Chem.*, [2] **17,** 97 (1878).
[119] Pictet, *Ber.*, **19,** 1063 (1886).
[120] Montague and Garry, *Compt. rend.*, **204,** 1659 (1937).
[121] Garry, *Ann. Chim.*, **17,** 5 (1942).

derivatives have also been obtained[122] from α,α-dibromo-2,2'-diaminostilbene through boiling with alcoholic picric acid. Long boiling with alcoholic potassium hydroxide, on the other hand, yields a compound called "di-indole" or "dindole" by Ruggli (this compound is not to be confused with the indole dimer which is usually called bi-indole). The same compound has been obtained by the slow addition of zinc dust to 2,2-dinitrobenzil in acetic acid.[123]

2-Methyl-5,6-dimethoxyindole has been prepared from 5-acetylamino-4-allylveratrol dibromide through treatment with alcoholic potassium hydroxide.[124]

$$(CH_3O)_2C_6H_2(CH_2CHBrCH_2Br)(NHCOCH_3) \xrightarrow{\text{alc. KOH}} (CH_3O)_2C_6H_2\langle CH{=}C(CH_3){-}NH\rangle$$

Dihydroindole and its homologs have been prepared through the catalytic dehydration of o-aminophenyl-β-ethanol and its derivatives.[125] If a dehydrogenating catalyst (Cu, Co, or Ag) is added indole is also obtained. Indole has been prepared from *N*-acetyldihydroindoxyl through dehydration and hydrolysis.[126] Indole has been prepared from indoxyl by heating with

$$C_6H_4\langle CHOH{-}CH_2{-}N(COCH_3)\rangle \xrightarrow{-H_2O} C_6H_4\langle CH{=}CH{-}N(COCH_3)\rangle \longrightarrow C_6H_4\langle CH{=}CH{-}NH\rangle$$

alkali in the presence of catalytically active metals at 250–325°.[127] Overheating of the indoxyl melt in the preparation of synthetic indigo gives indole[128] as does autoclaving of the indoxyl melt with water at 220–240° for three hours under pressure. Skatole has been isolated from the mixture

[122] Ruggli, *Ber.*, **50**, 883 (1917).

[123] Heller, *ibid.*, **50**, 1202 (1917).

[124] Lions, *J. Proc. Roy. Soc. N. S. Wales*, **63**, 168 (1930); *Chem. Abstr.*, **24**, 5299 (1930).

[125] German patent 606,027; *Chem. Abstr.*, **29**, 1439 (1935).

[126] German patent 516,675; *Chem. Abstr.*, **25**, 3016 (1931). German patent 518,675; *Chem. Abstr.*, **25**, 3364 (1931).

[127] Vorlander and Apelt, *Ber.*, **37**, 1134 (1904). U. S. patent 2,365,966; *Chem. Abstr.*, **40**, 609 (1946).

[128] German patent 260,327; *Chem. Abstr.*, **7**, 3236 (1913). Stepanof and Polyakova, *Sintezy Dushistykh Veshchesto, Sbornik Statei*, **1939**, 127–130; *Khim. Referat. Zhur.*, **1940**, No. 4, 114; *Chem. Abstr.*, **36**, 3802 (1942).

resulting from heating together aniline, zinc chloride, and glycerol.[129] 2-Phenylindole is obtained when 2-phenyl-4-quinolinecarboxylic acid is treated with potassium amide and potassium nitrite in liquid ammonia.[130]

An interesting synthesis of indole derivatives results from the condensing of quinone with ethyl β-aminocrotonate.[131] Similarly quinone and glycine-

Quinone (O, O) + $H_2N(CH_3)C{=}CHCOOC_2H_5$ ⟶ HO–indole: C–$COOC_2H_5$, C–CH_3, N–H; m.p. 205°

$(CH_3)_2SO_4$ ⟶ CH_3O–indole: C–$COOC_2H_5$, C–CH_3, N–H; m.p. 161°

NaOH ⟶ HO–indole: CH, C–CH_3, N–H; m.p. 188°

NaOH ⟶ CH_3O–indole: C–COOH, C–CH_3, N–H; m.p. 208°

$-CO_2$ ⟶ CH_3O–indole: CH, C–CH_3, N–H; m.p. 89–90°

acetoacetic ester give ethyl (2-methyl-3-carbethoxy-5-hydroxyindyl-1-)-

HO–indole: C–$COOC_2H_5$, C–CH_3, N–$CH_2COOC_2H_5$

acetate, m.p. 148°. A number of indole derivatives[132] have been prepared

Cyclohexane-1,2-dione monoxime (CH_2, CH_2, CH_2, CH_2, C=O, C=NOH) ⟶ ($CH_3COCH_2COOC_2H_5$) tetrahydroindole: C–$COOC_2H_5$, C–CH_3, N–H; m.p. 134° ⟶ Free acid m.p. 170°

$-CO_2H$ ⟶ tetrahydroindole: CH, C–CH_3, N–H

[129] Fischer and German, *Ber.*, **16,** 710 (1883).

[130] White and Bergstrom, *J. Org. Chem.*, **7,** 497 (1942).

[131] Nenitzescu, *Bull. Soc. Chim. România*, **11,** 37–43 (1929); *Chem. Abstr.*, **24,** 110 (1930).

[132] Triebs and Dinelli, *Ann.*, **517,** 152 (1935). Triebs, Dornberger, Schröder, Albrecht, Reinecke, and Emmerich, *ibid.*, **524,** 285 (1936).

by condensing the monoxime of cyclohexane-1,2-dione with β-ketone esters. Ethyl pyruvate oxime and cyclohexanone similarly condensed to give an isomer of the product obtained above. Dehydrogenation with selenium gave skatole. Condensation of cyclohexane-1,2-dione monoxime with acetonyl-

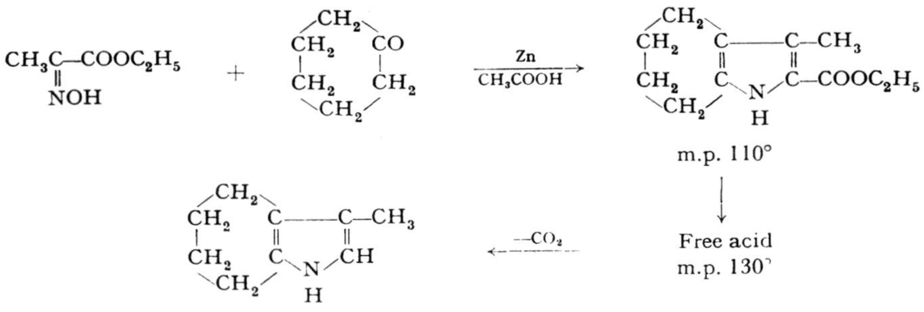

```
                         CH2                                       CH2
                       /     \                                   /     \
CH3C—COOC2H5          CH2     CO          Zn                    CH2     C ——— C—CH3
   ‖             +    |       |      ——————————→                |       ‖     ‖
   NOH                CH2     CH2     CH3COOH                   CH2     C     C—COOC2H5
                       \     /                                   \     / \   /
                         CH2                                       CH2     N
                                                                           H
                                                                    m.p. 110°
                                                                        |
        CH2                                                             ↓
      /     \                                                      Free acid
    CH2      C ——— C—CH3                                           m.p. 130°
    |        ‖     ‖               —CO2
    CH2      C     CH          ←————————
      \     / \   /
        CH2    N
               H
```

acetone similarly gave an indole derivative. Indole derivatives have also been prepared from cyclohexene pseudonitrosite and from chlorocyclohexanone as shown in the following scheme:[132]

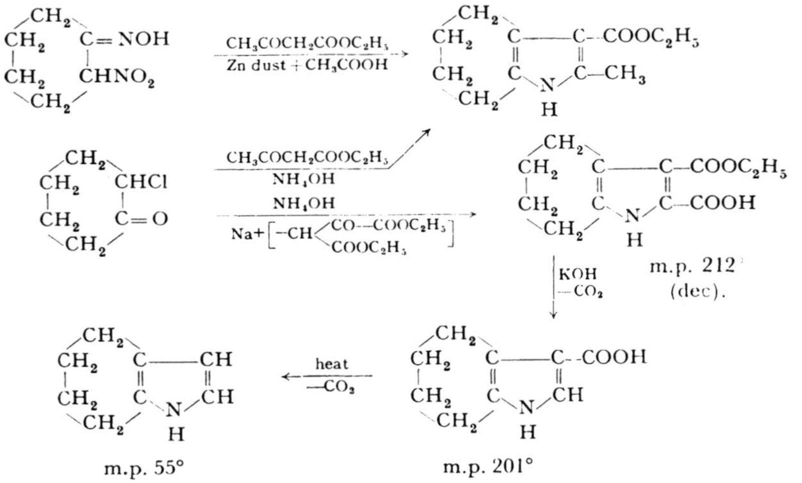

```
      CH2                                                    CH2
    /     \                                                /     \
  CH2      C=NOH      CH3COCH2COOC2H5                    CH2      C ——— C—COOC2H5
  |        |         ————————————————————→               |        ‖     ‖
  CH2      CHNO2      Zn dust + CH3COOH                  CH2      C     C—CH3
    \     /                                                \     / \   /
      CH2                                                    CH2     N
                                                                     H
                                              ↗
      CH2             CH3COCH2COOC2H5                        CH2
    /     \          ——————————————————                    /     \
  CH2      CHCl           NH4OH                          CH2      C ——— C—COOC2H5
  |        |              NH4OH                          |        ‖     ‖
  CH2      C=O       ————————————————————————→           CH2      C     C—COOH
    \     /          Na+[—CH<(CO—COOC2H5)(COOC2H5)]        \     / \   /
      CH2                                                    CH2     N
                                                                     H
                                                                          m.p. 212°
                                                            | KOH          (dec).
                                                            | —CO2
                                                            ↓
      CH2                                                CH2
    /     \                                            /     \
  CH2      C ——— CH            heat                  CH2      C ——— C—COOH
  |        ‖     ‖          ←————————                |        ‖     ‖
  CH2      C     CH            —CO2                  CH2      C     CH
    \     / \   /                                      \     / \   /
      CH2    N                                           CH2     N
             H                                                   H
       m.p. 55°                                           m.p. 201°
```

Dennstedt[133] found that certain pyrroles dimerize when treated with mineral acids in the absence of water and that subsequent treatment with sulfuric acid brought about the loss of ammonia and formation of an indole. The conditions requisite for indole formation in this series have been reported

[133] Dennstedt, *Ber.*, **21,** 3429 (1888); **22,** 1920 (1889); **24,** 2562 (1891). Dennstedt and Zimmermann, *ibid.*, **20,** 850 (1887; **21,** 1478 (1888).

by Allen and coworkers[134]: (*1*) from the pyrroles by long heating with zinc salts and acetic acid, (*2*) from the dipyrrole and sulfuric acid, and (*3*) from acetonylacetone and pyrroles. All three procedures give poor yields. Dennstedt started with the dimer of 2-methylpyrrole and obtained a dimethylindole which he thought was 2,6-dimethylindole. The same dimethylindole was obtained from the *m*-tolylhydrazone of acetone.[135] This compound has

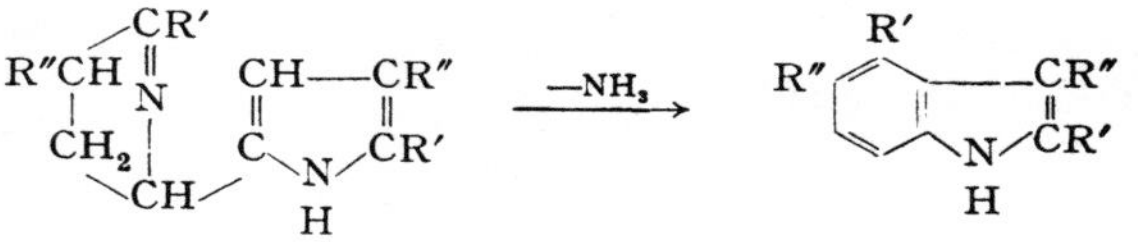

$$\xrightarrow{-NH_3}$$

subsequently been shown to be 2,4-dimethylindole ($R' = CH_3$, $R'' = H$).

The catalytic dehydrogenation of 3,4-diaryl-5,6,7,8-tetrahydrocinnolines yields as the principal product 2,3-diarylindoles.[135a]

Oxidation of Indole

Indole and indole derivatives are cleaved by perbenzoic acid, ozone, and in some cases peracetic acid, the pyrrole ring being opened at the double aond.[136]

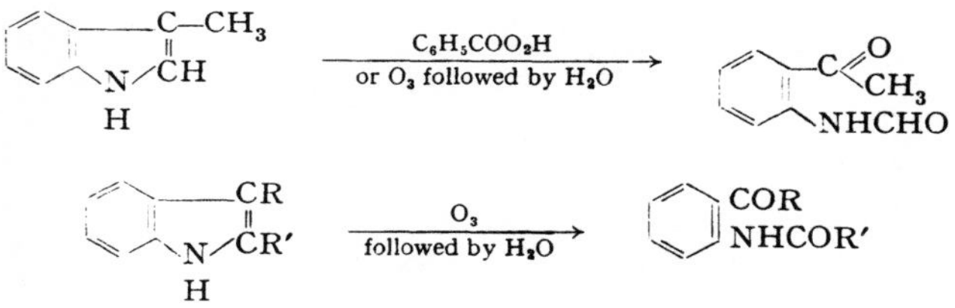

$$\xrightarrow[\text{or } O_3 \text{ followed by } H_2O]{C_6H_5COO_2H}$$

$$\xrightarrow[\text{followed by } H_2O]{O_3}$$

Oxidation of methylketole by oxygen,[137] hydrogen peroxide,[138] air and

[134] Allen, Young, and Gilbert, *J. Org. Chem.*, **2**, 235 (1937). Allen, *ibid.*, **2**, 400 (1937); *Can. J. Res.*, **25**, 1 (1947).

[135] Plancher and Ciusa, *Atti accad. Lincei*, [5] **15**, ii, 447 (1906).

[135a] Allen and Van Allen, *J. Am. Chem. Soc.*, **73**, 5850 (1951).

[136] Witkop and Graser, *Ann.*, **556**, 103 (1944). Witkop and Fiedler, *ibid.*, **558**, 91 (1947). Witkop, *ibid.*, **558**, 98 (1947). Witkop, *J. Am. Chem. Soc.*, **72**, 1428 (1950). Witkop and Patrick, *ibid.*, **73**, 2196 (1951). Mentzer, Molho, and Berger, *Compt. rend.*, **229**, 1937 (1949); *Bull. soc. chim. France*, **1950**, 555. Bergner, Molho, and Mentzer, *Compt. rend.*, **230**, 760 (1950). Witkop and Patrick, *J. Am. Chem. Soc.*, **74**, 3855, 3861 (1952).

[137] Oddo, *Gazz. chim. ital.*, **46**, I, 323 (1916).

[138] Plancher and Colacicchi, *Atti accad. Lincei*, [5] **20**, I, 453 (1911).

light,[139] autoxidation,[140] or peracetic acid[141] all convert the compound into a yellow substance $C_{18}H_{16}N_2O$ (I) formulated by Witkop as:

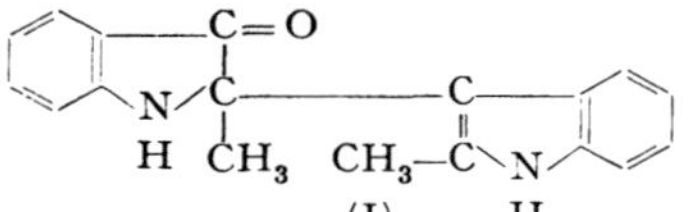

(I)

Reduction of this compound with $LiAlH_4$ gives some (II) and through a molecular rearrangement (III).

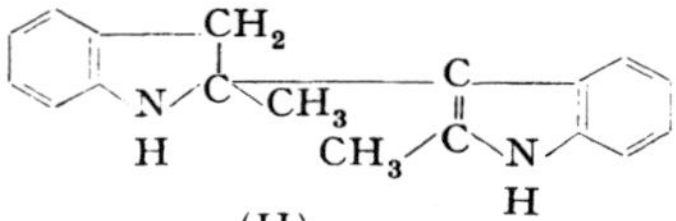

(II) 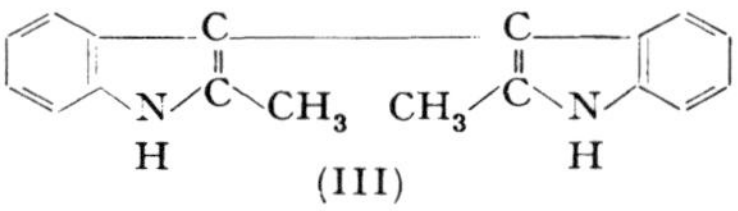 (III)

Oxidation of indole itself by air and light[137, 139, 142] or by peracetic acid[141] leads to indoxyl which can react further to yield either indigo or the trimeric condensation product which is probably formed via leucoindoxyl red and indoxyl red.[143] The latter compound is known to be capable of adding a molecule of indole to give VI.[143] VI has also been prepared through treatment

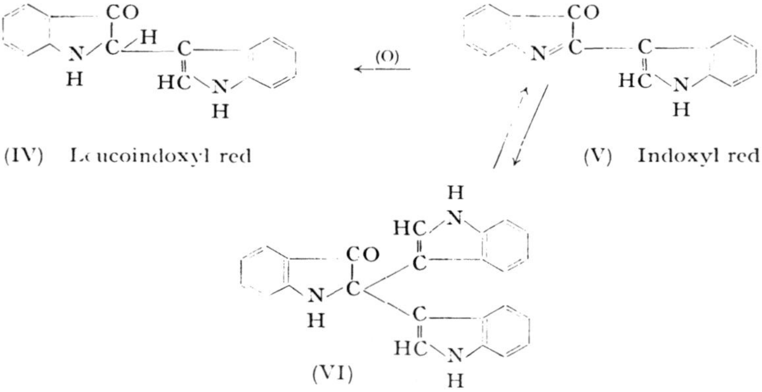

(IV) Leucoindoxyl red (V) Indoxyl red

(VI)

of indole with sulfuric acid and one-third mole sodium nitrite.[141, 143] Compound VI is dimorphous, m.p. 204°, 245°.[141]

[139] Baudisch and Hoschek, *Ber.*, **49**, 453, 2579 (1916).

[140] Toffolo, *Atti X⁰ cong. intern. chim.*, **3**, 369 (1939). Toffoli, *Rend. ist. sanita publ.*, **2**, II, 565 (1939).

[141] Witkop, *Ann.*, **558**, 98 (1947). Witkop and Patrick, *J. Am. Chem. Soc.*, **73**, 713 1951).

[142] Oddo, *Gazz. chim. ital.*, **50**, II, 276 (1920).

[143] Seidel, *Ber.*, **77**, 788, 797 (1944). Seidel, *Chem. Ber.*, **83**, 20 (1950).

Tautomerism of Indole

Indole and the 2-substituted indoles are tautomeric substances behaving as though they had both structures I and II. While these two tautomeric

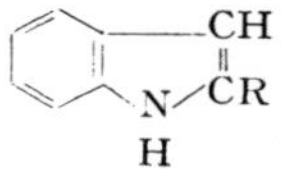

(I) (R.I. 821) 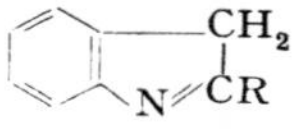 (II) (R.I. 823)

structures have not been isolated, derivatives corresponding to both structures are well known.[144] Indoles substituted in position 3 or in positions 2 and 3 give true nitroso derivatives (III) when treated with nitrous acid while those indoles substituted in position 2 only give nitroso derivatives which are regarded as derived from II. These oximes of structure IV were also obtained when the indole was treated with amyl nitrile and sodium ethoxide.

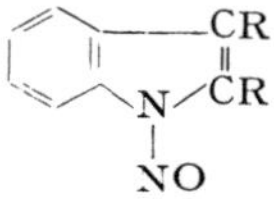

(III) 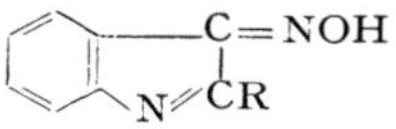 (IV)

Indole itself yields with nitrous acid or with amyl nitrite and sodium ethoxide the nitroso derivative[145] previously prepared by the action of sodium nitrite and acetic acid on indole.[146] This nitroso derivative is thought

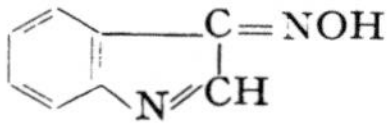

to correspond to structure IV.[145] In addition to this simple nitrosoindole the action of nitrous acid on indole leads to the formation of indole red (V) and

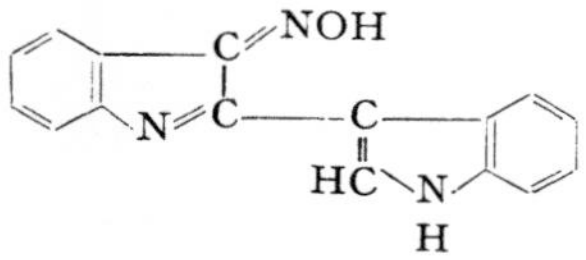

(V) 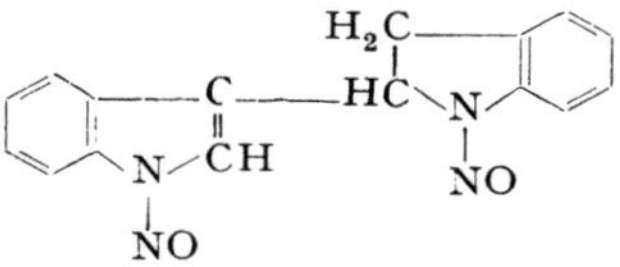 (VI)

[144] Angeli and Spica, *Gazz. chim. ital.*, **29,** i, 500 (1899). Spica and Angelico, *ibid.*, **29,** ii, 49 (1899). Angeli and Angelico, *Atti accad. Lincei*, [5] **12,** i, 344 (1903). Angeli and Marchetti, *ibid.*, [5], **16,** ii, 790 (1907). Angeli and Morelli, *ibid.*, [5], **17,** i, 697 (1908). König, *J. prakt. Chem.*, **84,** 194 (1911). For a comprehensive review see Angeli, *Ahren's Sammlung*, **17,** 311–332 (1912).

[145] Angeli and Marchetti, *Atti accad. Lincei*, [5] **16,** i, 381 (1907).

VI.[146,147] In support of the structure IV is the fact that these indoles (II) condense with nitrosobenzene and with *p*-nitrosodimethylaniline to give the derivatives:

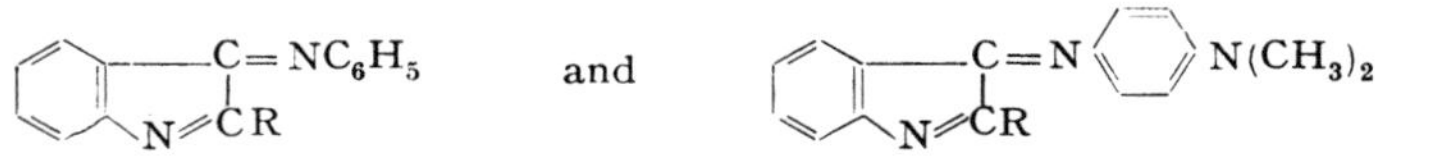

$C{=}NC_6H_5$, $N{=}CR$ and $C{=}N{-}C_6H_4{-}N(CH_3)_2$, $N{=}CR$

In this connection the suggestion has been made[148] that 2-phenylindole be used as a reagent for the identification of nitrosobenzene and its derivatives. The formula IV for the nitroso derivatives of the indoles has been supported by the work of a number of investigators.[149]

1-Hydroxy-2-phenylindole condenses with amyl nitrite[150] to give a compound VII or VIII identical with that obtained from phenylisatogen and hydroxylamine.[151] This derivative (VII) or (VIII) gives acetyl, benzoyl,

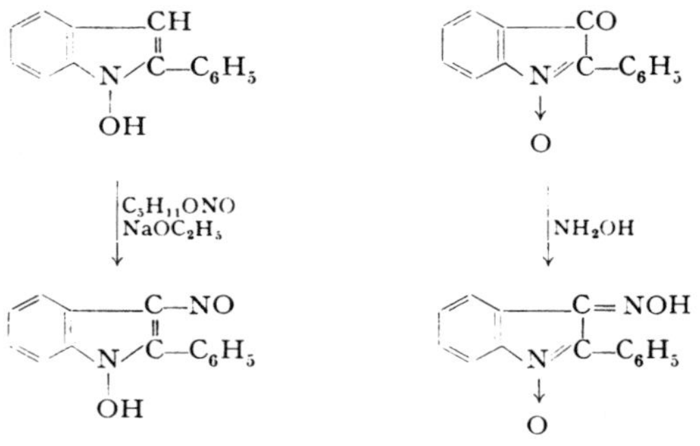

CH, $N{-}C{-}C_6H_5$, OH → ($C_5H_{11}ONO$, $NaOC_2H_5$) → $C{-}NO$, $N{-}C{-}C_6H_5$, OH (VII)

CO, $N{=}C{-}C_6H_5$, $\downarrow O$ → (NH_2OH) → $C{=}NOH$, $N{=}C{-}C_6H_5$, $\downarrow O$ (VIII)

and ethyl derivatives and is reduced by zinc and acetic acid to 3-amino-2-phenylindole.[152]

[146] Zatti and Ferratini, *Ber.*, **23,** 2299 (1890); *Gazz. chim. ital.*, **21,** ii, 19 (1891).

[147] Nencki, *Ber.*, **7,** 1597 (1874); **8,** 337, 772, 1519 (1875). Schmitz-Dumont, Hamann, and Geller, *Ann.*, **504,** 1 (1933). Seidel, *Ber.*, **B 77,** 797 (1944). La Parola, *Gazz. chim. ital.*, **75,** 157 (1945).

[148] Levy and Campbell, *J. Chem. Soc.*, **1939,** 1442.

[149] Verkade, Lieste, and Werner, *Rec. trav. chim.*, **64,** 289 (1945). Mohlau, *B er.*, **15,** 2487 (1882); **18,** 166 (1885). Campbell and Cooper, *J. Chem. Soc.*, **1935,** 1208. Illari and Cattalori, *Gazz. chim. ital.*, **69,** 31 (1939).

[150] Angeli and Angelico, *Atti accad. Lincei*, [5], **13,** i, 255 (1904).

[151] Pfeiffer, Braude, Fritsch, Halberstadt, Kirshoff, Kleber, and Witkop, *Ann.*, **411,** 72 (1916). Ajello, *Gazz. chim. ital.*, **68,** 646 (1939); **70,** 401 (1940).

[152] Angeli and Angelico, *Atti accad. Lincei*, [5], **15,** ii, 761 (1906).

The 2-substituted indoles react with benzenediazonium chloride to yield the corresponding phenylhydrazones.[153]

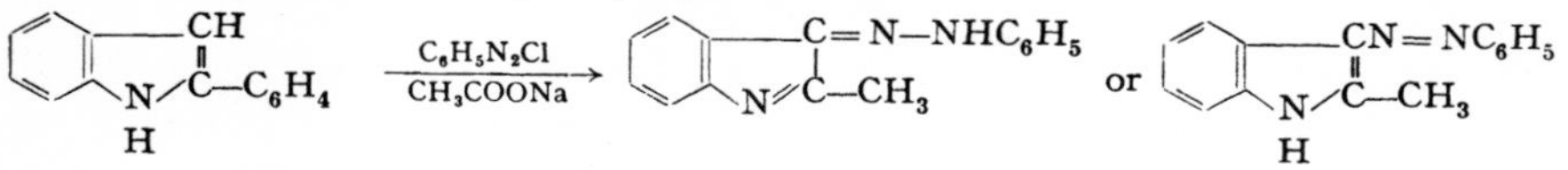

3-Nitroindole, m.p. 210°, cannot be prepared by direct nitration but can be prepared through condensation of ethyl nitrate with indole in the presence of sodium ethoxide.[154] 3-Nitromethylketole is obtained similarly from methylketole or by the oxidation of the nitroso derivative.[155] A number of other 3-nitroindole derivatives have been prepared similarly.[156] Thus 2-phenylindole in acetic acid treated first with sodium nitrite and then with an excess of concentrated nitric acid yields a dinitroderivative[156] which has been shown to be 3,5-dinitro-2-phenylindole[157] through oxidation to 5-nitro-*N*-benzoylanthranilic acid. This nitration of the indole nucleus in position 5 parallels the nitration of 2,3-dimethylindole which yields the same nitroindole obtained through ring closure from the *p*-nitrophenylhydrazone of

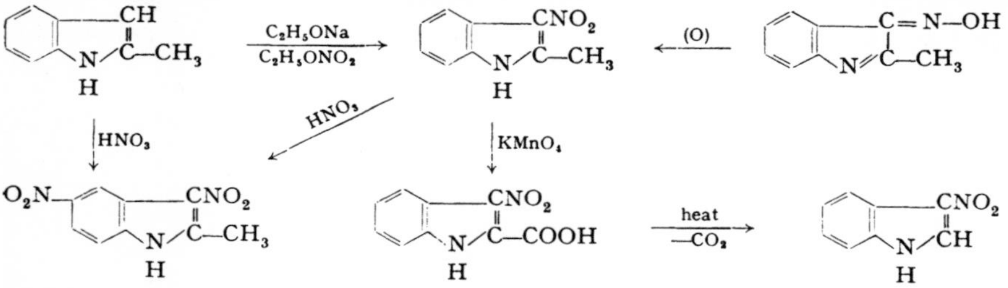

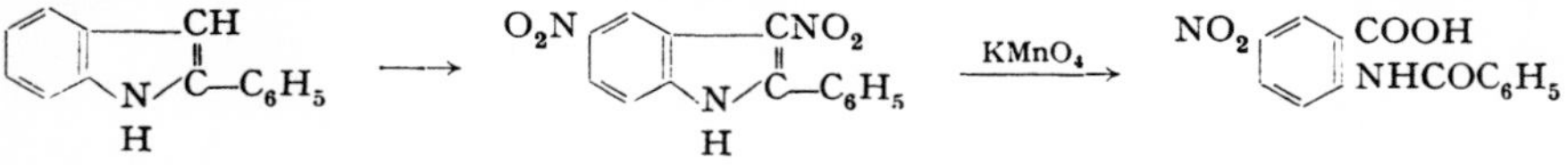

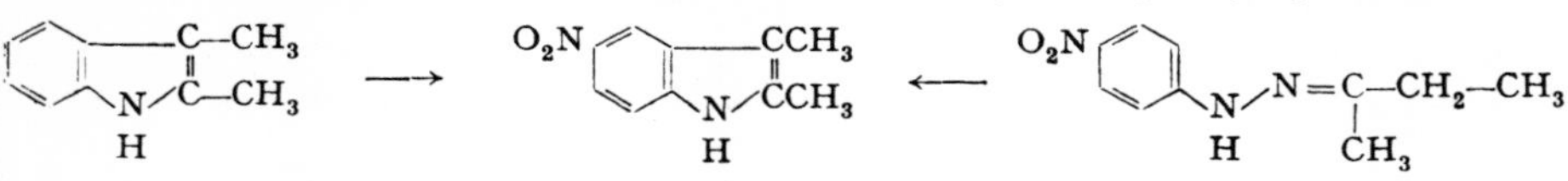

m.p. 188°

[153] Plancher and Soncini, *Gazz. chim. ital.*, **32,** ii, 436 (1902). Wagner, *Ann.*, **242,** 384 (1887).

[154] Angelico and Velardi, *Atti accad. Lincei*, [5], **13,** i, 241, (1904).

[155] Zatti, *Gazz. chim. ital.*, **19,** 260 (1889). Walther and Clemen, *J. prakt. Chem.*, [2] **61,** 268 (1900).

[156] Angeli and Angelico, *Gazz. chim. ital.*, **30,** ii, 268 (1900). Hsinmin and Mann, *J. Chem. Soc.*, **1949,** 2903.

[157] Womack, Campbell, and Dodds, *ibid.*, **1938,** 1402.

methyl ethyl ketone.[158] On the other hand, the nitration of 1-acetyl-2,3-dimethylindole with nitric acid in acetic acid solution gives a nitro derivative,[158] m.p. 170°, which on hydrolysis yields 6-nitro-2,3-dimethylindole, m.p. 142°, identical with one of the products obtained from the *m*-nitrophenylhydrazone of methyl ethyl ketone.[159] The nitration of 1-acetyl-2,3-diphenylindole takes place similarly in position 6.[160] Both dihydroskatole and benzoyldihydromethylketole have been found to nitrate in position 6.[161] The orientation observed here is in line with that found in the carbazole series where carbazole is nitrated in position 3, tetrahydrocarbazole in position 6 (corresponding to 5 in indole) and the *N*-acyl-tetrahydrocarbazoles in position 7 (corresponding to position 6 in indole). Ethyl indole-3-car-

$C—COOC_2H_5$ / $N—CH$ / H $\xrightarrow{HNO_3}$ O_2N $C—COOC_2H_5$ / $N—CH$ / H

O_2N CH / $N—CH$ / H $\xleftarrow{-CO_2}$ O_2N $C—COOH$ / $N—C—H$ / H $\xrightarrow{KMnO_4}$ O_2N COOH / NH_2

boxylate has been reported to undergo nitration in position 6.[162] Other indole derivatives have been found to undergo nitration in positions 5 and 7.[163]

Sulfonation of Indole

Sulfonation of indole with pyridine-sulfur trioxide complex at temperatures below 50° gives the unstable indole-1-sulfonic acid which is easily hydrolyzed giving indole. Sulfonation with the same reagent at higher

[158] Bauer and Straus, *Ber.*, **65**, 308 (1932). Plant and Tomlinson, *J. Chem. Soc.*, **1933**, 955.
[159] Schofield and Theobold, *J. Chem. Soc.*, **1949**, 796; **1950**, 1505. Plant and Whitaker, *ibid.*, **1940**, 283–286.
[160] Fennel and Plant, *ibid.*, **1932**, 2872.
[161] v. Braun, Gabowski, and Rowicz, *Ber.*, **46**, 3169 (1913).
[162] Majima and Kotake, *ibid.*, **B 63**, 2237 (1930).
[163] Menon, Perkin, and Robinson, *J. Chem. Soc.* ,**1932**, 780. Malthur and Robinson, *ibid.*, **1934**, 1415. Hill and Robinson, *ibid.*, **1933**, 486.

temperature (120°) gives indole-2-sulfonic acid. Skatole similarly gives 3-methylindole-2-sulfonic acid while 2-methylindole is not sulfonated.[164]

Halogen Derivatives of Indole

The introduction of an atom of halogen into the pyrrole ring of the indole molecule takes place especially easily through the action of the appropriate halogen on an *N*-acylindole in carbon disulfide. The 3-halogen-*N*-acyl derivatives are easily hydrolyzed to the corresponding 3-halogenindole.[165] 3-Chloroindole, m.p. 91.5°, is obtained also through the action of sulfuryl chloride on indole. The use of two molecular proportions of sulfuryl chloride yields 2,3-dichloroindole, m.p. 103–104°, which has also been prepared by the action of phosphorus pentachloride on either oxindole or dioxindole.[166] 3-Bromoindole, m.p. 67° (dec.), was prepared as indicated above. *N*-Bromosuccinimide has been employed as the brominating agent in converting 1-benzoylindole into 1-benzoyl-3-bromoindole, m.p. 97°,[167] while *N*-bromophthalimide has been used to convert skatole into a dibromoskatole, m.p. 100–102°.[168] Dioxane dibromide has been employed to convert indole into 3-bromoindole.[169] The action of iodine on indole and on 2-methylindole gives the corresponding 3-iodo derivatives.[170] Treatment of 1-acetyl-2,3-

$$\text{(C—}C_5H_5\text{; C—}C_6H_5\text{; N—}COCH_3\text{)} \xrightarrow[\text{in } CH_3COOH \text{ or } CS_2]{Br_2} \text{Br (C—}C_6H_5\text{; C—}C_6H_5\text{; N—}COCH_3\text{)}$$

[164] Terent'ev and Golubeva, *Compt. rend. acad. sci. U. R. S. S.*, **51**, 689 (1946); *Chem. Abstr.*, **41**, 2023 (1947). Terent'ev and Tsymbal, *Compt. rend. acad. sci. U. R. S. S.*, **55**, 833 (1947); *Chem. Abstr.*, **42**, 558 (1948). Terent'ev and Vestnik, *Moskov. Univ.*, **1947**, No. 6, 9–32; *Chem. Abstr.*, **44**, 1480 (1950). Terent'ev, Golubeva, and Tsymbal, *J. Gen. Chem.*, **19**, 763, 781 (1949); *Chem. Abstr.*, **44**, 1095 (1950). Terent'ev and Yanovskaya, *J. Gen. Chem. U. S. S. R.*, **21**, 1295 (1951); *Chem. Abstr.*, **46**, 2048 (1952).

[165] Mazarra and Borgo, *Gazz. chim. ital.*, **35**, ii, 100, 326, 563 (1905). Weissgerber, *Ber.*, **46**, 652 (1913).

[166] Baeyer, *Ber.*, **12**, 457 (1879); **15**, 786 (1882).

[167] Buu-Hoi, *Ann.*, **556**, 1 (1944).

[168] Putokhin, *J. Gen. Chem. (U. S. S. R.)*, **15**, 332 (1945); *Chem. Abstr.*, **40**, 3741 (1946). Borodkin, *J. Applied Chem. U.S.S.R.*, **23**, 1173, (1950); *Chem. Abstr.*, **46**, 10149 (1952).

[169] Yanovskaya, *Doklady Akad. Nauk S. S. S. R.*, **71**, 693–695 (1950); *Chem. Abstr.*, **44**, 8354 (1950).

[170] Oswald, *Z. physiol. Chem.*, **60**, 289 (1909); **73**, 128 (1911). Weissgerber, *Ber.*, **46**, 654 (1913).

diphenylindole with bromine gives 1-acetyl-2,3-diphenyl-6-bromoindole.[171] The treatment of 2,3-dimethylindole with bromine yielded quite different products as is shown in the following scheme:[171]

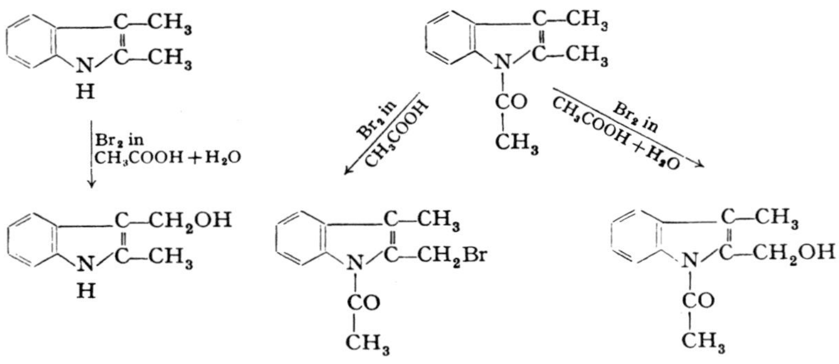

The bromination of ethyl indole-3-carboxylate has been reported to yield either the 6-bromo- or the 5,6-dibromo derivative depending on the conditions of bromination.[172] Reaction of indole with mercuric acetate gives a

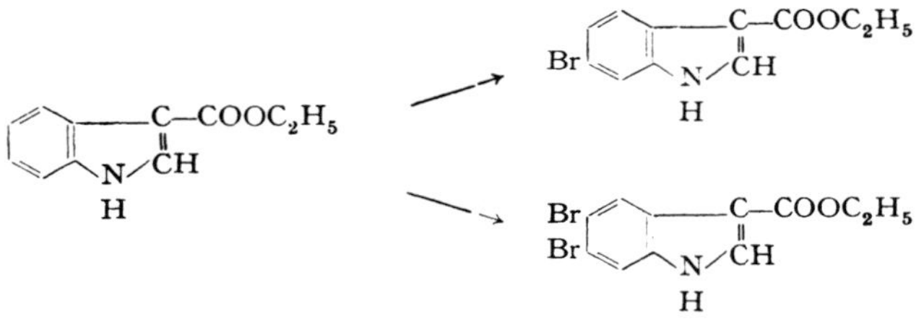

derivative which can be converted conveniently into 2,3-diiodoindole, m.p. 220° (dec.). Similarly 2-iodoskatole, m.p. 197–198° and 3-iodo-2-methylindole have been prepared from skatole and methylketole, respectively.[173]

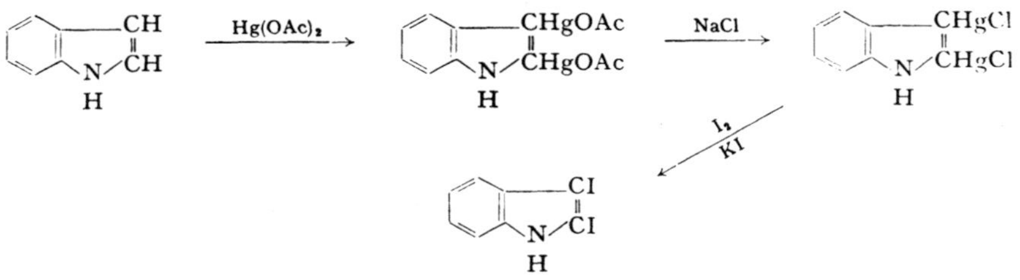

[171] Plant and Tomlinson, *J. Chem. Soc.*, **1933**, 955. Koelsch, *J. Am. Chem. Soc.*, **66**, 1983 (1944). Compare, however, Plant and Thompson, *J. Chem. Soc.*, **1950**, 1065.

[172] Majima and Kotake, *Ber.*, **B 63**, 2237 (1930).

[173] Mingoia, *Gazz. chim. ital.*, **60**, 509 (1930).

4-Chloroindole has been prepared through decarboxylation of 4-chloroindole-2-carboxylic acid which was in turn prepared through reduction of 2-nitro-6-chlorophenylpyruvic acid.[174]

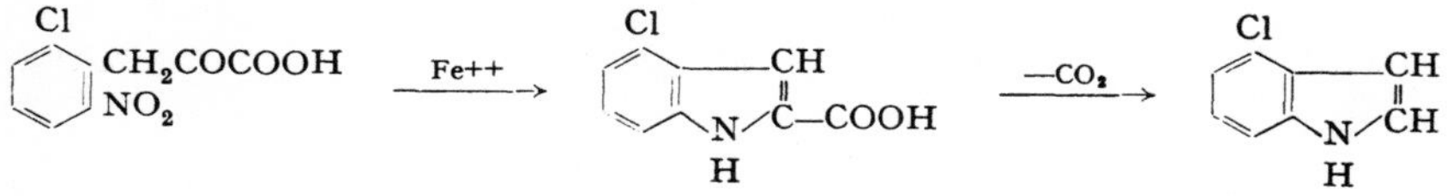

Cl $CH_2COCOOH$ NO_2 $\xrightarrow{Fe^{++}}$ Cl CH C—COOH N H $\xrightarrow{-CO_2}$ Cl CH CH N H

5- and 7-Chloroindole have been prepared through cyclization of the *p*- and *o*-chlorophenylhydrazones, respectively, of ethyl pyruvate, followed by decarboxylation of the resulting chloroindole-2-carboxylic acids. 6-Chloroindole has been prepared as shown in the following scheme.[175]

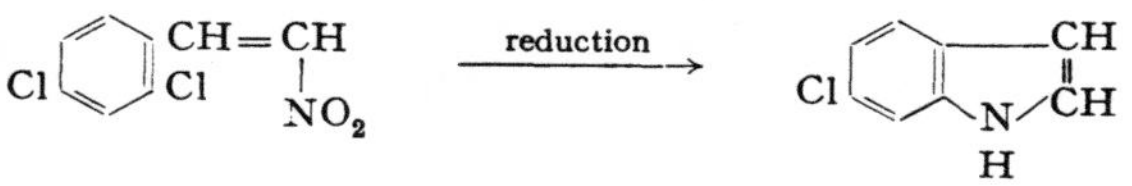

Cl CH=CH Cl NO_2 $\xrightarrow{\text{reduction}}$ Cl CH CH N H

The action of arsenic acid on members of the indole series in aqueous or organic solvents yields the corresponding β-indole arsonic acids.[176] Only the hydrogen atom in position 3 is replaced by the action of arsenic acid. The trimer of indole was obtained as a by-product in the arsonation of indole. 3-Indolyl-azo-*p*-phenylarsonic acid was prepared by coupling diazotized

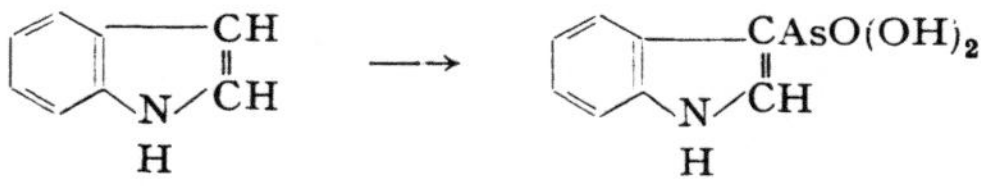

CH CH N H $\longrightarrow$ $CAsO(OH)_2$ CH N H

atoxyl with indole.[177] Skatole and methylketole couple similarly. The corresponding antimony derivatives have been prepared similarly.

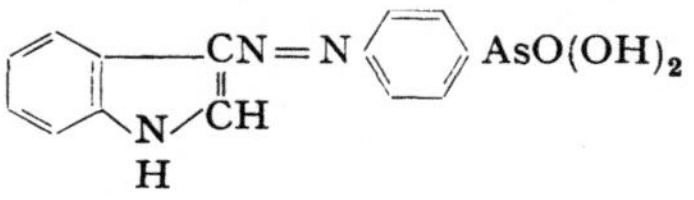

CN=N $AsO(OH)_2$ CH N H

Alkyl Derivatives of Indole

The alkylation of indole has been the subject of careful study. A complete survey of the literature to 1904 has been given by Ciamician.[178] The

[174] Uhle, *J. Am. Chem. Soc.*, **71**, 761 (1949).
[175] Rydon and Long, *Nature*, **164**, 575 (1949).
[176] Funakubo, *J. Chem. Soc. Japan*, **48**, 526 (1927); *Chem. Abstr.*, **22**, 1775 (1928).
[177] Mingoia, *Gazz. chim. ital.*, **60**, 124 (1930); *Chem. Abstr.*, **24**, 3781 (1930). Mingoia, *Gazz. chim. ital.*, **62**, 343 (1932).
[178] Ciamician, *Ber.*, **37**, 4227 (1904).

earliest alkylation studies were those conducted by Emil Fischer.[179] The somewhat complicated behavior of indoles in alkylation reactions was cleared up largely by the careful and thorough studies of Brunner[180] and of Plancher.[181] Through the action of methyl iodide on indole, 2-methylindole, 3-methylindole, or 1-methylindole the following changes are effected, it being difficult to stop the reaction at the first two stages:

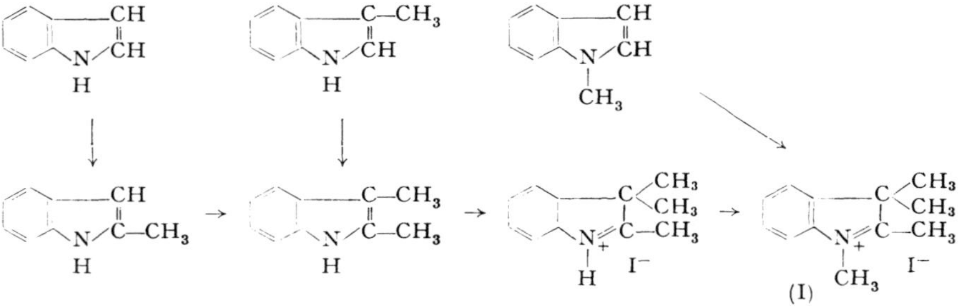

(I)

From the quaternary compound I a tertiary compound II is set free by the action of alkalis. This tertiary base II is the α-methylene derivative of 1,2,3-trimethyldihydroindole. The methylene group is active and can be alkylated

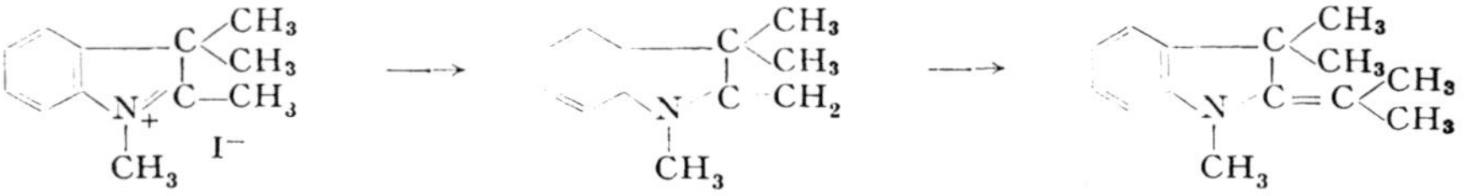

(I) (II) (III)

[179] Fischer and Meyer, *ibid.*, **23**, 2629 (1890). Fischer and Steche, *Ann.*, **242**, 348 (1887); *Ber.*, **20**, 818, 2199 (1887).

[180] Brunner, *Monatsh.*, **17**, 253, 479 (1896); **18**, 95, 527 (1897); **21**, 156 (1900); **27**, 1183 (1906).

[181] Ciamician and Plancher, *Ber.*, **29**, 2475 (1896). Plancher, *Gazz. chim. ital.*, **28**, ii, 374, 418 (1898); **30**, ii, 564 (1900); **31**, i, 280 (1901); **32**, ii, 398 (1902). Plancher, *Ber.*, **31**, 1488 (1898). Plancher, *Atti accad. Lincei*, [5] **11**, ii, 182 (1902).

[182] For other references to alkylation studies see: Ciamician and Zatti, *Ber.*, **22**, 1980 (1899). Zatti and Ferratini, *ibid.*, **23**, 2302 (1890); *Gazz. chim. ital.*, **21**, ii, 309 (1891); *Atti accad. Lincei*, [5] **6**, i, 463 (1897). Ciamician and Boeris, *Gazz. chim. ital.*, **24**, ii, 299 (1894); **27**, i, 77 (1897); *Ber.*, **29**, 2472 (1896). Ciamician, *ibid.*, **29**, 2460 (1896). Ciamician and Piccinini, *ibid.*, **29**, 2465 (1896). Plancher, *Chem. Ztg.*, **32**, 37 (1898); *Gazz. chim. ital.*, **28**, ii, 333, 374, 418 (1898); **31**, i, 280 (1901); **32**, ii, 398 (1902); **31**, 1488 (1898). Brunner, *ibid.*, **31**, 612, 1943 (1898); *Gazz. chim. ital.*. **31**, i, 181 (1901). Piccinini, *ibid.*, **28**, i, 187 (1898); **28**, ii, 40, 51, 87 (1898). Plancher and Bonavia, *ibid.*, **32**, ii, 414 (1902). Plancher and Carrasco, *Atti accad. Lincei*, [5] **14**, ii, 31 (1905). Hantzsch and Horn, *Ber.*, **35**, 883 (1902). Decker and Klauser, *ibid.*, **37**, 524 (1904).

further.[182] Compound III can be prepared also as indicated in the following scheme:[183]

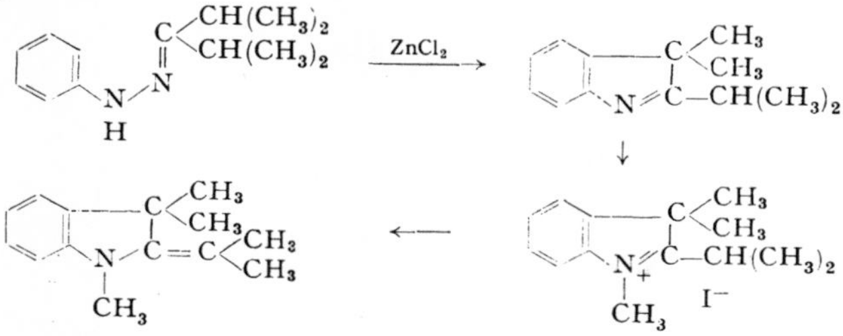

$$C_6H_5NH{-}N{=}C[CH(CH_3)_2]_2 \xrightarrow{ZnCl_2} \text{2-isopropyl-3,3-dimethylindolenine} \rightarrow \text{methiodide} \rightarrow \text{(III)}$$

(III)

The quaternary compound (I) is also obtained in good yield when the methylphenylhydrazone of methyl isopropyl ketone is allowed to stand for many days at room temperature with alcoholic hydrogen iodide. On heating, the compound I loses methyl iodide yielding 1,2,3-trimethylindole which

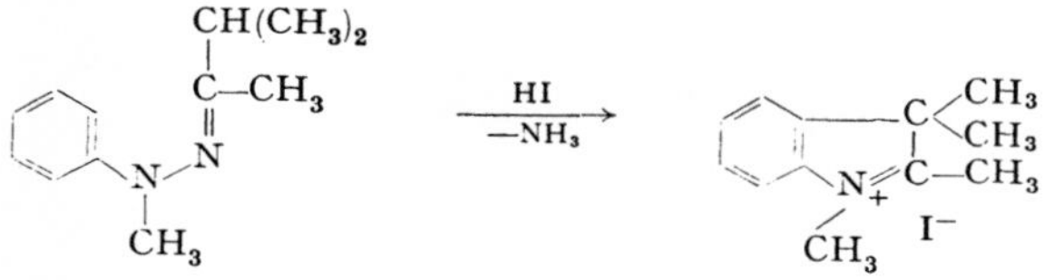

$$C_6H_5N(CH_3){-}N{=}C(CH_3)CH(CH_3)_2 \xrightarrow[-NH_3]{HI} \text{(I)}$$

can combine reversibly with methyl iodide to yield I again.

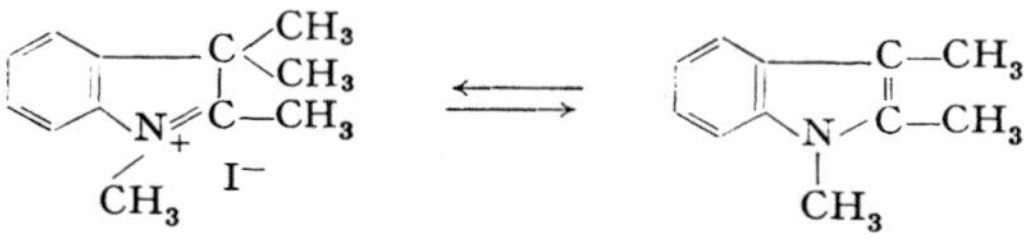

An interesting molecular rearrangement takes place when the methiodide of 2-ethyl-3,3-dimethylindolenine is heated to melting, the ethyl group migrating to position 3, changing places with a methyl group. Similarly

2-isopropyl-3,3-dimethylindolenine rearranges with the isopropyl group exchanging positions with a methyl group.[184] A reverse migration is observed

[183] Plancher, *Ber.*, **31,** 1496 (1898). Plancher and Bettinelli, *Gazz. chim. ital.*, **29,** i, 106 (1899). Plancher and Ravenna, *Atti accad. Lincei,* [5] **15,** ii, 557 (1905). Plancher and Bettinelli, *ibid.*, **7,** i, 367 (1898).

[184] Plancher, *Atti accad. Lincei,* [5] **9,** i, 115 (1900); [5] **11,** ii, 182 (1902).

when 3-phenylindole is heated with zinc chloride, the phenyl group migrating to position 2.[185] Other molecular rearrangements have been observed in the indole series[186] including the following reactions:

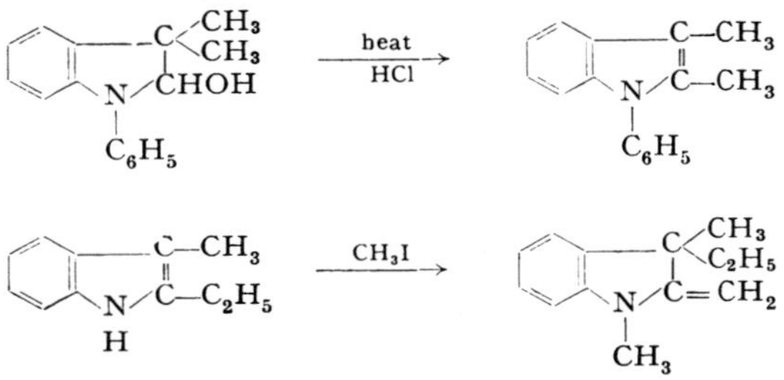

C CH_3 CH_3 N CHOH C_6H_5 — heat, HCl → C—CH_3 N C—CH_3 C_6H_5

C—CH_3 N C—C_2H_5 H — CH_3I → C CH_3 C_2H_5 N C=CH_2 CH_3

The tertiary base II on oxidation with potassium permanganate or chromic anhydride gives 1,3,3-trimethyloxindole which is also obtained through the oxidation of III (page 33).

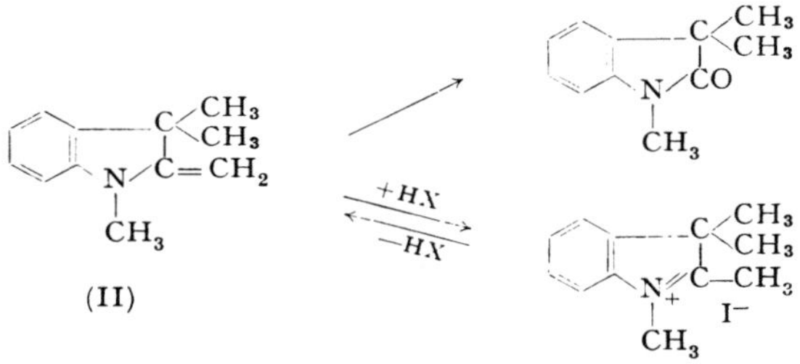

C CH_3 CH_3 N C=CH_2 CH_3 (II) → C CH_3 CH_3 N CO CH_3

+HX / −HX ⇄ C CH_3 CH_3 N+ C—CH_3 I^- CH_3

The analog of II without a substituent group on N (IV) can be prepared through the Fischer synthesis from the phenylhydrazone of methyl isopropyl ketone. Alternative formula (V) must also be taken into account for this substance. With potassium nitrite and acetic acid the oxime (VI) is obtained.[187]

C CH_3 CH_3 N C=CH_2 H (IV) C CH_3 CH_3 N C—CH_3 (V)

The formation of the oxime along with the stability of the compound against cold permanganate renders the formula V more acceptable than IV. Dehydration of the oxime (VI) yields the nitrile (VII). 2-Methyl-3,3-di-

[185] Fischer and Schmidt, *Ber.*, **21**, 1071, 1811 (1888). Ince, *Ann.*, **253**, 35 (1889).

[186] Brunner, *Monatsh.*, **17**, 253 (1896); **21**, 156 (1900). Knorr, *Ber.*, **36**, 1272 (1903). Plancher and Bonavia, *Gazz. chim. ital.*, **32**, ii, 414 (1902).

[187] Plancher, *Gazz. chim. ital.*, **28**, ii, 405 (1898). Plancher and Bettinelli, *ibid.*, **29**, i, 108 (1899).

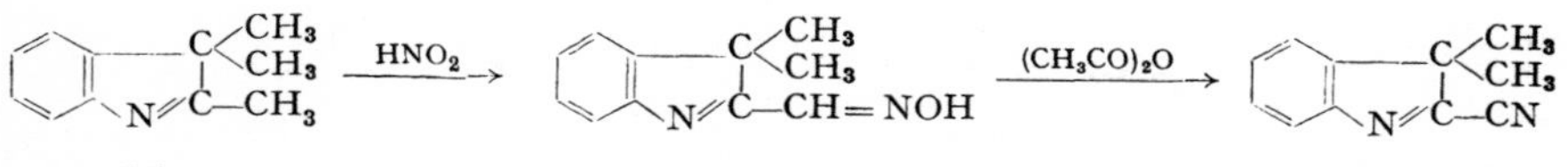

$$\text{(V)} \xrightarrow{HNO_2} \text{(VI)} \xrightarrow{(CH_3CO)_2O} \text{(VII)}$$

benzylindolenine behaves toward nitrous acid in the same manner as does compound V.[188] The methiodide of compound II has been found to couple with benzenediazonium chloride to give the phenylhydrazone.[189] The nitrile (VII) can be converted into the ketone (VIII) through interaction with methylmagnesium iodide.[190] Treatment with phenylmagnesium bromide gives the phenylketone in similar fashion.

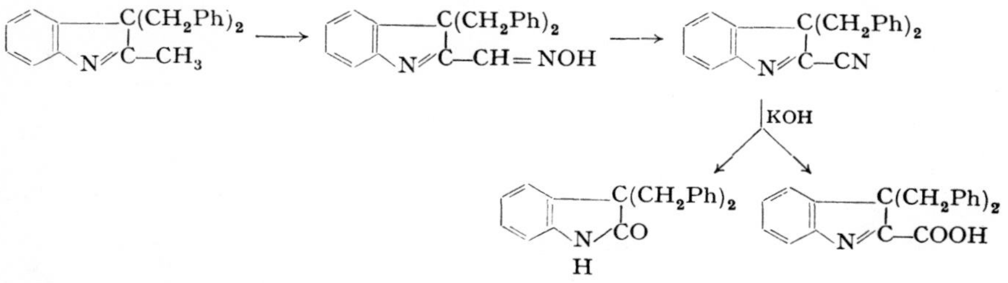

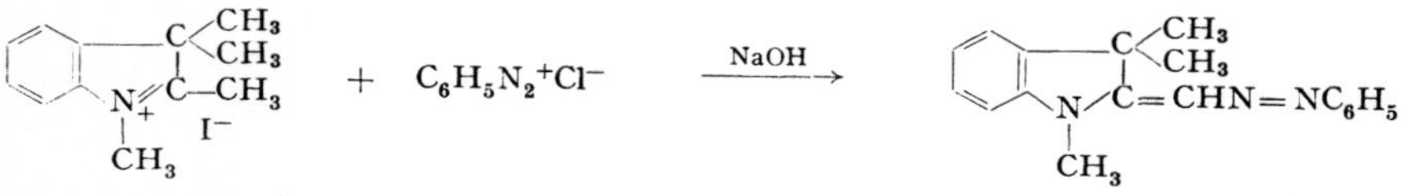

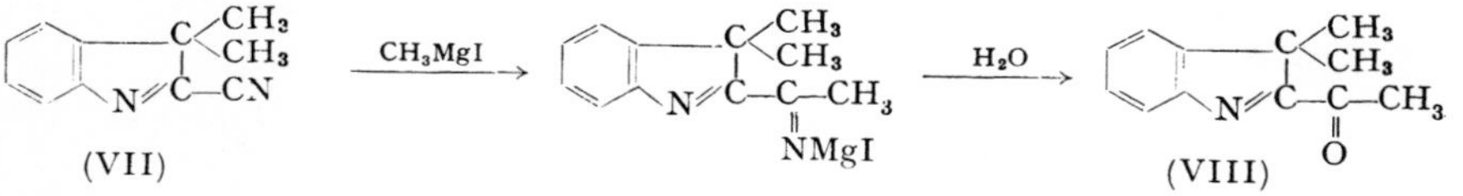

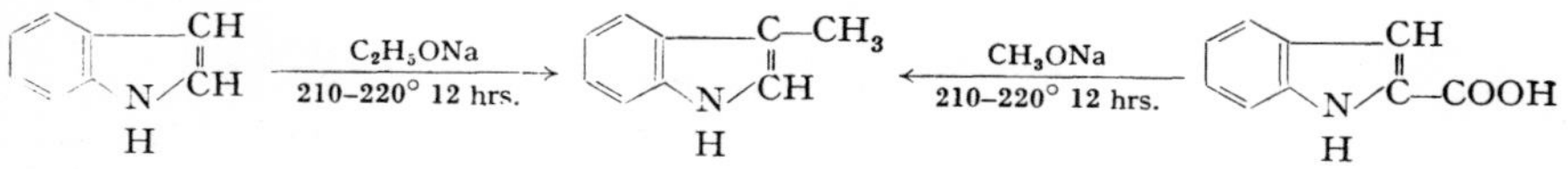

$$\text{(VII)} \xrightarrow{CH_3MgI} \xrightarrow{H_2O} \text{(VIII)}$$

Indole and various indole derivatives are alkylated by means of heating with sodium alkoxides or with the alkoxides of other alkali or alkaline earth metals.[191] While the 2-alkylindoles are similarly alkylated in position

[188] Leuches and Overberg, *Ber.*, **B 64,** 1896 (1931).

[189] König and Müller, *ibid.*, **B 57,** 144 (1924). König, *ibid.*, **B 57,** 891 (1924). Rosenhauer, *ibid.*, **B 57,** 1192 (1924). Ghigi, *Gazz. chim. ital.*, **63,** 791 (1933); *Chem. Abstr.*, **28,** 2354 (1934).

[190] Plancher and Guimelli, *Atti accad. Lincei*, **18,** ii, 393 (1909).

[191] Cornforth and Robinson, *J. Chem. Soc.*, **1942,** 680. British patent 550,060.

3 those indoles already substituted in position 3 are not alkylated in this manner.[192] 3-Acetylindole and 2-methyl-3-acetylindole exchange the acetyl group for alkyl when heated with the alcohol and the appropriate sodium alkoxide. Those indole ketones with the acyl group in position 2 do not undergo this reaction.[193] It has also been reported that indole and its simple derivatives are alkylated by treatment with an alcohol in the presence of hydrochloric, hydrobromic, or an arylsulfonic acid.[194] 1-Monoalkylindoles can be prepared by alkylating indole with an alkyl sulfate and alkali.[195] 1-[2-(1-Pyrrolidyl)ethyl]indole has been prepared through the action of 1-[(1-pyrrolidyl)ethyl] chloride on the sodium salt of indole.[196]

1-Methylgramine methiodide reacts with methylmagnesium iodide and with phenylmagnesium iodide to give 1-methyl-3-ethylindole and 1-methyl-

$$\text{1-methylindole-3-}CH_2\overset{+}{N}(CH_3)_3\ I^- \xrightarrow{RMgBr} \text{1-methylindole-3-}CH_2R$$

3-benzylindole. Gramine methiodide similarly reacts with Grignard reagents to give 3-alkylindoles.[197]

Indoline

The dihydro derivatives of the indoles are called indolines. The indolines are obtained from the corresponding indoles by reduction with zinc dust or with tin and hydrochloric acid or through electrolytic reduction in acid media.[198] The yield is lessened because of polymerization of the indole in the

[192] U. S. patent 2,407,452; *Chem. Abstr.*, **41**, 488 (1947). Oddo and Alberti, *Gazz. chim. ital.*, **63**, 236 (1933).

[193] Alberti, *Gazz. chim. ital.*, **67**, 238 (1937); **69**, 568 (1934).

[194] British patent 341,554; *Chem. Abstr.*, **25**, 4894 (1931). French patent 703,838; *Chem. Abstr.*, **25**, 4558 (1931). German patent 534,552; *Chem. Abstr.*, **26**, 995 (1932).

[195] U. S. patent 2,460,745; *Chem. Abstr.*, **43**, 4302 (1949).

[196] Wright, *J. Am. Chem. Soc.*, **71**, 1028 (1949).

[197] Snyder, Eliel, and Carnahan, *ibid.*, **73**, 970 (1951).

[198] v. Braun, *Ber.*, **37**, 2915, 3210 (1904). v. Braun, and Sobecki, *ibid.*, **44**, 2158 (1911). Jackson, *ibid.*, **14**, 883 (1881). Fischer, *Ann.*, **236**, 123 (1886). Carrasco, *Gazz. chim. ital.*, **38**, ii, 301 (1908). Carrasco and Padoa, *ibid.*, **36**, ii, 512 (1906). Plancher and Ravenna, *Atti accad. Lincei*, [5] **14**, i, 632 (1905). Benedicenti, *Z. physiol. Chem.*, **53**, 181 (1907). Wenzing, *Ann.*, **239**, 239 (1887). Hayashi, *Sci. Papers Inst. Phys. Chem. Research Tokyo*, **39**, 94 (1941); *Chem. Abstr.*, **41**, 6561 (1947).

$$\text{(1-methylindole: CH=CH, N–CH}_3\text{)} \xrightarrow{2\,(H)} \text{(1-methylindoline: CH}_2\text{–CH}_2\text{, N–CH}_3\text{)}$$

presence of acid. Indoline is a colorless stable liquid, b.p. 230°, heavier than water, $d_4^{20} = 1.069$, slightly soluble in water. The *N*-nitroso derivative melts at 83–84° and the benzenesulfonyl derivative at 133°.

The indolines are also obtained by the catalytic reduction of the indoles. The catalysts most commonly used are platinum, nickel, nickel salts, and copper and copper salts. The indolines on distillation with silver sulfate are oxidized to the corresponding indoles.[199] Catalytic dehydrogenation of indoline gives indole.[200] Dehydrogenation of octahydro-2-methylindole with selenium at 310–335° for fifteen hours gives 2-methylindole.[201] While catalytic hydrogenation of indole gives first the dihydroindole the absorption of hydrogen may continue yielding either an octahydroindole or through rupture of the pyrrole ring an aniline derivative.[202] With a nickel catalyst at 225° hydrogenation past the dihydrostage gives 1-ethyl-2-aminocyclohexane[203] which had previously been thought to be octahydroindole.[204] Octahydroindole has been prepared through hydrogenation of indole using a platinum catalyst.[205] Under proper conditions 2-methylindole, skatole, and 1,2-dimethylindole can be catalytically reduced to the octahydro derivatives.[206] Dihydroindole is obtained in good yields through the use of copper catalysts.[207] Treatment of melted indole with hydrogen in the presence of reduced nickel at 200° gives *o*-toluidine.[208] Catalytic reduction of indoxyl gives dihydroindole.[209]

[199] Kann and Tafel, *Ber.*, **27**, 826 (1894).
[200] Akabori and Suzuki, *Proc. Imp. Acad. (Japan)*, **5**, 255 (1929).
[201] Sin-itero, Huzise, and Kozoteba, *Bull. Chem. Soc. Japan*, **14**, 478 (1939); *Chem. Abstr.*, **34**, 1987 (1940).
[202] King, Bartrop, and Walley, *J. Chem. Soc.*, **1945**, 277.
[203] Willstätter, Seitz, and v. Braun, *Ber.*, *B* **58**, 385 (1925).
[204] v. Braun, Baeyer, and Blessing, *ibid.*, *B* **57**, 392 (1924).
[205] Willstätter, and Jaquet, *ibid.*, **51**, 767 (1918).
[206] v. Braun and Baeyer, *ibid.*, *B* **58**, 387 (1925).
[207] German patent 623,693; *Chem. Abstr.*, **30**, 4874 (1936). British patent 44,631; *Chem. Abstr.*, **30**, 5592 (1936). French patent 792,064; *Chem. Abstr.*, **30**, 4181 (1936).
[208] Carrasco and Padoa, *Gazz. chim. ital.*, **36**, ii, 512 (1906); *Atti accad. Lincei*, [5] **15**, i, 699 (1906).
[209] Schumann, Muench, and Christ, U. S. patent 1,891,057; *Chem. Abstr.*, **27**, 1892 (1933).

Indoline has been synthesized from *o*-aminophenyl-*β*-ethanol through conversion to the chlorine and subsequent ring closure.[210]

$$C_6H_4(NH_2)CH_2CH_2OH \xrightarrow{HCl} C_6H_4(NH_2)CH_2CH_2Cl \xrightarrow{NaOH} C_6H_4\langle CH_2CH_2NH\rangle$$

An elaborate study of the catalytic hydrogenation of indoles and carbazoles was conducted by Adkins and Coonradt.[211] In general they found that copper chromite catalysts gave an equilibrium between indole and hydrogen and dihydroindole.[212] On the other hand, Raney nickel gave octahydro derivatives. The catalytic hydrogenation of indoles has also been studied by several other investigators.[213] Indoline has been prepared through the electrolytic reduction of thio-oxindole[214] which was prepared in turn from oxindole through the agency of phosphorus trisulfide.

The indolines are typical secondary amines and undergo reactions typical of these compounds.[215] Nitrous acid converts indoline into the *N*-nitroso derivative which on reduction with zinc and acetic acid gives the corresponding hydrazine[211,216] and which rearranges under the influence of alcoholic hydrogen chloride, the nitroso group entering the para position in the benzene ring.[217] The analogy with aromatic secondary amines is

$$\text{N-NH}_2\text{ (2-methylindoline)} \xleftarrow{\text{reduction}} \text{N-NO (2-methylindoline)} \xrightarrow{HCl} \text{ON-(2-methylindoline, N-H)}$$

further shown in the reaction of the indolines with benzenediazonium chloride to the diazoamino derivatives which rearrange in the presence of mineral acids.[217] The indoles give *N*-benzoyl derivatives in the Schotten-

[210] Ferber, *Ber.*, **B 62**, 183 (1929). Plancher, *Atti accad. Lincei*, [5] **14**, i, 632 (1905); *Ber.*, **B 62**, 1088 (1929). Bennett and Hafez, *J. Chem. Soc.*, **1941**, 287, 652.

[211] Adkins and Coonradt, *J. Am. Chem. Soc.*, **63**, 1563 (1941).

[212] Adkins and Burks, *ibid.*, **70**, 4174 (1948).

[213] Fujise, *Sci. Papers Inst. Phys. Chem. Research Tokyo*, **9**, 91 (1928); *Chem. Abstr.*, **23**, 144 (1929).

[214] Sugasawa, Satoda, and Yamagisawa, *Sci. Papers Inst. Phys. Chem. Research Tokyo*, **58**, 139 (1938); *Chem. Abstr.*, **32**, 4161 (1938). Japanese patent 132,391; *Chem. Abstr.*, **35**, 3270 (1941).

[215] Bamberger and Sternitzki, *Ber.*, **26**, 1291 (1893).

[216] Wenzing, *Ann.*, **239**, 239 (1887).

[217] Bamberger and Sternitzki, *Ber.*, **26**, 1291 (1883). Bamberger and Zumbo, *ibid.*, **26**, 1285 (1893).

$C_6H_5N{=}N$ on N of 2-methylindoline ($N{=}NC_6H_5$) $\longrightarrow$ $C_6H_5N{=}N$ on the ring of 2-methylindoline (N–H)

Baumann reaction. The opening of the dihydropyrrole ring in the benzoylindolines is effected by the action of phosphorus pentachloride[218] and also in certain cases by reduction.[219] The cleavage of the heterocyclic ring of dihydroindole is also effected by the action of cyanogen bromide.[220]

N-methylindoline (N–CH_3) $\xrightarrow{BrCN}$ o-(CH_2CH_2Br)–C_6H_4–$N(CH_3)CN$ (about 70%) and N-cyanoindoline (N–CN) (about 30%)

N-Methylindoline[221] has been prepared through the reduction of *N*-methylindole. It boils at 216° at 728 mm. and is slightly soluble in water. It is converted by hydriodic acid and phosphorus to indoline with removal of the methyl group. 2-Methylindoline has been the subject of many studies.[222] Distillation of 2-methylindoline hydrochloride with zinc dust gives 2-methylindole along with a little quinoline. 2-Methylindoline is converted by phosphorus and hydriodic acid into *o*-amino-*n*-propylbenzene. Since the compound contains an asymmetric carbon atom it has been obtained in optically active forms. The *N*-nitroso derivative melts at 55° and is reduced by zinc dust and acetic acid to the corresponding hydrazine, 1-amino-2-methylindoline; m.p. 40–41°.[216]

Reduction of indole with Raney nickel in presence of methanol gives methyloctahydroindole through reductive alkylation.[223]

[218] v. Braun, Grabowski, and Rawicz, *ibid.*, **46**, 3169 (1913). v. Braun and Steindorff, *ibid.*, **37**, 4581 (1904). v. Braun and Sobecki, *ibid.*, **44**, 2158 (1911). v. Braun and Korschmann, *ibid.*, **45**, 1263 (1912).

[219] v. Braun, Heider, and Neumann, *ibid.*, **49**, 2613 (1916). v. Braun, and Neumann, *ibid.*, **49**, 1283 (1916); **50**, 50 (1917).

[220] v. Braun, *ibid.*, **51**, 96 (1918).

[221] Wenzing, *Ann.*, **239**, 246 (1887). Plancher and Ravenna, *Atti accad. Lincei*, [5] **14**, i, 633 (1905). Benedicenti, *Z. physiol. Chem.*, **53**, 181 (1907). Carrasco, *Gazz. chim. ital.*, **38**, ii, 306 (1908).

[222] Jackson, *Ber.*, **14**, 833 (1881). Wenzing, *Ann.*, **239**, 244 (1887). Bamberger and Zumbo, *Ber.*, 26, 1285 (1893). Bamberger and Sternitzki, *ibid.*, **26**, 1292 (1893). Piccinini, *Gazz. chim. ital.*, **28**, ii, 58 (1898). Piccinini and Camozzi, *ibid.*, **28**, ii, 91 (1898). Stoemmer, *Ber.*, **31**, 2540 (1898). Pope and Clark, *J. Chem. Soc.*, **85**, 1330 (1904). v. Braun and Steindorf, *Ber.*, **37**, 4581, 4723 (1904). Carrasco, *Gazz. chim. ital.*, **38**, ii, 305 (1908). König, *J. prakt. Chem.*, [2] **88**, 218 (1913).

[223] Metayer, *Bull. soc. chim. France*, **1948**, 1093.

2,2-Dimethylindoline results from *o*-isopropylaminobenzyl alcohol through distillation with splitting out of water.[224] The compound, b.p. 210°, has a weak turpentine-like odor and is almost insoluble in water. 3,3-Dimethylindoline, m.p. 34–35°, has also been described.[225]

The preparation of an indoline through an internal Michael reaction has been described.[226]

$$O_2N\text{-}C_6H_3(CH{=}CHCOOCH_3)\text{-}N(CH_3)CH_2COOCH_3 \longrightarrow O_2N\text{-}C_6H_3\text{-}CHCH_2COOCH_3\text{-}CHCOOCH_3\text{-}N(CH_3)$$

Metal Salts of Indole

Like pyrrole, indole yields alkali metal salts.[227] The sodium salt is prepared by the addition of sodium or of sodium amalgam to molten indole, the potassium salt through the interaction of potassium hydroxide and indole at 125–130°. On treatment with methyl iodide the sodium salt of indole gives *N*-methylindole as the principal product along with some 2-methylindole and some skatole. The formation of the alkali metal salt is sometimes used to separate indole from the other constituents in the coal-tar fraction boiling between 240–260°.[228] The preparation of mercury salts of indole has been described.[229] Salts of several members of the indole series with stannous chloride, titanium chloride, aluminium chloride, and other metal salts have been described.[230]

Indole Aldehydes

Indole reacts with chloroform and alkali in a Reimer-Tiemann reaction to give indole-3-aldehyde[231, 232] along with β-chloroquinoline. Indole-3-

[224] Paal and Landenheimer, *Ber.*, **25**, 2974 (1892).

[225] Brunner, *Monatsh.*, **18**, 97 (1897). Ciamician and Piccinini, *Ber.*, **29**, 2740 (1896).

[226] Koelsch and Stephens, *J. Am. Chem. Soc.*, **72**, 2209 (1950).

[227] Weissgerber, *Ber.*, **43**, 3521 (1910).

[228] German patent 223,304; *Chem. Zentr.*, **1910**, ii, 349. German patent 515,543; *Chem. Abstr.*, **25**, 2443 (1931).

[229] German patent 236,893; *Chem. Abstr.*, **6**, 1600 (1912).

[230] Schmitz-Dumont and Motzkus, *Ber.*, **B 62**, 466 (1929). Franklin, *J. Phys. Chem.*, **24**, 81 (1920). Delavigne, *Gazz. chim. ital.*, **68**, 271 (1938).

[231] Ellinger, *Ber.*, **39**, 2515 (1906).

[232] Boyd and Robson, *Biochem. J.*, **29**, 555 (1935). Rudinov and Veselovskaya, *Zhur. Obshchei Khim.* (*J. Gen. Chem.*), **20**, 2202 (1950); *Chem. Abstr.*, **45**, 7106 (1951).

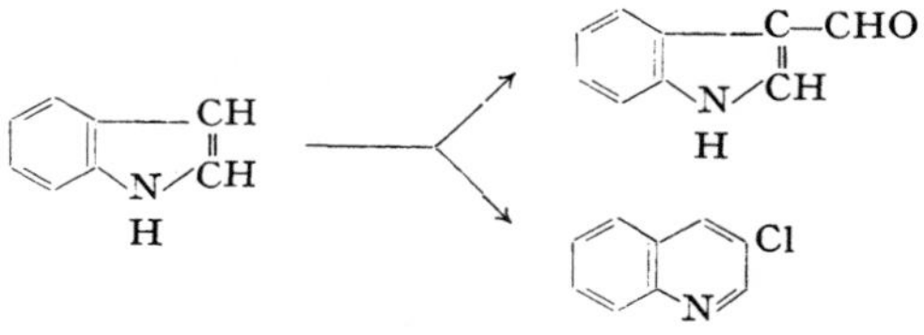

aldehyde has also been prepared in better yield through the action of zinc cyanide and hydrogen chloride on ethyl indole-2-carboxylate.[232] The

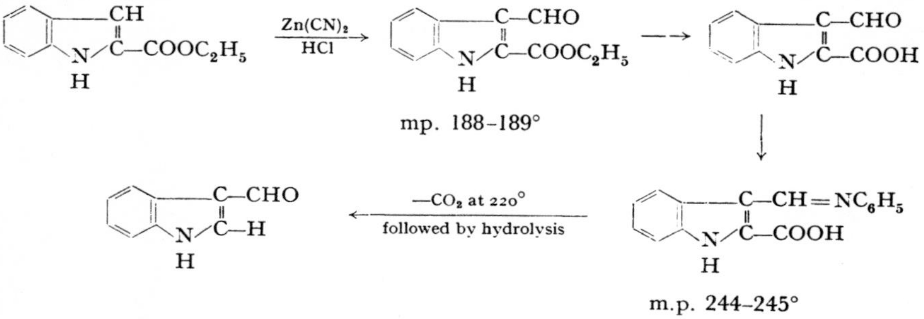

$\xrightarrow[HCl]{Zn(CN)_2}$

mp. 188–189°

$—CO_2$ at 220° followed by hydrolysis

m.p. 244–245°

compound was first prepared from tryptophan through oxidation with ferric chloride.[233] The compound was also prepared from indolylmagnesium iodide through the action of ethyl formate,[234] a reaction which also yields the 1-formyl derivative.[235] In similar fashion the 2-substituted indoles react with chloroform and alkali to give the corresponding indole-3-aldehydes and β-chloroquinolines.[236] The Adams modification of the Gattermann synthesis

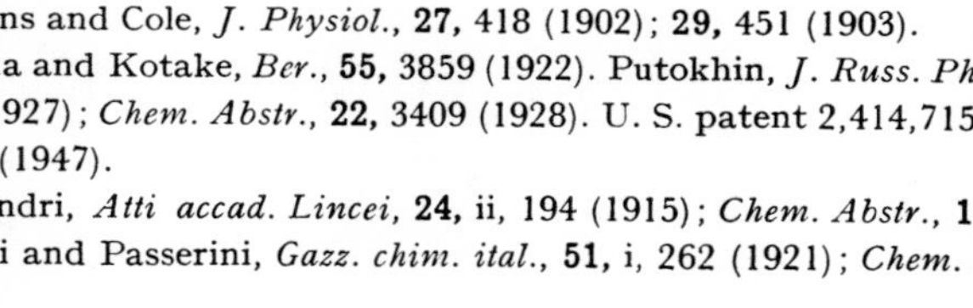

[233] Hopkins and Cole, *J. Physiol.*, **27**, 418 (1902); **29**, 451 (1903).

[234] Majima and Kotake, *Ber.*, **55**, 3859 (1922). Putokhin, *J. Russ. Phys.-Chem. Soc.*, **59**, 761 (1927); *Chem. Abstr.*, **22**, 3409 (1928). U. S. patent 2,414,715; *Chem. Abstr.*, **41**, 3129 (1947).

[235] Allesandri, *Atti accad. Lincei*, **24**, ii, 194 (1915); *Chem. Abstr.*, **10**, 1350 (1916). Allesandri and Passerini, *Gazz. chim. ital.*, **51**, i, 262 (1921); *Chem. Abstr.*, **16**, 94 (1922).

[236] Plancher and Ponti, *Atti accad. Lincei*, **16**, i, 130 (1907). Kermack, Perkin, and Robinson, *J. Chem. Soc.*, **121**, 1872 (1922). Blume and Lindwall, *J. Org. Chem.*, **10**, 255 (1945).

has also been used to prepare 2-methylindole-3-aldehyde.[237] The Houben-Hoesch synthesis has similarly been employed for the synthesis of indole

$$\text{2-methylindole (CH, C–CH}_3\text{, NH)} \xrightarrow[\text{CH}_3\text{CN}]{\text{HCl}} \text{3-acetyl-2-methylindole (C–COCH}_3\text{, C–CH}_3\text{, NH)}$$

ketones. 2-Methylindole condenses with amyl formate in the presence of sodium amoxide yielding the sodium salt of the hydroxymethylene form of 2-methylindole-3-aldehyde. The question of whether these indole-3-aldehydes should be formulated as true aldehydes (I) or as hydroxymethylene deriva-

$$\text{2-methylindole (CH, C–CH}_3\text{, NH)} \xrightarrow[\text{NaOC}_5\text{H}_{11}]{\text{HCOOC}_5\text{H}_{11}} \text{C=CHOH, N=C–CH}_3$$

(I): C–CHO, C–CH$_3$, N–H (II): C=CHOH, N=C–CH$_3$

tives (II) has been raised by Angeli.[238,239] The compounds yield oximes, phenylhydrazones, and semicarbazones.[240] On the other hand, the 2-substituted indole-3-aldehydes fail to give bisulfite addition products or positive fuchsin tests.[241] 2-Phenylindole-3-aldehyde does not give a silver mirror with Tollen's reagent and does not reduce Fehling's solution. Benzoin condensation and Cannizzaro reactions attempted with indole-3-aldehydes were failures.[241] Attempts to effect Perkin condensations with these aldehydes likewise failed.[241,242] Oxidation of 2-phenylindole-3-aldehyde with 30% hydrogen peroxide gave not 2-phenylindole-3-carboxylic acid but *N*-benzoylanthranilic acid. Attempts to effect the oxidation with less vigorous oxidizing agents gave only smaller yields of *N*-benzoylanthranilic acid.[241] It had previously been reported that 3-nitro-2-phenylindole, 3-nitroso-2-phenylindole,[243] and 1-hydroxy-2-phenylindole[244] are oxidized by alkaline permanganate to give *N*-benzoylanthranilic acid. 2-Methylindole-3-aldehyde

[237] Seka, *Ber.*, **B 56,** 2058 (1923).
[238] Angeli and Marchetti, *Atti accad. Lincei,* [5] 16, i, 381 (1907).
[239] Angeli and Allesandri, *ibid.*, 23, ii, 93 (1914).
[240] Angeli and Marchetti, *ibid.*, [5] 16, ii, 790 (1907). Blume and Lindwall, *J. Org. Chem.*, **10,** 255 (1945).
[241] Van Order and Lindwall, *J. Org. Chem.*, **10,** 128 (1945). Blume and Lindwall, *ibid.*, **11,** 185 (1945).
[242] Majima and Kotake, *Ber.*, **58,** 2037 (1925).
[243] Angeli and Angelico, *Gazz. chim. ital.*, **30,** ii, 277 (1901).
[244] Fischer, *Ber.*, **29,** 2063 (1896).

under similar conditions yields *N*-acetylanthranilic acid.[245] Angeli[246] thought that these results indicated that the compounds possessed the hydroxymethylene structure. On the other hand, it has been found that 2-methylindole-3-aldehyde can be oxidized by potassium permanganate in acetone to 2-methylindole-3-carboxylic acid[247] while the condensation reactions with hydantoin,[248] hippuric acid,[249] barbituric acid,[250] and other compounds containing active methylene groups are normal-aldehyde reactions.[232, 251, 252]

The best preparation of indole-3-aldehydes has been accomplished through the action of *N*-methylformanilide and phosphorus oxychloride on the appropriate indole.[251, 253, 254] The cyanoethylation of certain indoles and indole-3-aldehydes has been effected through the addition of acrylonitrile.[251, 255]

$CH_2{=}CH_2CN$

$POCl_3$, $C_6H_5N(CHO)CH_3$

$POCl_3$, $C_6H_5N(CH_3)CHO$

$CH_2{=}CHCN$

CH, C—C_6H_5, N, H; CH, C—C_6H_5, N, CH_2CH_2CN; C—CHO, C—C_6H_5, N, H; C—CHO, C—C_6H_5, N, CH_2CH_2CN

[245] Plancher and Ponti, *Atti accad. Lincei*, **16,** i, 130 (1907); *Chem. Abstr.*, **2,** 1147 (1908).

[246] Angeli and Allesandri, *Atti accad. Lincei*, **23,** ii, 93 (1914); *Chem. Abstr.*, **9,** 1322 (1915).

[247] Allesandri, *Atti accad. Lincei*, **24,** ii, 194 (1915).

[248] Boyd and Robson, *Biochem. J.*, **29,** 2256 (1935). Majima and Kotake, *Ber.*, **55,** 3859 (1922). Miller and Robson, *J. Chem. Soc.*, **1938,** 1910.

[249] Ellinger and Flamand, *Ber.*, **40,** 3029 (1907). Restelli, *Anales asoc. quim. argentina*, **23,** 58 (1935); *Chem. Abstr.*, **30,** 1054 (1936).

[250] Akabori, *Proc. Imp. Acad. (Japan)*, **3,** 342 (1927); *C.A.*, **21,** 3185 (1927).

[251] Blume and Lindwall, *J. Org. Chem.*, **10,** 255 (1945).

[252] Seka, *Ber.*, **B 57,** 1868 (1924).

[253] German patent 614,325; *Chem. Abstr.*, **29,** 5861 (1935). French patent 773,259; *Chem. Abstr.*, **29,** 1431 (1935). French patent 49,986, *Chem. Abstr.*, **36,** 3188 (1942). French patent 44,865; *Chem. Abstr.*, **29,** 6246 (1935). German patent 615,130; *Chem. Abstr.*, **29,** 6248 (1935). German patent 677,207; *Chem. Abstr.*, **33,** 6880 (1939). British patent 438,278; *Chem. Abstr.*, **34,** 447 (1940).

[254] Shabica, Howe, Zeigler, and Tishler, *J. Am. Chem. Soc.*, **68,** 1156 (1946).

[255] German patent 641,597; *Chem. Abstr.*, **31,** 5813 (1937). British patent 438,278; *Chem. Abstr.*, **30,** 2577 (1936). British patent 466,316; *Chem. Abstr.*, **31,** 7887 (1937). French patent 47,563; *Chem. Abstr.*, **34,** 447 (1940).

Recent syntheses of indole-3-aldehyde have been accomplished through treatment of gramine with hexamethylenetetramine,[255a] by the action of dimethylformamide on indole in the presence of phosphorus oxychloride[255b] and of carbon monoxide on the potassium salt of indole.[255b]

Catalytic reduction of indole-3-aldehyde yields 3-hydroxymethylindole, m.p. 90°.[256]

Indole-2-aldehyde (2,4-dinitrophenylhydrazone, m.p. 315–320° (dec.)) and certain of its derivatives have been prepared through oxidation of 2-

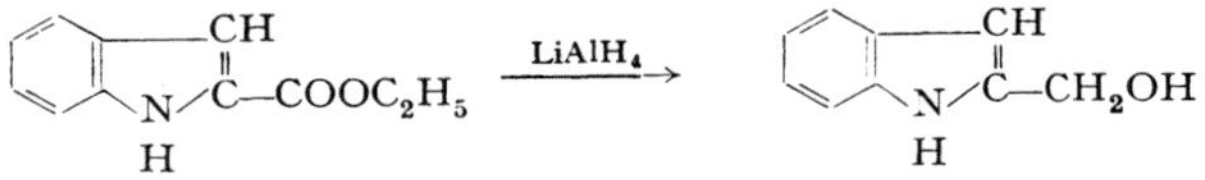

CH, N, H, C—$COOC_2H_5$ $\xrightarrow{LiAlH_4}$ CH, N, H, C—CH_2OH

hydroxymethylindole which in turn was prepared through reduction of ethylindole-2-carboxylate with $LiAlH_4$.[257]

Acyl Derivatives of Indole

Treatment of indole with acetic anhydride at 180–200° yields 1-acetylindole along with 1,3-diacetylindole.[258] The 1-acetylindole is volatile with

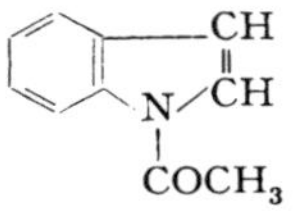

CH, CH, N, $COCH_3$

steam while 3-acetylindole and 1,3-diacetylindole are not. 1-Acetylindole is a liquid, boiling at 152–153° at 14 mm., which is hydrolyzed to indole by heating with aqueous alkali. 1,3-Diacetylindole is the principal product of the action of acetic anhydride on indole at higher temperatures. The compound melts at 150–151° and is hydrolyzed by boiling with either water

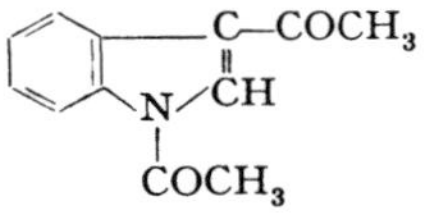

C—$COCH_3$, CH, N, $COCH_3$

[255a] Snyder, Swaminathan, and Sims, *J. Am. Chem. Soc.*, **74**, 5110 (1952).

[255b] Tyson and Shaw, *ibid.*, **74**, 2273 (1952).

[256] Madinaveitia, *J. Chem. Soc.*, **1937**, 1927.

[257] Goutarel, Janot, Prelog, and Taylor, *Helv. Chim. Acta*, **33**, 150 (1950). Taylor, *ibid.*, **33**, 164 (1950).

or dilute alkali yielding 3-acetylindole.[258,259] Skatole yields 2-acetyl-3-methylindole when treated with acetyl chloride.

β-Indolyl alkyl ketones are obtained easily through the reaction of acid chlorides with indolylmagnesium iodide.[260] Small yields of the *N*-acyl derivatives are obtained at the same time.

1-(MgI)indole —CH_3COCl→ 3-acetylindole (C—$COCH_3$, N—H)

1-(MgBr)-3-methylindole —$ClCH_2COCl$→ 3-methyl-2-($COCH_2Cl$)indole (N—H)

1-Benzoylindole has been prepared through the action of benzoyl chloride on a suspension of the sodium salt of indole in benzene.[261] The compound melts at 67–68° and boils at 213° at 16 mm. Members of the indole series are acylated by acetyl chloride or acetic anhydride at 100° in positions 2 and 3 and if these positions are occupied the compounds react with acetyl chloride and aluminium chloride in carbon disulfide to introduce the acetyl group in the benzene ring.[262]

1,2-dimethylindole —$(CH_3CO)_2O$→ 3-$COCH_3$-1,2-dimethylindole, m.p. 103-104°

1,2-dimethylindole —CH_3COCl, $AlCl_3$→ CH_3CO-(benzene ring)-1,2-dimethylindole, m.p. 157°

On the other hand, 1-acyl-2,3-dialkylindoles are acylated in the Friedel-Crafts reaction in position 6.[263]

[258] Zatti and Ferratini, *Ber.*, **23,** 1359 (1890). Ciamician and Zatti, *ibid.*, **22,** 1977 (1889).

[259] Oddo and Sessa, *Gazz. chim. ital.*, **41,** i, 237 (1911).

[260] Oddo and Sessa, *ibid.*, **41,** i, 234 (1911). Salway, *J. Chem. Soc.*, **103,** 353 (1913). Oddo, *Gazz. chim. ital.*, **43,** ii, 190, 362 (1913). Mingoia, *ibid.*, **59,** 105 (1929); **61,** 646 (1931).

[261] Weissgerber, *Ber.*, **43,** 3523 (1910).

[262] Borsche and Groth, *Ann.*, **549,** 238 (1941).

[263] Gaudion, Hook, and Plant, *J. Chem. Soc.*, **1947,** 1631. Plant and Thompson, *ibid.*, **1950,** 1069.

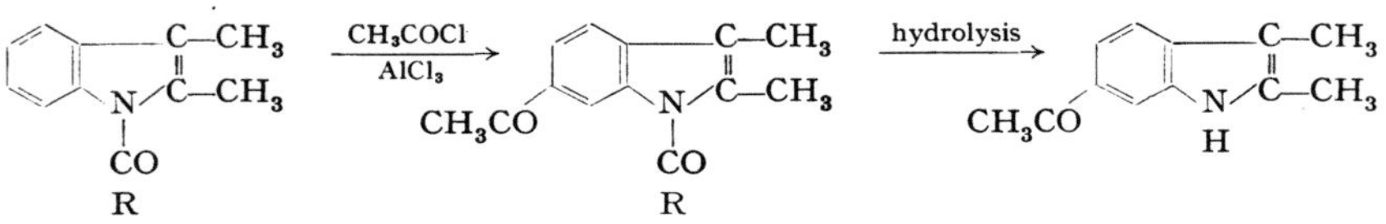

Indole ketones have been synthesized[264] from the appropriate hydrazones through the Fischer synthesis.

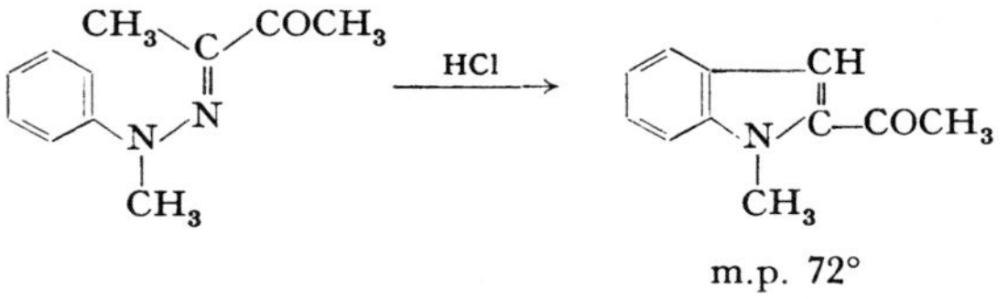

m.p. 72°

3,3-Dimethyl-2-acetylindolenine has been prepared[265] through the action of methylmagnesium iodide on 3,3-dimethyl-2-cyanoindolenine.

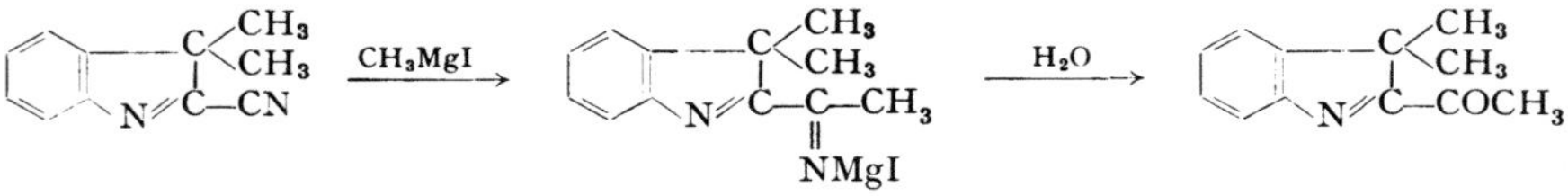

The Houben synthesis has also been employed to obtain ketones of the indole series.[266] Indole ketones have been prepared also from the 1,2-disubstituted indoles through the action of the anilides of substituted benzoic acids on the indole under the influence of phosphorus oxychloride. Thus 1-methyl-2-phenylindole yields 1-methyl-2-phenyl-3-*p*-chlorobenzoylindole[267] on treatment with *p*-chlorobenzanilide.

Extensive studies have been made[268] of the acylation of the indolenines.

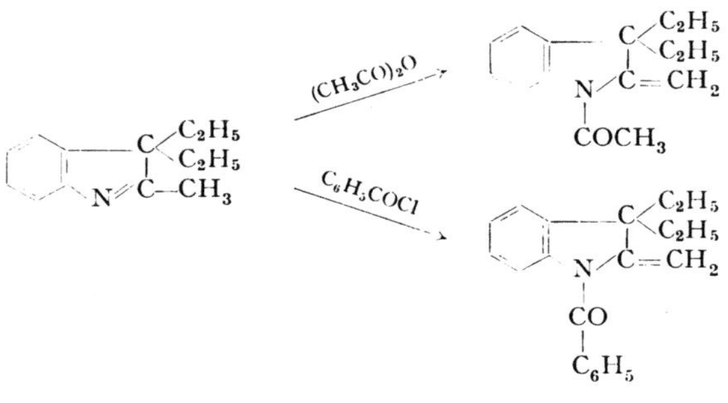

[264] Diels and Köllisch, *Ber.*, **44**, 264 (1911). Diels and Dürst, *ibid.*, **47**, 284 (1914).

[265] Plancher and Giumelli, *Atti accad. Lincei*, **18**, i, 393 (1909).

[266] Houben and Fischer, *Ber.*, **B 64**, 2645 (1931).

[267] German patent 614,326; *Chem. Abstr.*, **29**, 5861 (1935).

[268] Leuchs, Heller and Hoffman, *Ber.*, **B 62**, 871 (1929). Leuchs, Philipott, Sander, Heller, and Kohler, *Ann.*, **461**, 27 (1928).

The 2-methyl-3,3-dialkylindolenines yield 2-methylene derivatives on treating with acylating agents (see eq. bottom p. 46). Acetic anhydride adds to 3,3-dialkylindolenines to give diacyl derivatives of the indolinol. Similarly when the substituent in position 2 is a phenyl group, precluding formation of a methylene derivative, an indolinol derivative is obtained through the action of benzoyl chloride. Certain of the indolenines react with acid

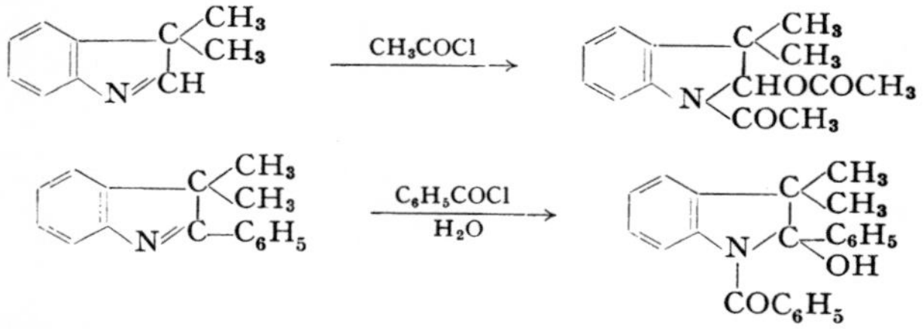

CH_3COCl →

C_6H_5COCl, H_2O →

chlorides to give 1-acyl-2-chloroindolines.[269] Some of the transformations which may be accomplished through the agency of these 2-chloro derivatives are shown in the accompanying scheme.

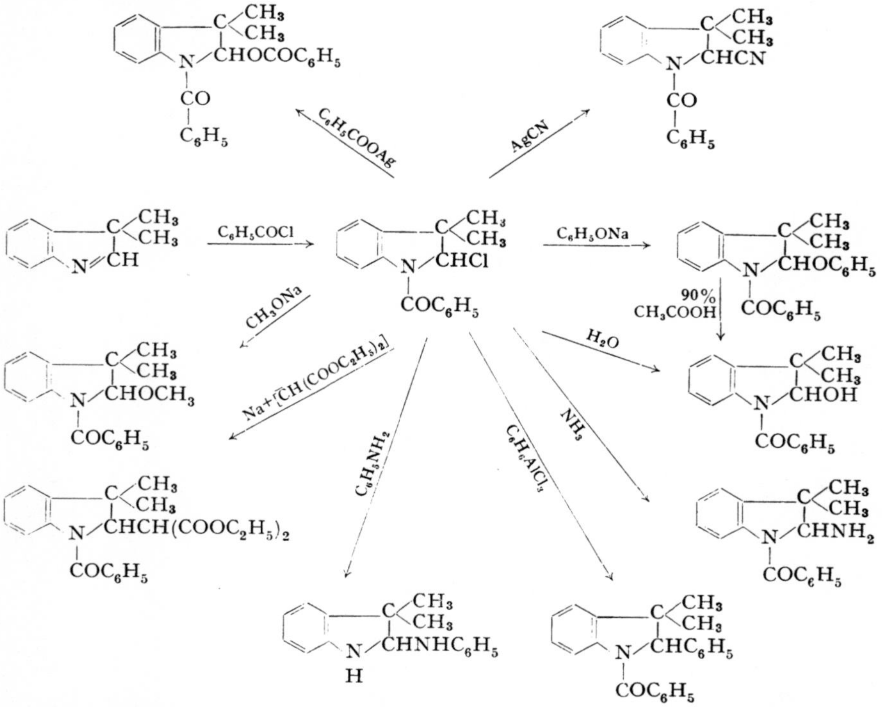

C_6H_5COOAg; AgCN; C_6H_5COCl; C_6H_5ONa; CH_3ONa; 90% CH_3COOH; H_2O; $Na^+[CH(COOC_2H_5)_2]^-$; $C_6H_5NH_2$; C_6H_6, $AlCl_3$; NH_3

[269] Leuchs and Schlolzer, *Ber.*, **B 67,** 1572 (1934). Leuchs, Wulkow, and Gerland, *ibid.*, **B 65,** 1586 (1932).

Condensation of Indoles with Aldehydes

Aldehydes condense with 2-substituted indoles to give products analogous to the triphenylmethanes.[270] Thus benzaldehyde condenses with methyl ketole to give the leuco base which on oxidation followed by treatment with a halogen acid yields a dye known as a "rosindole." Compounds identical with the leuco bases of the rosindoles have been obtained through

$$C_6H_5CHO + 2\,C_6H_4NHC(CH_3)=CH \longrightarrow \left(C_6H_4NHC(CH_3)=C\right)_2 CHC_6H_5$$

$$\xrightarrow[+HX]{\text{oxidation}}$$

$$\left[\begin{array}{c} C_6H_4NHC(CH_3)=C-C(C_6H_5)=C-C(CH_3^+)=NC_6H_4 \end{array}\right] X^-$$

action of indolylmagnesium halides on aldehydes.[271] Similar condensation products are obtained through the reaction of skatole with aldehydes.[272] Thus skatole and benzaldehyde give phenylbis(3-methyl-2-indolyl)methane. Methylketole can also be condensed with aldehydes in molecular proportions to yield products formulated as follows:[273]

$$RCH=C\begin{array}{c}\diagup C_6H_4 \diagdown \\ \diagdown C(CH_3) \diagup\!\!\diagup \end{array}N$$

Formaldehyde reacts with indole derivatives to give compounds of the type

$$CH_2\left[C\begin{array}{c}\diagup\!\!\diagup C(CH_3) \diagdown \\ \diagdown C_6H_4 \diagup \end{array}NH\right]_2$$

which are colorless crystals when freshly prepared and become orange red on exposure to air and light. Indole, skatole, and methyl ketole all condense with glyoxal to give derivatives formulated as tetraindolylethylenes.[274]

[270] Fischer, *ibid.*, **19,** 2988 (1886); *Ann.*, **242,** 372 (1887). Compare Walther and Clemen, *J. prakt. Chem.*, [2] **61,** 256 (1930). Renz and Löw, *Ber.*, **36,** 4326 (1903). Voisenet, *Bull. soc. chim.*, [4] **5,** 736 (1909).

[271] Mingoia, *Gazz. chim. ital.*, **56,** 772 (1926); *Chem. Abstr.*, **21,** 1117 (1927).

[272] Etienne and Heymes, *Bull. soc. chim. France*, **1948,** 841.

[273] Freund and Lebach, *Ber.*, **36,** 308 (1903); **37,** 332 (1904); **38,** 2640 (1905). Scholtz, *ibid.*, **38,** 2138 (1905).

[274] Hasselstrom, *Ann. Acad. Sci. Fennicae,* **Ser. A, 30. No. 13,** 18 pp. (1930); *Chem. Abstr.*, **25,** 4880 (1931).

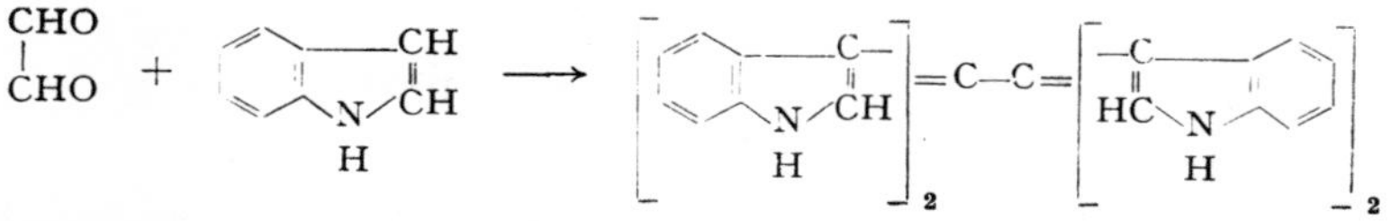

Ehrlich[275] found that a red color is developed when hydrochloric acid is added to an alcoholic solution of 2-methylindole and *p*-dimethylaminobenzaldehyde. This color, which is the basis of a colorimetric procedure for the estimation of indole, was found to be due to the formation of a rosindole.[273, 276] This characteristic color test is given by many indoles.

2-Methylindole-3-aldehyde on treatment with sulfuric acid yields an orange-yellow dye (A), m.p. 234–237°[277] also obtained by the reaction of 2-methylindole with formic acid in the presence of sulfuric acid. The same compound is obtained through the condensation of 2-methylindole with 2-methylindole-3-aldehyde. The substance was first formulated as di(2-methyl-3-indyl)-2-methyl-3-indolidene methane (I). A similar product was obtained when indole-3-aldehyde was treated with sulfuric acid or when

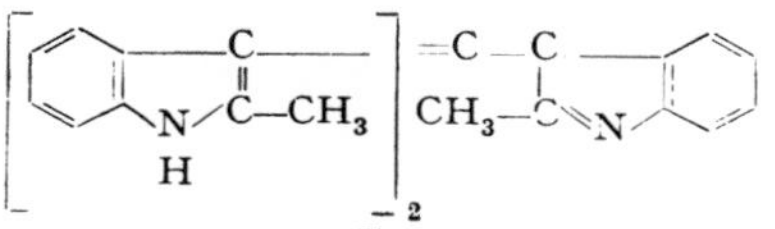

(I)

indole was condensed with formic acid in the presence of sulfuric acid. Ethyl orthoformate has been employed in a similar condensation in place of formic acid.[278] On the other hand, 2-methylindole condenses with formic acid in the

[275] Ehrlich, *Deut. med. Wochschr.* (April, 1901). Chernoff, *Ind. Eng. Chem., Anal. Ed.*, **12,** 273 (1940).

[276] For additional condensations of aldehydes with indoles compare: Hadano, *J. Pharm. Soc. Japan*, **48,** 919 (1928); *Chem. Abstr.*, **23,** 1635 (1929). Dost'al, *Chem. Listy*, **31,** 250 (1937); *Chem. Abstr.*, **31,** 7873 (1937). Dost'al, *Chem. Listy*, **32,** 161 (1938); *Chem. Abstr.*, **32,** 6242 (1938). Ludwig and Tache, *Soc. Chim. România Sect. Soc. Romane Slünte, Bull. Chim. Pura Apl.*, [2] **3A,** 3–15 (1941–1942); *Chem. Zentr.*, **1943,** II, 1275; *Chem. Abstr.*, **38,** 5499 (1944). Cook and Majer, *J. Chem. Soc.*, **1944,** 486. Homer, *Biochem. J.*, **7,** 101 (1913). Scholtz, *Arch. Pharm.*, **253,** 629 (1915). Hoschek, *Ber.*, **49,** 2584 (1916). Plancher, *Giorn. chim. ind. ed appl.*, **2,** 458 (1920); *Chem. Abstr.*, **15,** 3103 (1921). Brooker and Sprague, *J. Am. Chem. Soc.*, **63,** 3203 (1941). Passerini and Bonciani, *Gazz. chim. ital.*, **63,** 138 (1933); *Chem. Abstr.*, **27,** 3473 (1933). Passerini and Albani, *Gazz. chim. ital.*, **65,** 933 (1935); *Chem. Abstr.*, **30,** 3817 (1936). Wenzing, *Ann.*, **239,** 239 (1887).

[277] Ellinger and Flamand, *Z. physiol. Chem.*, **62,** 276 (1909); **71,** 7 (1911); **78,** 365 (1912); **91,** 15 (1914).

[278] Nenitzescu, *Bul. Soc. Chim. România*, **11,** 37 (1929); *Chem. Abstr.*, **24,** 110 (1930).

presence of hydrochloric acid to give tri-(2-methyl-3-indyl)methane (B), m.p. 319°. The pigment (A) is now regarded[279, 280] as having the formula II

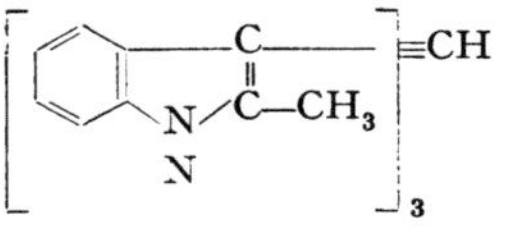

$$\left[C_6H_4 \text{—C—} ; \text{N—C—}CH_3 ; \text{N} \right]_3 \equiv CH$$

(B)

rather than I which was originally assigned to it. The compound has also been prepared through the action of 2-methylindolemagnesium bromide with either iodoform or carbon tetrachloride.[281]

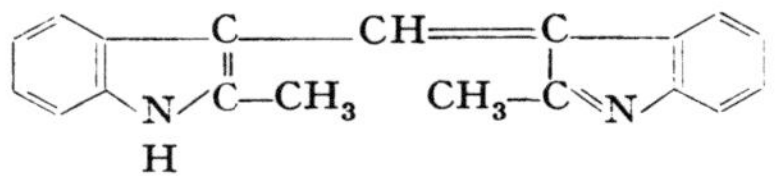

C——CH═══C
N(H)—C—CH_3 CH_3—C═N

(II)

Indocyanine dyes are obtained through the condensation of the methiodides of 2,3,3-trialkylindolenines with ethyl orthoformate.[282]

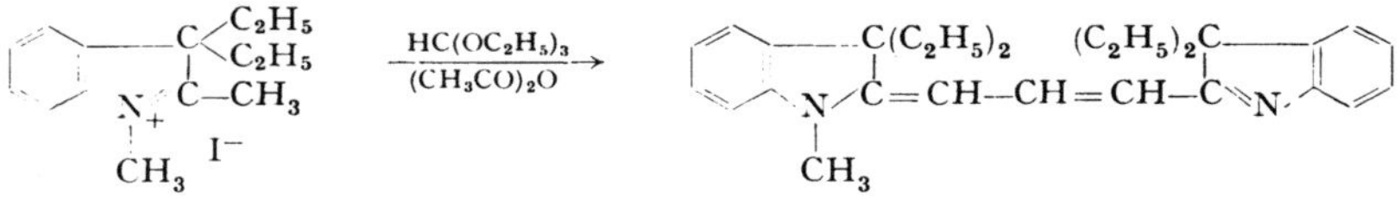

$C(C_2H_5)(C_2H_5)$; $\overset{+}{N}$—C—CH_3; CH_3 I^- $\xrightarrow[(CH_3CO)_2O]{HC(OC_2H_5)_3}$ $C(C_2H_5)_2$ $(C_2H_5)_2C$; N—C=CH—CH=CH—C═N; CH_3

Indolylmagnesium Halides

Indole reacts with Grignard reagents to give indolylmagnesium halides.[283] These indolylmagnesium complexes can be analyzed as the pyridine addition compounds (*e.g.*, $C_8H_6NMgI \cdot 2C_5H_5N$). The indolylmagnesium halide derivatives are usually regarded as being N–MgX derivatives[284] although the alternative C–MgX structure has been suggested.[285] The fact that the pyrrolemagnesium halides (regarded as C–MgX derivatives) give a

[279] König, *J. prakt. Chem.*, **84**, 194 (1911). Scholtz, *Ber.*, **46**, 1082, 2138, 2539 (1913).

[280] Fischer and Pistor, *Ber.*, **56**, 2316 (1923).

[281] Oddo and Sanna, *Gazz. chim. ital.*, **54**, 682 (1924). Oddo and Tognacchini, *ibid.*, **53**, 271 (1923).

[282] Ghigi, *ibid.*, **63**, 698 (1933); *Chem. Abstr.*, **28**, 2003.

[283] Oddo, *Ber.*, **43**, 1012 (1910); *Gazz. chim. ital.*, **41**, i, 222 (1911).

[284] Chelintzev and Tronov, *J. Russ. Phys.-Chem. Soc.*, **46**, 1876 (1914); *Chem. Abstr.*, **9**, 2071 (1915).

[285] Nenitzescu, *Bul. Soc. Chim. România*, **11**, 130 (1930); *Chem. Abstr.*, **24**, 2458 (1930).

violet color with Michler's ketone while the indole derivatives do not is regarded as evidence that the latter are N–MgX derivatives.

Indolylmagnesium iodide reacts with methyl iodide to give a mixture of skatole, 1-methylindole, and 1,3-dimethylindole. Indolylmagnesium iodide, after removal of the ether, then heated fifteen hours with methyl iodide gave skatole. Shorter heating periods gave 1-methylindole and a little 1,3-dimethylindole.[286] Skatolylmagnesium iodide similarly gave with methyl iodide a mixture of 1,3-dimethylindole and 3,3-dimethylindolenine.[287] Methylketolylmagnesium iodide and methyl iodide gave 2,3-dimethylindole and 2,3,3-trimethylindolenine. The latter compound was also obtained by the action of methyl iodide on 2,3-dimethylindolylmagnesium iodide. Benzyl chloride and methylketolylmagnesium iodide similarly gave 2-methyl-3-benzylindole and 2-methyl-3,3-dibenzylindolenine.

It has been reported that indolylmagnesium bromide and triphenylmethyl chloride react to give 1-triphenylmethylindole, m.p. 211.5–212°,[288] which was also obtained through the action of triphenylmethyl chloride on indole in pyridine solution. On the contrary it has also been reported[289] that indolylmagnesium iodide and triphenylmethyl chloride yield 2-triphenylmethylindole which was also obtained from the sodium salt of indole and triphenylmethyl chloride. The discrepancies between these two reports have not been resolved. Indolylmagnesium iodide reacts readily with acid chlorides yielding 3-acyl derivatives of indole along with small yields of 1,3-diacyl derivatives.[290]

The indolylmagnesium halides in general react with carbon dioxide at low temperatures to give the rather unstable indole-1-carboxylic acids. At higher temperatures the 3-carboxylic acids are formed. For example, indolylmagnesium iodide has been reported to yield indole-1-carboxylic acid[290] as well as indole-3-carboxylic acid[291] on treatment with carbon dioxide. Indolylmagnesium iodide and ethyl chlorocarbonate react to give ethyl

[286] Oddo, *Gazz. chim. ital.*, **41**, i, 222 (1911); *Chem. Abstr.*, **5**, 2638 (1911). Oddo, *Gazz. chim. ital.*, **63**, 234 (1933); *Chem. Abstr.*, **27**, 3933 (1933).

[287] Hoshino, *Proc. Imp. Acad.* (*Tokyo*), **8**, 171 (1932); *Chem. Abstr.*, **26**, 2814 (1932). Hoshino, *Abstracts of Japan. Chem. Lit.*, **6**, 390 (1932); *Chem. Abstr.*, **27**, 291 (1933).

[288] Funakubo and Hirotani, *Ber.*, **B 69**, 2123 (1936).

[289] Kubota, *J. Chem. Soc. Japan*, **59**, 399, 407 (1938); *Chem. Abstr.*, **32**, 9080 (1938). Kubota and Mila, *J. Chem. Soc. Japan*, **59**, 409–411 (1938); *Chem. Abstr.*, **32**, 9080 (1938).

[290] Oddo and Sessa, *Gazz. chim. ital.*, **41**, i, 234 (1911). Salway, *J. Chem. Soc.*, **103**, 351 (1913).

[291] Majima and Kotake, *Ber.*, **B 55**, 3865 (1922); **B 63**, 2237 (1930).

indole-3-carboxylate along with some diethyl indole-1,3-dicarboxylate. Skatolylmagnesium bromide and carbon dioxide give 3-methylindole-1-carboxylic acid while the same Grignard reagent with ethyl chlorocarbonate gives ethyl 3-methylindole-1-carboxylate.[292]

2-Methylindolylmagnesium bromide in ether yields with carbon dioxide the unstable 2-methylindole-1-carboxylic acid. At higher temperatures the absorption of carbon dioxide yields 2-methylindole-3-carboxylic acid. Similarly, reaction of 2-methylindolylmagnesium bromide with ethyl chlorocarbonate gives ethyl 2-methylindole-3-carboxylate, m.p. 135°.[292]

Skatolylmagnesium bromide reacts with acyl halides in dry ether at the boiling temperature to give only the *C*-acyl derivatives (2-derivatives), the 1-acyl derivatives being obtained in traces only. At lower temperatures mixtures of the *C*- and *N*-derivatives are obtained.[293] With methylketole the tendency to formation of *C*-acyl derivatives is more pronounced, with benzoyl chloride only small yields of the *N*-benzoyl derivative are obtained.

The indolylmagnesium halides react with formaldehyde in the usual manner.[294] Oxidation of the alcohols so obtained with potassium permanganate yields the corresponding acids. Indolylmagnesium iodide reacts with

$$\text{Indole(N-MgBr)} + CH_2O \xrightarrow[\text{by } H_2O]{\text{followed}} \text{Indole(NH)-3-}CH_2OH \xrightarrow{KMnO_4} \text{Indole(NH)-3-}COOH$$

ethylene oxide to give β-indolylethanol (tryptophol).[295] Methylketolylmagnesium iodide reacts similarly. Indolylmagnesium iodide reacts with γ-butyrolactone yielding a complex which on decomposition gives indole-3-γ-butyric acid.[296] 2,3-Dimethylindolylmagnesium bromide reacts with

$$\text{Indole(N-MgI)} + \underset{\text{(γ-butyrolactone)}}{CH_2CH_2CH_2CO\text{—}O} \xrightarrow[HCl + H_2O]{\text{followed by}} \text{Indole(NH)-3-}CH_2CH_2CH_2COOH$$

ethylene bromide to give an indolenine derivative.[297] Studies have been

[292] Oddo, *Mem. accad. Lincei*, [5] **14,** 510 (1923); *Chem. Abstr.*, **19,** 2492 (1925). Oddo, *Gazz. chim. ital.*, **42,** i, 361 (1912). Wislicenus and Arnold, *Ber.*, **20,** 3395 (1887).

[293] Oddo, *Gazz. chim. ital.*, **43,** ii, 190 (1913).

[294] Mingoia, *ibid.*, **62,** 844 (1932).

[295] Oddo and Cambieri, *ibid.*, **69,** 19 (1939).

[296] Stepanov, U. S. S. R. patent 66,681; *Chem. Abstr.*, **41,** 2087 (1941).

[297] Kobayasi, *Ann.*, **539,** 213 (1939).

made[298] of the action of oxygen on 2-methylindolylmagnesium bromide.

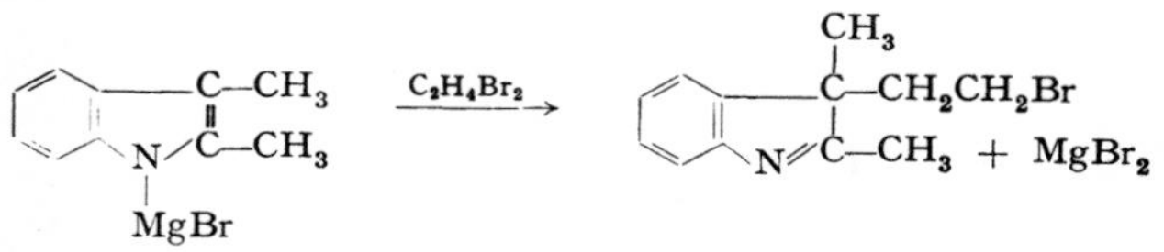

$$\text{2,3-dimethylindolyl-MgBr} \xrightarrow{C_2H_4Br_2} \text{3-methyl-3-(}CH_2CH_2Br\text{)-2-methyl-3H-indole} + MgBr_2$$

Treatment of indolylmagnesium bromide with sulfur gives 3,3′-diindolyl sulfide.[299] Methylketolylmagnesium bromide similarly gave 3,3′-di-(2-methylindolyl) sulfide. Indolylmagnesium bromide reacts with sulfur dioxide to give 3,3′-diindolyl sulfoxide mixed with the sulfide.[300] In similar fashion indolylmagnesium bromide reacts with sulfuryl chloride to give 3,3′-diindolylsulfone. 2-Methylindolylmagnesium bromide reacts in corresponding

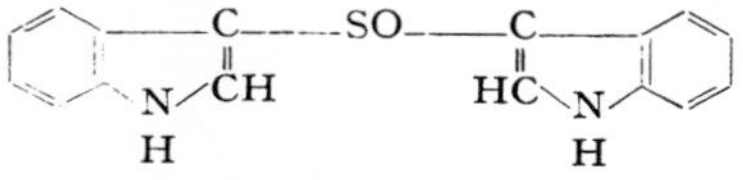

3,3′-diindolyl sulfoxide

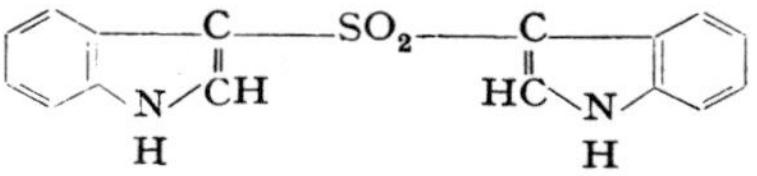

3,3′-diindolylsulfone

manner with sulfur dioxide and with sulfuryl chloride. Skatolylmagnesium bromide reacts to give the corresponding 2,2′-derivatives. Treatment of

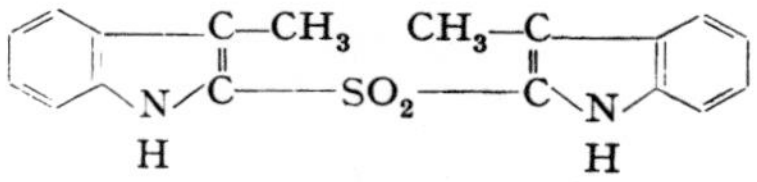

indolylmagnesium bromide with sulfur and then with benzoyl chloride gives benzoylthioindoxyl, m.p. 157°, and α-benzoylthioindole, m.p. 128°.[301] With acetyl chloride 1,3-diacetylthioindoxyl was the product. Saponification with potassium hydroxide gave acetylthioindoxyl, m.p. 208°, and 2-acetylthioindole, m.p. 184°. Thioindoxyl was obtained through further saponification

of benzoylthioindoxyl or of acetylthioindoxyl while α-thioindole, m.p. 148–150°, was obtained by further saponification of α-benzoylthioindole or of

[298] Toffoli, *Rend. ist. sanità publ.*, **2**, 565 (1939); *Chem. Abstr.*, **34**, 4733 (1940). Oddo, *Gazz. chim. ital.*, **50**, ii, 268 (1920); *Chem. Abstr.*, **15**, 2272 (1921).

[299] Madelung and Tencer, *Ber.*, **48**, 949 (1915). Oddo, *Mem. accad. Lincei*, [5] **14**, 510–623 (1923); *Chem. Abstr.*, **19**, 2492 (1925).

[300] Oddo and Mingoia, *Gazz. chim. ital.*, **56**, 782 (1926).

[301] Oddo and Mingoia, *ibid.*, **62**, 299–317 (1932).

Thioindoxyl

α-acetylthioindole. Acyl derivatives of 2-methylthioindoxyl have been

α-Thioindole

prepared similarly from 2-methylindolemagnesium bromide while similar derivatives have been obtained from skatolylmagnesium bromide.[302] Indolyl-

followed by CH_3COCl

magnesium iodide reacts with phosphorus trichloride to give a mixture of tri(3-indolyl)phosphine, m.p. 195–196°, and tri(1-indolyl)phosphine, m.p. 223–225°. 2-Methylindolemagnesium bromide reacts similarly to give tri(2-methylindolyl-1)phosphine, m.p. 180°, while 3-methylindolylmagnesium bromide gives tri-(3-methylindolyl-2-)phosphine, m.p. 156–158°. 2-Methylindolylmagnesium bromide reacts with phosphorus oxychloride to give tri(2-methylindolyl-3)phosphoxide,[303] $(C_6H_4NH\text{-}C(CH_3){:}C)_3PO$, m.p. 170°, tri(2-methylindolyl-1)phosphoxide, $(CH_3C{:}CHC_6H_4N)_3{:}PO$, m.p. 140–142°, and di(2-methylindolyl-3)phosphinic acid, $(C_6H_4NH\text{-}C(CH_3){:}C)_2POOH$, m.p. 159–160°. Indolylmagnesium iodide reacts similarly with phosphorus oxychloride to give tri(indolyl-3)phosphoxide, m.p. 138–140°, and diindolyl-3-phosphinic acid, m.p. 190°.

Treatment of indolylmagnesium bromide with ethyl formate yields 1-formylindole (indole-1-aldehyde) in 90% yield.[304] Heating favors the formation of the 3-formyl derivative (indole-3-aldehyde).[305] 2-Methylindolylmagnesium iodide reacts similarly with ethyl formate and with isoamyl formate giving a mixture of 1-formyl-2-methylindole and 2-methyl-3-

[302] Oddo and Raffa, *ibid.*, **71**, 242 (1941); *Chem. Abstr.*, **36**, 2854 (1942).

[303] Mingoia, *Gazz. chim. ital.*, **60**, 144 (1930); *Chem. Abstr.*, **24**, 3783 (1930). Mingoia, *Gazz. chim. ital.*, **62**, 333 (1932); *Chem. Abstr.*, **26**, 4813 (1932).

[304] Putokhin, *Ber.*, **B 59**, 1987 (1926). Putokhin, *J. Russ. Phys. Chem. Soc.*, **59**, 761 (1927); *Chem. Abstr.*, **22**, 3409 (1928).

[305] U. S. patent 2,414,715; *Chem. Abstr.*, **41**, 3129 (1947).

formylindole. The 1-formyl-2-methylindole rearranges to the 3-formyl derivative on heating with zinc chloride.[306] Treating indolylmagnesium bromide with ethyl acetate gave a mixture of 1-acetyl and 3-acetylindole. Treatment of 2-methyl-indolylmagnesium bromide with ethyl nitrate yielded 2-methyl-3-nitro-indole.[306, 307]

The action of the acid chlorides of several dibasic acids on indolyl-, skatolyl-, and methylketolylmagnesium halides has been the object of several studies. Oxalyl chloride[308] reacts with indolylmagnesium bromide to give compound I, 1,3-bisindil (I) along with some 2,2-indil (IV).[309] Treatment of I with alkali gives 3,3-indil (II) along with some 1,1-indil (III). Methylketolylmagnesium bromide and oxalyl chloride react to give V.

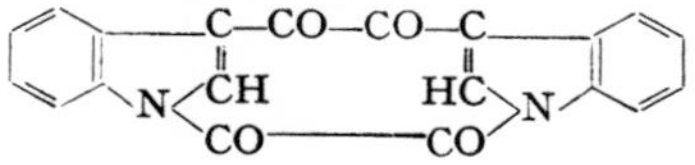

C—CO—CO—C
CH HC
N N
CO CO
(I)

Malonyl chloride reacts with indolylmagnesium bromide and with methylketolylmagnesium bromide to give compounds VI and VII, re-

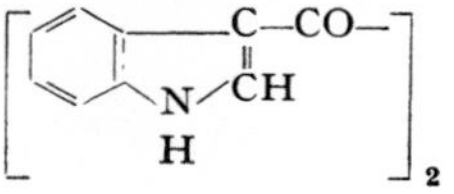

[C—CO— / CH / N / H]$_2$ (II) 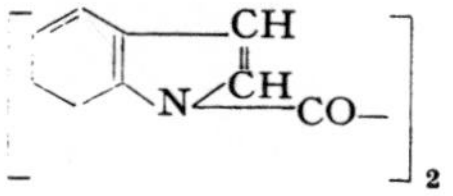
[CH / CH / N / CO—]$_2$ (III) 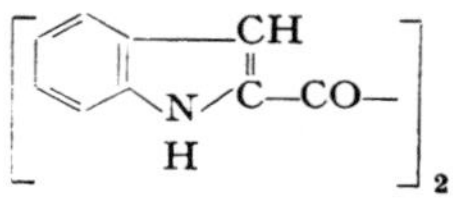
[CH / C—CO— / N / H]$_2$ (IV)

spectively,[309] while succinyl chloride similarly yields compounds VIII and IX. (It should be stated that the evidence supporting the structures assigned

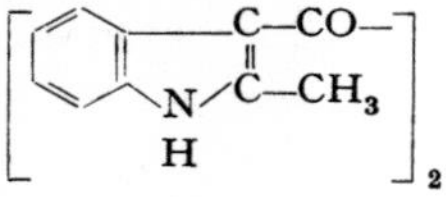

[C—CO— / C—CH_3 / N / H]$_2$ (V) 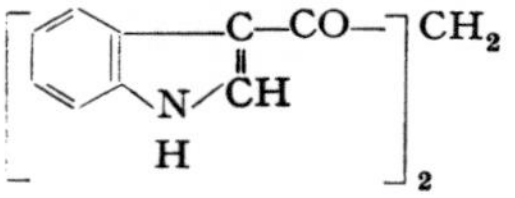
[C—CO— / CH / N / H]$_2$ CH_2 (VI) 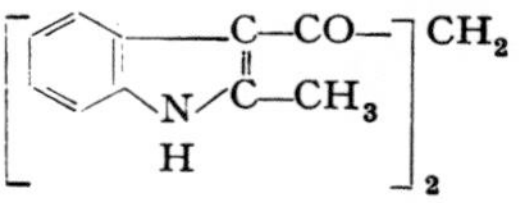
[C—CO— / C—CH_3 / N / H]$_2$ CH_2 (VII)

some of these derivatives is not completely convincing). The interaction of indolylmagnesium bromide and phosgene[310] gave a mixture of three products

(VIII) [C—$COCH_2$— / CH / N / H]$_2$
[C—CO—CH_2— / C—CH_3 / N / H]$_2$ (IX)

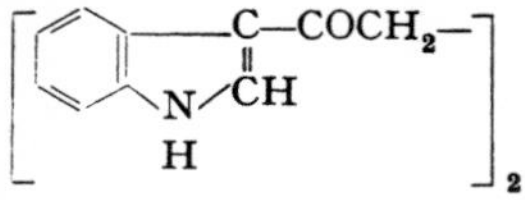

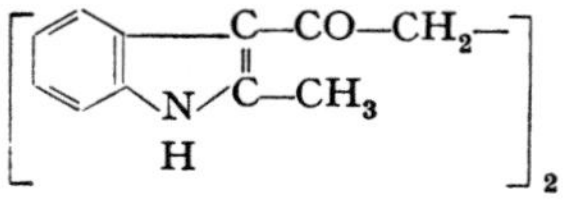

[306] Allesandri and Passerini, *Gazz. chim. ital.*, **51**, i, 262 (1921).

[307] Oddo, *Mem. accad. Lincei*, [5] **14**, 510 (1923); *Chem. Abstr.*, **19**, 2492 (1925).

[308] Oddo and Sanna, *Gazz. chim. ital.*, **51**, ii, 337 (1921). Majima and Shigematsu, *Ber.*, **B 57**, 1449 (1924). Hishida, *J. Chem. Soc. Japan, Pure Chem. Sect.*, **72**, 312–314 (1951); *Chem. Abstr.*, **46**, 5038 (1952).

[309] Sanna, *Gazz. chim. ital.*, **52**, ii, 165 (1922).

[310] Oddo and Mingoia, *ibid.*, **57**, 473 (1927).

X, XI, and XII, while methylketolylmagnesium bromide and phosgene gave similarly the 3,3-, the 1,1-, and the 1,3-derivatives (XIII, XIV, and XV).

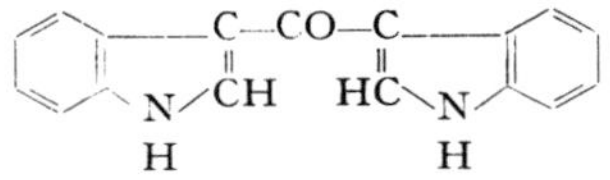

(X) m.p. 280°

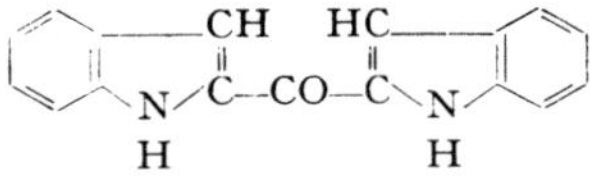

(XI) m.p. 198°

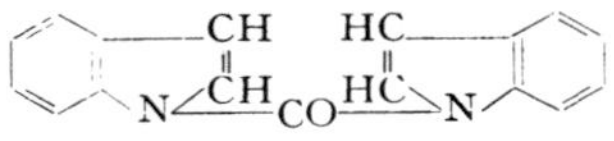

(XII) m.p. 245°

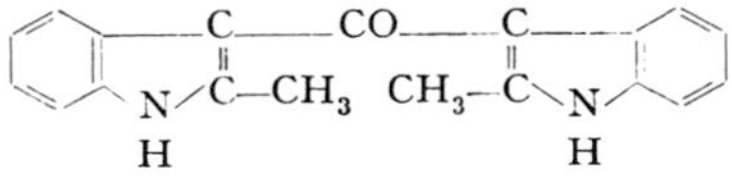

(XIII) m.p. 290°

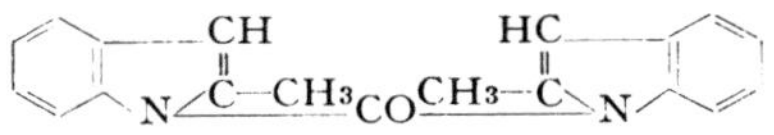

(XIV) m.p. 180°

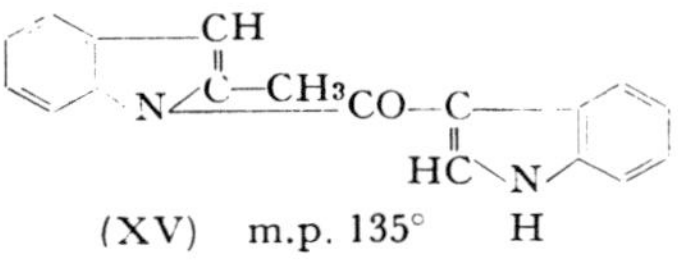

(XV) m.p. 135°

Acetylsalicyl chloride and indolylmagnesium bromide react to give a complex which on hydrolysis gives principally 3-salicylindole, m.p. 171°.[311] Methylketolylmagnesium bromide similarly gives 2-salicyl-3-salicylindole, m.p. 167°, while skatolylmagnesium bromide gives 1-salicyl-3-methylindole, m.p. 151°.

The reactions of the chlorides of the acid esters of dibasic acids with the indolylmagnesium halides have been the subject of several studies.[312, 313, 314, 315] Indolylmagnesium bromide reacts with ethyl oxalyl chloride to yield ethyl indole-3-glyoxalate,[312, 313] while methylketolylmagnesium bromide yields the corresponding 2-methyl derivative. The free

$\longrightarrow$ C—$COCOOC_2H_5$

acid (indole-3-glyoxylic acid) decomposes on heating yielding indole-3-aldehyde. Indolylmagnesium bromide likewise reacts with the chloride of

[311] Toffoli, *ibid.*, **64,** 364 (1934); *Chem. Abstr.*, **28,** 6437 (1934). Toffoli, *Gazz. chim. ital.*, **65,** 487 (1935); *Chem. Abstr.*, **30,** 455 (1936).

[312] Oddo and Albanese, *Gazz. chim. ital.*, **57,** 827 (1927); *Chem. Abstr.*, **22,** 1775 (1928).

[313] Baker, *J. Chem. Soc.*, **1940,** 455; **1946,** 461; **1947,** 558.

[314] Albanese, *Gazz. chim. ital.*, **60,** 21 (1930); *Chem. Abstr.*, **24,** 4029 (1930).

[315] Majima, Shigematsu, and Rokkaku, *Ber.*, **B 57,** 1453 (1924).

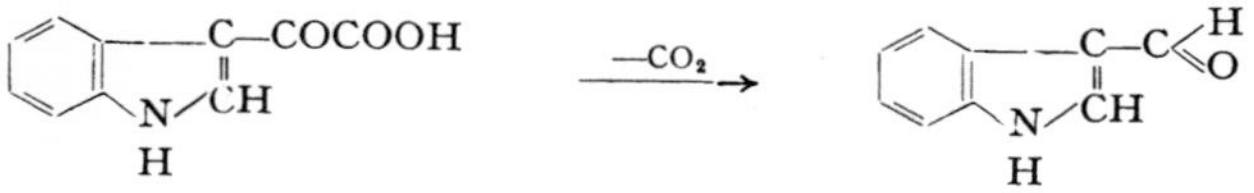

$$\text{(indole)-3-COCOOH} \xrightarrow{-CO_2} \text{(indole)-3-CHO}$$

the acid ester of malonic acid[312, 313] yielding principally the β-derivative along with a small amount of the α-derivative. This β-ketone ester (XVI) on being subjected to "ketone cleavage" gives 3-acetylindole. The 2-methyl deriva-

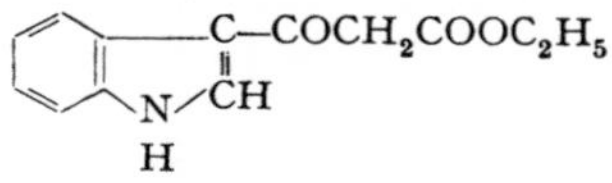

$$\text{(indole)-3-}COCH_2COOC_2H_5$$

(XVI)

tive of XVI is obtained through the interaction of methylketolylmagnesium bromide with the same ester chloride.

3-Chloroacetylindole has been prepared through the interaction of chloroacetyl chloride and indolylmagnesium bromide.[316] Methylketolylmagnesium bromide in similar fashion gives the 3-chloroacetyl derivative while skatolylmagnesium bromide gives the 2-chloroacetyl derivative. Bromoacetyl and iodoacetyl derivatives of indole and of 2-methylindole have been prepared similarly as have the dichloroacetyl and trichloroacetyl derivatives.[317]

Indolylmagnesium iodide reacts with diethylcarbamyl chloride to give the diethylamide of indole-3-carboxylic acid.[318] Chloroacetonitrile and indolylmagnesium iodide yield indole-3-acetonitrile[319] while indolylmag-

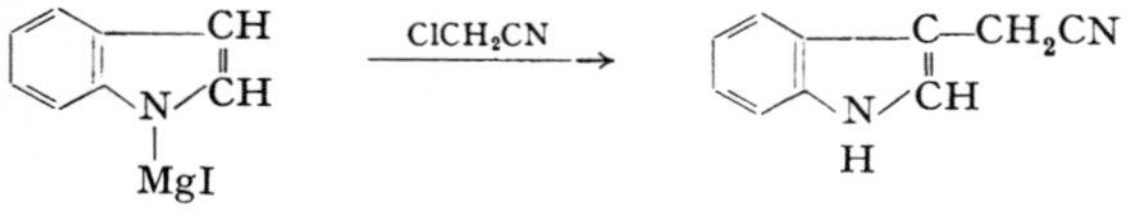

$$\text{(indole)-N-MgI} \xrightarrow{ClCH_2CN} \text{(indole)-3-}CH_2CN$$

nesium bromide and cyanogen chloride yield indole-3-nitrile and methylketolylmagnesium bromide similarly gives 2-methylindole-3-nitrile. Indole-

[316] Salway, *J. Chem. Soc.*, **103**, 351 (1913). Majima and Kotake, *Ber.*, **B 55**, 3865 (1922). Mingoia, *Gazz. chim. ital.*, **61**, 646 (1931). Sanna, *ibid.*, **59**, 838 (1929).

[317] Sanna, *Rend. seminar. fac. sci. univ. Cagliari*, **4**, 28–33 (1934); *Chem. Zentr.*, **1935**, II, 367; *Chem. Abstr.*, **30**, 6363 (1936). Sanna and Massidda, *Gazz. chim. ital.*, **61**, 60 (1931); *Chem. Abstr.*, **25**, 2720 (1931). Sanna and Athene, *Rend. seminar. fac. sci. univ. Cagliari*, **4**, 62 (1934); *Chem. Zentr.*, **1935**, ii, 219; *Chem. Abstr.*, **30**, 6363 (1936).

[318] Weyler and Binder, *Arch. Pharm.*, **276**, 506–516 (1937); *Chem. Abstr.*, **32**, 939 (1938).

[319] Majima and Hoshino, *Ber.*, **B 58**, 2042 (1925). Wieland, Konz, and Mittasch, *Ann.*, **513**, 1 (1934). Hoshino and Kotake, *ibid.*, **516**, 76 (1935).

3-γ-butyric acid has been prepared[320] by the hydrolysis of the nitrile obtained through the action of indolylmagnesium bromide with γ-chlorobutyronitrile.

Gramine has been prepared through the interaction of indolylmagnesium bromide with dimethylaminoacetonitrile.[321]

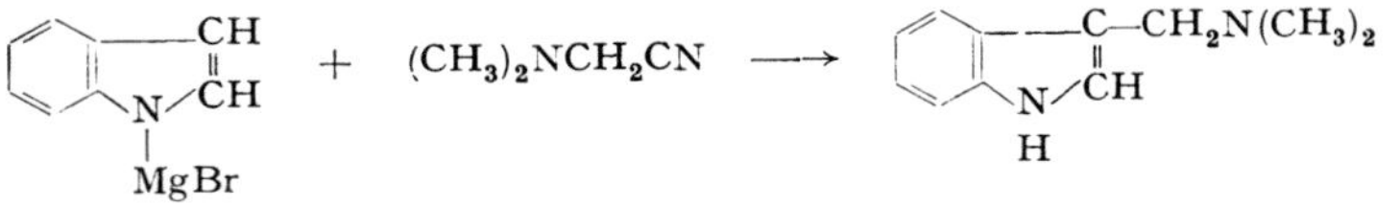

$$\text{Indolyl-MgBr} + (CH_3)_2NCH_2CN \longrightarrow \text{3-}(CH_2N(CH_3)_2)\text{-indole}$$

Gramine, m.p. 134°

The reaction of phthalyl chloride and of phthalic anhydride with indolylmagnesium halides as well as with the corresponding derivatives of skatole and methylketole has been the object of a number of studies by Oddo and co-workers.[322] Methylketolylmagnesium bromide and phthalyl chloride react to give methylketolphthalein (XVII) as the principal product. An isomer of unestablished structure was also obtained (possibly from the symmetrical form of phthalyl chloride). The methylketolphthalein on treatment with potassium hydroxide yields the salt of the dye "methylketole

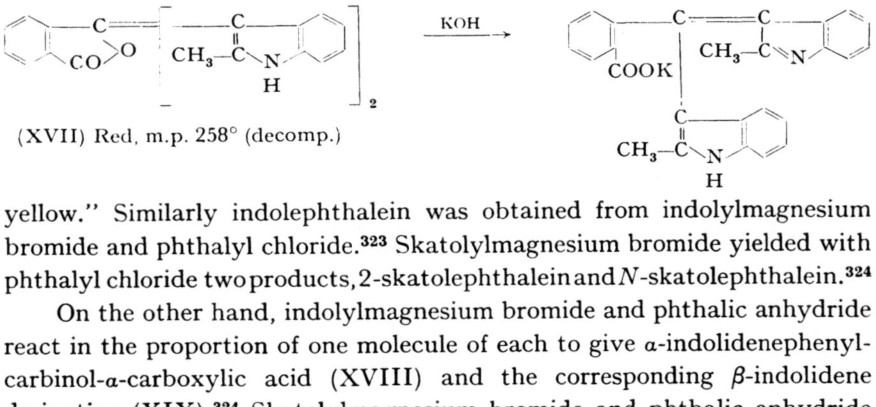

$$\text{(XVII)} \xrightarrow{KOH} \text{dye potassium salt (COOK)}$$

(XVII) Red, m.p. 258° (decomp.)

yellow." Similarly indolephthalein was obtained from indolylmagnesium bromide and phthalyl chloride.[323] Skatolylmagnesium bromide yielded with phthalyl chloride two products, 2-skatolephthalein and *N*-skatolephthalein.[324]

On the other hand, indolylmagnesium bromide and phthalic anhydride react in the proportion of one molecule of each to give α-indolidenephenylcarbinol-α-carboxylic acid (XVIII) and the corresponding β-indolidene derivative (XIX).[324] Skatolylmagnesium bromide and phthalic anhydride yield two products, *N*-skatolylhydroxyphthalide (XX) and the skatole

[320] Nametkin, Dzbenovskii, and Rudnev, *Compt. rend. acad. sci. U. S. S. R.*, **32**, 333 (1941); *Chem. Abstr.*, **37**, 3756 (1943).

[321] Wieland and Hsing, *Ann.*, **526**, 188 (1936).

[322] Oddo, *Atti accad. Lincei*, [6] **1**, 236–238 (1925); *Chem. Abstr.*, **19**, 2823 (1925). Oddo, *Gazz. chim. ital.*, **56**, 437 (1926); *Chem. Abstr.*, **21**, 211 (1927).

[323] Oddo, *Gazz. chim. ital.*, **58**, 569 (1928); *Chem. Abstr.*, **23**, 1634 (1929).

[324] Oddo and Toffoli, *Gazz. chim. ital.*, **60**, 3 (1930); *Chem. Abstr.*, **24**, 3784 (1930).

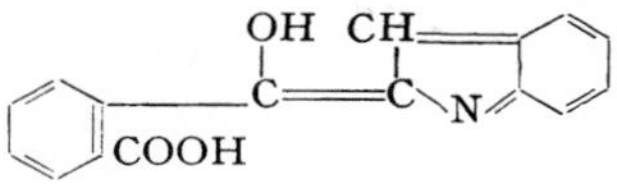

(XVIII) 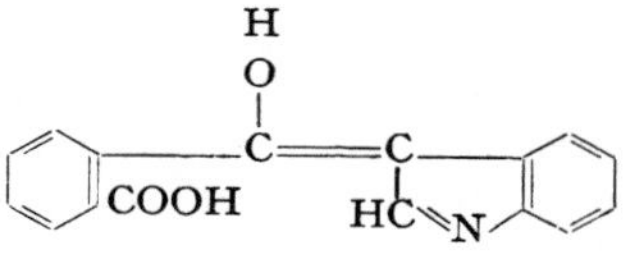 (XIX)

analog of XVIII (XXI). Phthalic anhydride and methylketolylmagnesium bromide react to give 2-methylindolidenephenylcarbinol-*o*-carboxylic acid (the 2-methyl analog of XIX).[325]

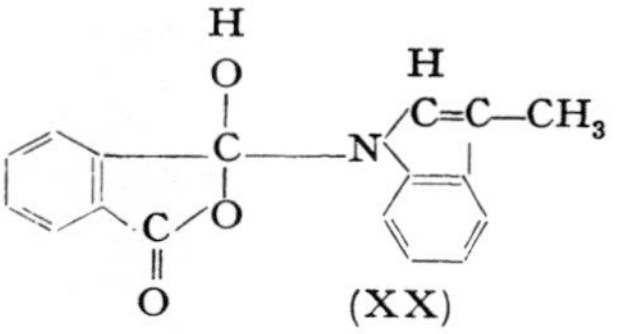

(XX) 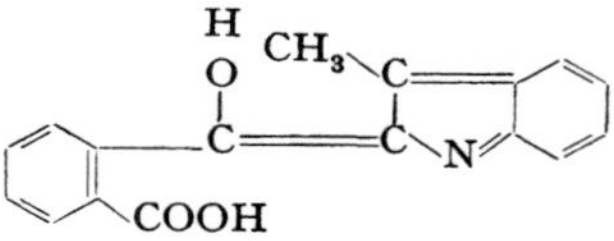 (XXI)

Methylketolylmagnesium bromide and acetophenone[326] react to yield di-α-methylindolylphenylmethylmethane (XXII) rather than the carbinol which might have been expected. Methylketolylmagnesium bromide and indolylmagnesium bromide also give similar products with acetone and other ketones.[326] The diindolyldimethylmethane (XXIII) obtained in this manner from acetone and indolylmagnesium bromide has also been prepared by direct condensation of indole and acetone in presence of hydrogen chloride.[327]

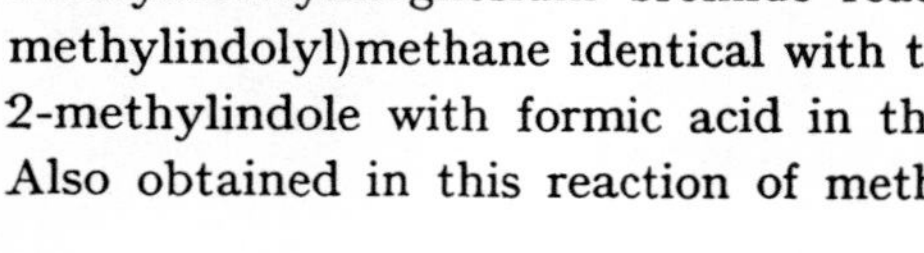

(XXII) 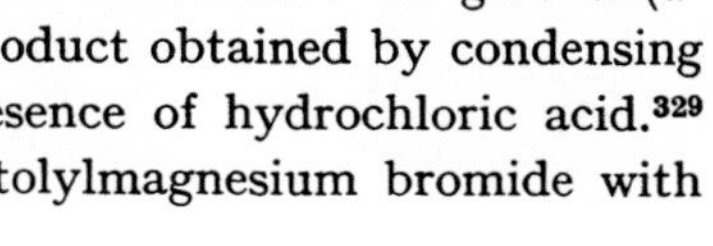 (XXIII)

Methylketolylmagnesium bromide reacts with iodoform[328] to give tri(α-methylindolyl)methane identical with the product obtained by condensing 2-methylindole with formic acid in the presence of hydrochloric acid.[329] Also obtained in this reaction of methylketolylmagnesium bromide with

[325] As to structure compare Fischer, *Ann.*, **242,** 381 (1882).

[326] Oddo and Perotti, *Gazz. chim. ital.*, **60,** 13 (1930); *Chem. Abstr.*, **24,** 3785 (1930). Majima and Kotake, *Ber.*, **B 55,** 3865 (1922).

[327] Scholtz, *Ber.*, **46,** 1082 (1913).

[328] Oddo and Tognaschini, *Gazz. chim. ital.*, **53,** 271 (1923); *Chem. Abstr.*, **17,** 2883 (1923).

[329] Ellinger and Flamand, *Z. physiol. Chem.*, **62,** 276 (1909); **71,** 6 (1911).

iodoform were α-methylindolyl-α-methylindolidenemethane (XXIV) and α-methylindolidene-di(α-methylindolyl)methane (XXV). The latter compound (XXV) is obtained also through the action of methylketolylmagnesium

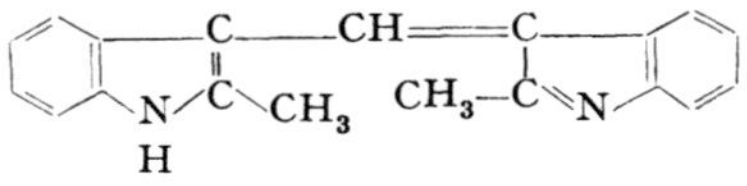

(XXIV) 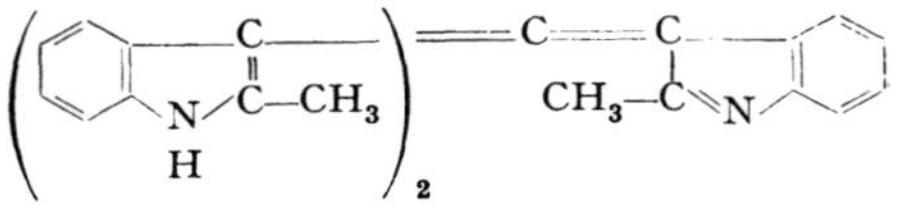 (XXV)

bromide on carbon tetrachloride.[330]

The leuco bases of the rosindoles which were first prepared by the action of aldehydes on methylketole[331] can be prepared in good yields also through the action of methylketolylmagnesium bromide on aldehydes.[332] The leuco

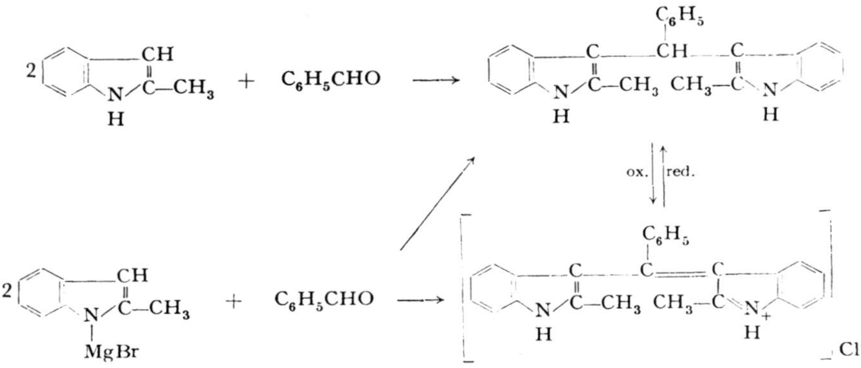

$$2\ \text{(methylketole)} + C_6H_5CHO \longrightarrow \text{leuco base}$$

$$2\ \text{(N-MgBr methylketole)} + C_6H_5CHO \longrightarrow [\text{rosindole}]^+\ Cl^-$$

ox. red.

bases on oxidation in the presence of acids give the rosindoles which can be reduced in turn to the leuco bases. Skatolylmagnesium bromide and indolylmagnesium bromide react similarly with aldehydes. On the other hand, when equimolecular proportions of indolylmagnesium bromide and acetaldehyde are heated together for ten hours the product is the ether XXVI.[333] The expected intermediate alcohol was not isolated. The reaction of methylketolylmagnesium bromide parallels that with indolylmagnesium bromide.

[330] Oddo and Sanna, *Gazz. chim. ital.*, **54**, 687 (1924); *Chem. Abstr.*, **19**, 829 (1925).

[331] Fischer, *Ber.*, **19**, 2988 (1886); *Ann.*, **242**, 373 (1887). Renz and Löw, *Ber.*, **36**, 4326 (1903). Walther and Clemen, *J. prakt. Chem.*, [2] **61**, 249 (1900).

[332] Mingoia, *Gazz. chim. ital.*, **56**, 772 (1926); *Chem. Abstr.*, **21**, 1117 (1927). Oddo and Toffoli, *Gazz. chim. ital.*, **64**, 359 (1934); *Chem. Abstr.*, **28**, 6436 (1934). Majima and Kotake, *Ber.*, **B 55**, 3865 (1922).

[333] Oddo and Cambieri, *Gazz. chim. ital.*, **70**, 559–566 (1940); *Chem. Abstr.*, **35**, 1050 (1941).

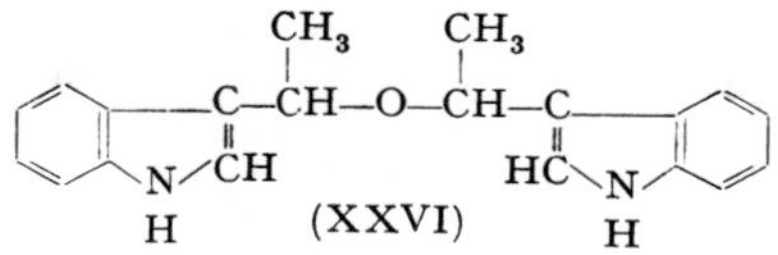

(XXVI)

The action of hydrogen peroxide on indolylmagnesium bromide, methylketolylmagnesium bromide, and skatolylmagnesium bromide gave 1-hydroxyindole, 2-methylindoxyl and 1-hydroxy-3-methylindole, respectively.[334]

The action of various substances with indolylmagnesium bromide in anisole has been compared with that in ethyl ether.[335] Carbonyl compounds are reported to give better yields of products when the reaction is carried out in anisole while acid chlorides are said to give better yields when they react in ether.

Indole Polymers

Indolenines of the type of 2,3,3-trimethylindolenine yield dimers on treatment with Grignard reagents,[335] the compound behaving as though it had the formula I. The dimer melts at 132° and is reconverted into the monomer on heating above its melting point.

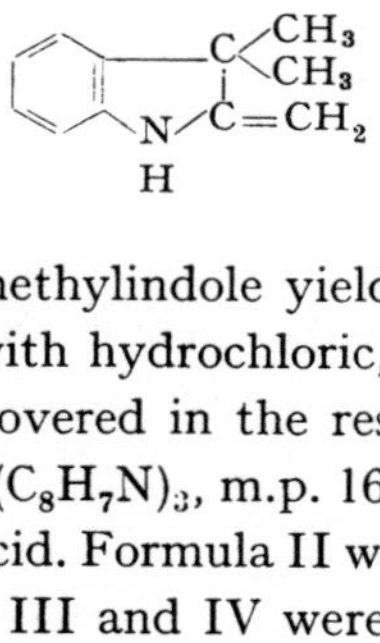

(I)

Indole, skatole, and 1-methylindole yield dimeric and trimeric forms quite readily on treatment with hydrochloric, hydrobromic, or phosphoric acid. Triindole was first discovered in the residue from the distillation of indole.[336] The same polymer, $(C_8H_7N)_3$, m.p. 169°, was obtained when indole was heated with phosphoric acid. Formula II was proposed for the compound by Keller[336] while structures III and IV were proposed by Oddo[337] for di-

[334] Ingraffia, *Gazz. chim. ital.*, **63,** 175 (1933); *Chem. Abstr.*, **27,** 3710 (1933).

[335] Majima and Kotake, *Ber.*, **B 55,** 3865 (1922); *Rept. Inst. Phys. Chem. Research Japan*, **2,** 82–91 (1922); *J. Chem. Soc. Japan*, **43,** No. 12 (1922); *Chem. Abstr.*, **17,** 2285 (1923).

[336] Keller, *Ber.*, **46,** 726 (1913).

[337] Oddo, *Gazz. chim. ital.*, **43,** I, 385 (1913). Oddo and Crippa, *ibid.*, **54,** 339 (1934). Oddo and Mingoia, *ibid.*, **57,** 380 (1927).

[338] Chelintzev *et al.*, *J. Russ. Phys. Chem. Soc.*, **47,** 1224 (1915); *Chem. Abstr.*, **9,** 3055 (1915).

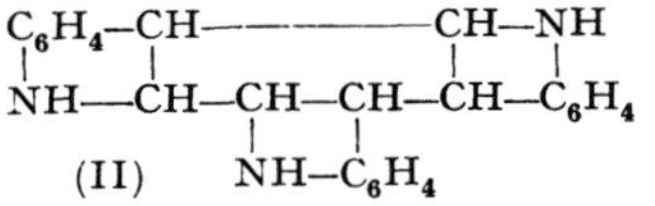

```
C6H4–CH————————CH–NH
|     |         |    |
NH—–CH–CH–CH–CH–C6H4
         |   |
(II)     NH–C6H4
```

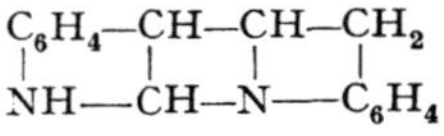

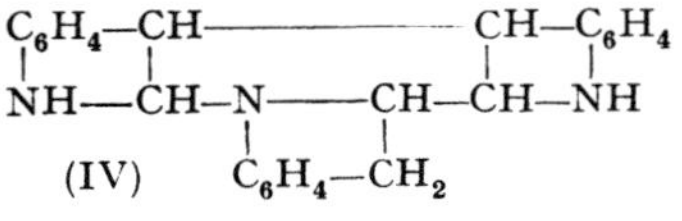

```
C6H4–CH–CH–CH2            C6H4–CH——————————CH–C6H4
|     |   |   |           |     |            |    |
NH—–CH–N—–C6H4            NH—–CH–N——–CH–CH–NH
                                  |      |
(III)                      (IV)  C6H4–CH2
```

indole and tri-indole, respectively. It was first reported[338] that tri-indole contains only one active hydrogen atom but subsequently shown that di-indole contains two and tri-indole three active hydrogen atoms.[339] The dimer and trimer were formulated by Schmitz-Dumont as the linear structures V and VI.[340] This formulation seems much more reasonable than the earlier ones. The kinetics of the polymerization of indole as well as of the

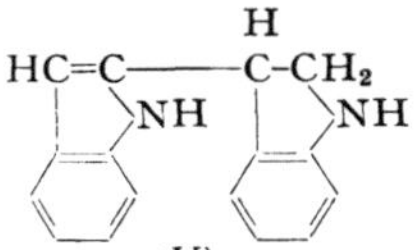

```
           H                   H             H
HC=C———C-CH2            H2C-C———C=C———C-CH2
 | \NH  | \NH              | \NH | \NH  | \NH
       (V)                         (VI)
```

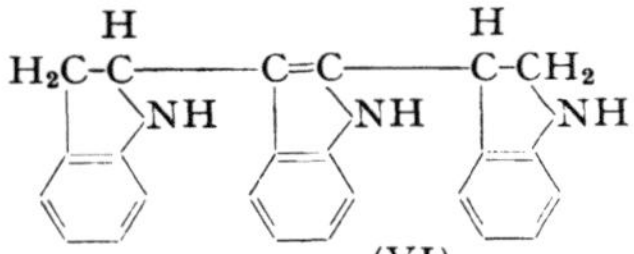

mixed polymerization of skatole and various ethylene derivatives were studied by Schmitz-Dumont and co-workers.[341]

Preparation and Reactions of Gramine

Gramine, donaxine, 3-dimethylaminomethylindole, was first isolated in 1932[342] from the germ of Swedish barley. Later, in 1935, the substance was isolated from the Asiatic reed, *Arundo donax* L.[343] The compound was first

[339] Schmitz-Dumont, Hamann, and Geller, *Ann.*, **504,** I (1933). Schmitz-Dumont and Hamann, *Ber.*, **B 66,** 71 (1933).

[340] Schmitz-Dumont, ter Horst, and Müller, *Ann.*, **538,** 261 (1939). Schmitz-Dumont, and ter Horst, *Ber.*, **B 58,** 250 (1933). Schmitz-Dumont and Hamann, *J. prakt. Chem.*, **139,** 167 (1934).

[341] Schmitz-Dumont, Hamann, and Diebold, *Ber.*, **B 71,** 205 (1928). Diebold and Thëmke, *ibid.*, **B 70,** 2189 (1937).

[342] Von Euhler and Hellström, *Z. physiol. Chem.*, **208,** 43 (1932). Von Euhler and Hellström, *ibid.*, **217,** 23 (1933).

[343] Orechoff and Norkina, *Ber.*, **66,** 436 (1935). Von Euhler, Erdtman, and Hellström, *ibid.*, **69,** 743 (1936).

synthesized through the reaction of indolylmagnesium iodide with dimethylaminoacetonitrile.[321] A later synthesis was effected through the application

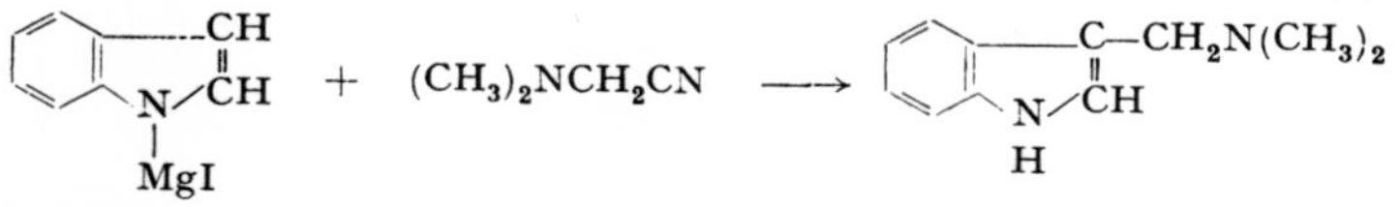

$$\text{Indole-}N\text{-MgI} + (CH_3)_2NCH_2CN \longrightarrow \text{3-}(CH_2N(CH_3)_2)\text{-indole (N-H)}$$

of the Mannich reaction using indole, formaldehyde, and dimethylamine.[344]

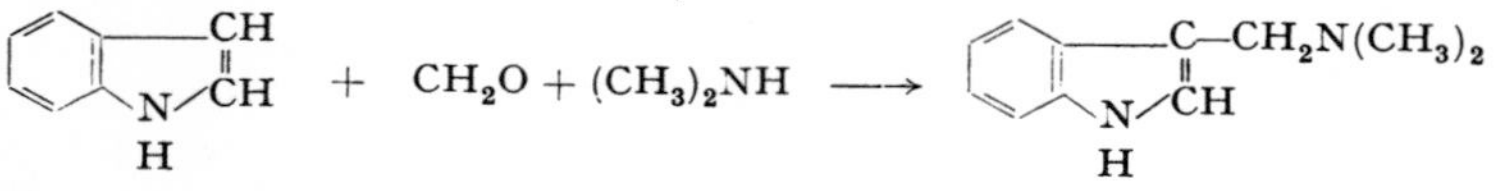

$$\text{Indole (N-H)} + CH_2O + (CH_3)_2NH \longrightarrow \text{3-}(CH_2N(CH_3)_2)\text{-indole (N-H)}$$

A number of homologs and analogs of gramine have been prepared similarly.[345] Recently gramine and its methiodide have proved to be quite useful agents in the synthesis of a number of interesting and valuable indole derivatives. Thus gramine methiodide reacts with potassium cyanide to give the nitrile of indole-3-acetic acid,[346] the latter being an important plant hormone. Gramine methiodide and gramine itself condense with many

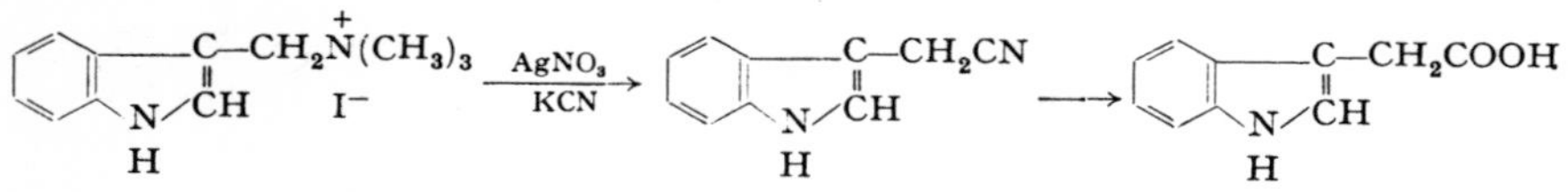

$$\text{3-}(CH_2\overset{+}{N}(CH_3)_3\ I^-)\text{-indole} \xrightarrow[KCN]{AgNO_3} \text{3-}(CH_2CN)\text{-indole} \longrightarrow \text{3-}(CH_2COOH)\text{-indole}$$

compounds containing active methylene groups (malonic ester, acetoacetic

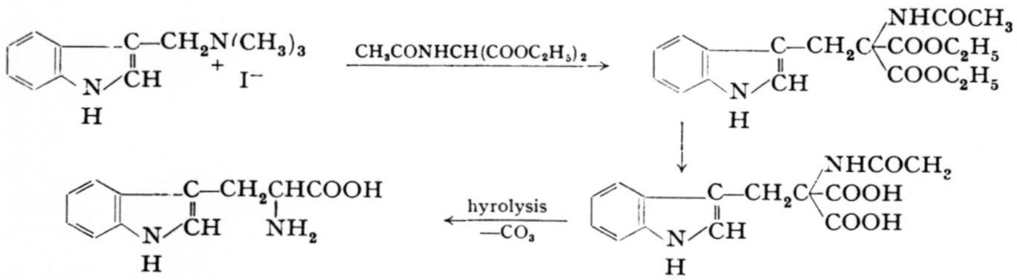

$$\text{3-}(CH_2\overset{+}{N}(CH_3)_3\ I^-)\text{-indole} \xrightarrow{CH_3CONHCH(COOC_2H_5)_2} \text{3-}(CH_2C(NHCOCH_3)(COOC_2H_5)_2)\text{-indole}$$

$$\downarrow$$

$$\text{3-}(CH_2CH(NH_2)COOH)\text{-indole} \xleftarrow[-CO_2]{\text{hyrolysis}} \text{3-}(CH_2C(NHCOCH_3)(COOH)_2)\text{-indole}$$

D,L-Tryptophan

[344] Kühn and Stein, *ibid.*, **B 70,** 567 (1937).

[345] Bell and Lindwall, *J. Org. Chem.*, **13,** 547 (1948). Brehm, *J. Am. Chem. Soc.*, **71,** 3541 (1949). Brehm and Lindwall, *J. Org. Chem.*, **15,** 685 (1950). Craig and Tarbell, *J. Am. Chem. Soc.*, **71,** 462 (1949).

[346] Snyder, Smith, and Stewart, *J. Am. Chem. Soc.*, **66,** 200 (1944). Saltzer, *U. S. Dept of Commerce, Office Technical Service*, P. B. Report No. 706 (1946). Wieland, Fischer, and Moewus, *Ann.*, **561,** 47 (1948). Japanese patent 161,544; *Chem. Abstr.*, **43,** 2236 (1949). Hausch, and Godfrey, *J. Am. Chem. Soc.*, **73,** 3518 (1951). Cf. also Eliel and Murphy, *ibid.*, **75,** 3589 (1953).

Gramine $\xrightarrow{O_2NCH_2COOC_2H_5}$ indol-3-yl–$CH_2CH(NO_2)COOC_2H_5$ $\xrightarrow{\text{catalytic reduction}}$ D,L-Tryptophan

Gramine $\xrightarrow{CH_3CH_2CH_2NO_2}$ indol-3-yl–$CH_2CH(NO_2)CH_2CH_3$

ester, ethyl cyanoacetate, etc.) giving thereby many important indole acids and other derivatives.[347]

Gramine methiodide was reported by Schramm[348] to be dimorphous yielding a stable form darkening at 175°, melting above 350°, and a metastable form melting at 172–173°, but this has recently been shown to be in error, pure gramine methiodide melting at 168–169°.[348a]

The analog of gramine with the dimethylaminomethyl group in position 2 has been prepared from indole-2-carboxylic acid as well as through ring closure from dimethylaminoacet-*o*-toluidide.[349] The reactions of 2-dimethyl-

Indole-2-COOH $\xrightarrow[(CH_3)_2NH]{SOCl_2}$ Indole-2-$CON(CH_3)_2$ $\xrightarrow{LiAlH_4}$ Indole-2-$CH_2N(CH_3)_2$

o-$CH_3C_6H_4NHCOCH_2N(CH_3)_2$ $\xrightarrow{NaNH_2}$ Indole-2-$CH_2N(CH_3)_2$

aminomethylindole and its salts are similar to those of gramine and its salts.

[347] Albertson and Tuller, *J. Am. Chem. Soc.*, **67,** 502 (1945). Snyder and Smith, *ibid.*, **66,** 350 (1944). British patent 599,046; *Chem. Abstr.*, **42,** 7341 (1948). Albertson, Archer and Suter, *J. Am. Chem. Soc.*, **66,** 500 (1944). Howe, Zambito, Snyder, and Tishler, *ibid.*, **67,** 58 (1945). Boekelheide and Ainsworth, *ibid.*, **72,** 2134 (1950). Hegedüs, *Helv. Chim. Acta*, **29,** 1599 (1946). Jackman and Archer, *J. Am. Chem. Soc.*, **68,** 2105 (1946). Albertson, Tuller, King, Fishburn and Archer, *ibid.*, **70,** 1150 (1948). Lyttle and Weisblatt, *ibid.*, **69,** 2118 (1947); **71,** 3079 (1949). Snyder and Katz, *ibid.*, **69,** 3140 (1947). Snyder and Eliel, *ibid.*, **70,** 1703, 1857, 3835 (1948); **71,** 663 (1949). Snyder and Pilgrim, *ibid.*, **70,** 3770, 3787, 4233 (1948). Stork and Singh, *ibid.*, **73,** 4742 (1951).

[348] Schramm, *J. Am. Chem. Soc.*, **63,** 2961 (1951).

[348a] Geissman and Armen, *ibid.*, **74,** 3916 (1952). Gray, *ibid.*, **75,** 1252 (1953).

[349] Kornfield, *J. Org. Chem.*, **16,** 806 (1951).

Carboxylic Acids of the Indole Series

The Fischer synthesis has been the means of the direct synthesis of many acids in this series. Illustrations of this method of synthesis have been given under the Fischer synthesis.

Direct introduction of the carboxyl group into the pyrrole part of the indole molecule has been effected by heating the alkali metal salts of indole in a stream of carbon dioxide.[350] Indole itself yields principally indole-3-carboxylic acid along with small amounts of indole-2-carboxylic acid. The preparation of the indole carboxylic acids from the magnesyl derivatives of indole, skatole, and 2-methylindole through the interaction of these derivatives with carbon dioxide and with chloroformic ester has been discussed under the heading of the indolemagnesium halides.

2-Methylindole is converted into indole-2-carboxylic acid by fusion with alkali in the presence of air.[351] 2-Methyl-3,3-diethylindolenine on heating with aqueous permanganate is converted to 3,3-diethylindolenine-2-carboxylic acid.[352] 2-Methyl-3,3-diethylindolenine is converted by nitrous acid to the oxime of 3,3-diethylindolenine-2-aldehyde.[352] The latter with acetic anhydride loses water yielding the nitrile which on saponification yields not only the corresponding carboxylic acid but (through elimination of hydrogen cyanide) an oxindole homolog as well. The same series of transformations

$$C_6H_4\left<\begin{matrix}C(C_2H_5)_2 \\ | \\ N{=}C{-}CH_3\end{matrix}\right. \xrightarrow{HNO_2} C_6H_4\left<\begin{matrix}C(C_2H_5)_2 \\ | \\ N{=}C{-}CH{=}NOH\end{matrix}\right. \longrightarrow C_6H_4\left<\begin{matrix}C(C_2H_5)_2 \\ | \\ N{=}C{-}CN\end{matrix}\right.$$

$$\xrightarrow{\text{Saponification}} C_6H_4\left<\begin{matrix}C(C_2H_5)_2 \\ | \\ N{=}C{-}COOH\end{matrix}\right. \text{ and } C_6H_4\left<\begin{matrix}C(C_2H_5)_2 \\ | \\ NH{-}CO\end{matrix}\right.$$

have been effected in the preparation of 3,3-dibenzylindolenine-2-carboxylic acid and 3,3-dibenzyloxindole from 2-methyl-3,3-dibenzylindolenine.[353]

The ionization constants of several indole and pyrrole carboxylic acids have been determined showing that the α-carboxylic acids are considerably stronger than the β-acids.[354]

[350] Zatti and Ferratini, *Ber.*, **23**, 2296 (1890). Weisgerber, *ibid.*, **43**, 3520 (1910). See also Shirley and Roussel, *J. Am. Chem. Soc.*, **75**, 375 (1953).

[351] Ciamician and Zatti, *Ber.*, **21**, 1929 (1888).

[352] Plancher, *Gazz. chim. ital.*, **28**, ii, 363 (1898).

[353] Leuchs and Overberg, *Ber.*, **B 64**, 1896 (1931).

[354] Angeli, *Gazz. chim. ital.*, **22**, ii, 16 (1892).

Pyrrole-α-carboxylic acid	$K = 0.00403$
Indole-α-carboxylic acid	$K = 0.0177$
Indole-β-carboxylic acid	$K = 0.00056$
3-Methylindole-α-carboxylic acid	$K = 0.0047$
2-Methylindole-β-carboxylic acid	$K = 0.00013$

Indole-2-carboxylic acid can be prepared conveniently through the application of procedures already described.[355] The acid crystallizes from hot water in colorless needles, m.p. 206–208°.[356] On heating above its melting point the compound loses carbon dioxide and gives indole. The *N*-hydroxy derivative was the first member of the series of *N*-hydroxyindoles to be prepared.[357] It is obtained when *o*-nitrobenzylmalonic acid is heated with sodium hydroxide and also through reduction of *o*-nitrophenylpyruvic acid with sodium amalgam. The compound is obtained as colorless prisms which melt with decomposition at 159.5° and is very unstable. It is reduced by zinc dust and acetic acid to indole-2-carboxylic acid.

A number of indole-2-carbonyl derivatives of amino acid esters have been prepared through the action of indole-2-carbonyl chloride on the esters.[358]

Indole-3-carboxylic acid[359] melts in a closed tube at about 218° with gas evolution. It loses carbon dioxide more readily than does the α-acid. The nitrile of this acid has been prepared[360] through the condensation of amyl formate with *o*-aminobenzyl cyanide. 2-Methylindole, potassium cyanide,

$$C_6H_4(CH_2CN)(NH_2) \xrightarrow[Na]{HCOOC_5H_{11}} \text{indole-3-carbonitrile (C—CN, CH, NH)}$$

[355] Fischer, *Ann.*, **236,** 141 (1886). Ciamician and Zatti, *Ber.*, **21,** 1930 (1888). Magnanini, *ibid.*, **21,** 1938 (1888). Zanetti, *ibid.*, **26,** 2007 (1893). Reissert, *ibid.*, **29,** 655 (1896); **30,** 1045 (1897). Piccinini and Salmoni, *Gazz. chim. ital.*, **32,** I, 252 (1902). Porcher and Hervieux, *Compt. rend.*, **145,** 345 (1907). Oddo and Sessa, *Gazz. chim. ital.*, **41,** i, 236 (1911). German patent 238,138; *Chem. Zentr.*, **1911,** ii, 1080. Madelung, *Ber.*, **45,** 3521 (1912). Gränacher, Mahal, and Gero, *Helv. Chim. Acta*, **7,** 579 (1924).

[356] Brehm, *J. Am. Chem. Soc.*, **71,** 3541 (1949).

[357] Reissert, *Ber.*, **29,** 639 (1896); **30,** 1035 (1897).

[358] Johnson, Andreen, and Holley, *J. Am. Chem. Soc.*, **67,** 423 (1945); **69,** 2370 (1947).

[359] Ciamician and Zatti, *Ber.*, **21,** 1929 (1888); **22,** 1976 (1889). Zatti and Ferratini, *ibid.*, **23,** 2296 (1890). Ellinger, *ibid.*, **39,** 2519 (1906). Porcher, *Compt. rend.*, **148,** 1211 (1909); *Bull. soc. chim.*, [4] **5,** 538 (1909). Pschorr and Hoppe, *Ber.*, **43,** 2544 (1910). Weissgerber, *ibid.*, **43,** 3526 (1910); **46,** 658 (1913).

[360] Gavrilov, *Isvestis Petrovosoi Selskochos Academii (Russia)*, **No. 1–4,** 14–15 (1919); *Chem. Abstr.*, **19,** 505 (1925). Pschorr and Hoppe, *Ber.*, **43,** 2543 (1910).

and mercury fulminate react to give 2-methylindole-3-carboxynitrile, m.p.

C—CN
‖
N—C—CH_3
H

207–210°. Hydrolysis of the nitrile gives the corresponding acid.

Compounds which contain the carboxyl group in a side chain and which are of considerable physiological interest are indole-3-acetic acid, indole-3-α-propionic acid, indole-3-β-propionic acid, and the α-amino derivative of the latter (tryptophan). Indole-3-acetic acid[361] was first observed in the products of the putrefaction of meat. Its structure was established through the synthesis of its methyl ester by the Fischer synthesis from the phenylhydrazone of methyl β-formylpropionate. The acid itself has been synthesized[362] through application of the Japp-Klingemann and Fischer syntheses from benzenediazonium chloride and ethyl α-acetoglutarate followed by hydrolysis and controlled decarboxylation of the diester which results. The compound has also been obtained from gramine methiodide.[346] Indole, formaldehyde, and hydrocyanic acid condense to give the nitrile of indole-

C—$CH_2\overset{+}{N}(CH_3)_3$ I^-
‖
N—CH
H

$\xrightarrow[KCN]{AgNO_3}$

C—CH_2CN
‖
N—CH
H

⟶

C—CH_2COOH
‖
N—CH
H

CH
‖
N—CH
H

$\xrightarrow[HCN]{CH_2O}$

3-acetic acid[363] which on hydrolysis yields the acid. Indole-3-acetic acid crystallizes from benzene in plates, melts at 164–165° and decomposes on strong heating into skatole and carbon dioxide. A very dilute solution of the acid made acidic with nitric acid is colored cherry red by nitrous acid and precipitates a red dye; the compound also yields an intense violet color with ferric chloride and hydrochloric acid (Salkowsky test).

Indole-3-acetic acid, also known as "heteroauxin," causes elongation of plant cells. Its unequal distribution in a plant causes phototropism and

[361] E. and H. Salkowski, *Ber.*, **13,** 192, 2217 (1880); *Z. physiol. Chem.*, **9,** 8 (1884). E. Salkowski, *ibid.*, **9,** 23 (1889). Ellinger, *Ber.*, **37,** 1801 (1904). Herter, *Chem. Zentr.*, **1908,** I, 1297, 1985. Hopkins and Cole, *ibid.*, **1903,** II, 1011.

[362] King and L'Ecuyer, *J. Chem. Soc.*, **1934,** 1901. Tanaka, *J. Pharm. Soc. Japan*, **60,** 74 (1940). Tanaka, *ibid.*, **60,** 219 (1940).

[363] French patent 841,676; *Chem. Abstr.*, **34,** 6947 (1940). British patent 517,692; *Chem. Abstr.*, **35,** 7107 (1941).

geotropism. It was the first of the so-called auxins of known structure to be discovered. The discovery of heteroauxin came about through the work of Kögl and co-workers.[364] The substance was isolated from urine as a crystalline material which stimulated cell elongation. Later the substance was found also in corn oil[365] and shown to be indole-3-acetic acid.[366] Indole-3-α-propionic acid and indole-3-pyruvic acid[367] and the 1-, 2-, and 5-methyl-indole-3-acetic acids possess similar activity while dihydroindole-3-acetic acid, 2-ethylindole-3-acetic acid, 2,5-dimethylindole-3-acetic acid, indole-2-carboxylic acid, indole-3-carboxylic acid, and indole-3-β-propionic acids are inactive.[368] Indole-3-β-propionic acid is one of the substances present in the putrefaction of proteins. It has been synthesized through the Fischer synthesis.[369] A number of derivatives of indole-3-acetic acid have been prepared through the Japp-Klingemann and Fischer syntheses.[370] Indole-3-acetic acid and its esters have been characterized as the 1,3,5-trinitrobenzene addition complexes.[371]

Cleavage of the indacylpyridinium salts gives acids of the indole series.[372]

$$\text{Indole-3-}COCH_2Br \xrightarrow{C_5H_5N} \left[\text{Indole-3-}COCH_2\overset{+}{N}C_5H_5\right] Br^- \xrightarrow{\text{alkali}} \text{Indole-3-}COOH$$

The same pyridinium salts are converted into ketoaldehydes by treatment with nitrosobenzene and hydrolysis of the resulting condensation product.[373]

[364] Kögl, Haagen-Smit, and Erxleben, *Z. physiol. Chem.*, **214**, 241 (1933); **216**, 31 (1933).

[365] Kögl, Erxleben, and Haagen-Smit, *ibid.*, **225**, 215 (1934).

[366] Kögl, Haagen-Smit, and Erxleben, *ibid.*, **288**, 90 (1934).

[367] Gränacher, Gero, and Schelling, *Helv. Chim. Acta*, **7**, 575 (1924).

[368] Pollard, *Ann. Repts.*, **32**, 425 (1936). Zimmerman, *Contrib. Boyce Thompson Inst.*, **7**, 439, 447 (1935).

[369] Ellinger, *Ber.*, **38**, 2884 (1905).

[370] Findlay and Dougherty, *J. Org. Chem.*, **13**, 560 (1948).
Stevens and Fox, *J. Am. Chem. Soc.*, **70**, 2263 (1948). Fox and Bullock, *ibid.*, **73**, 2754, 2756, 5155 (1951).

[371] Redemann, Wittwer, and Sell, *J. Am. Chem. Soc.*, **73**, 2957 (1951).

[372] Sanna, *Gazz. chim. ital.*, **72**, 357 (1942); *Chem. Abstr.*, **37**, 6662 (1943).

[373] Sanna, *Gazz. chim. ital.*, **72**, 363 (1942); *Chem. Abstr.*, **37**, 6662 (1943).

$$\left[\text{C}_8\text{H}_5\text{N(H)}\text{—C—COCH}_2\overset{+}{\text{N}}\text{C}_5\text{H}_5\right]\text{Br}^- \xrightarrow{\text{C}_6\text{H}_5\text{NO}} \text{C}_8\text{H}_5\text{N(H)}\text{—C—COCH}{=}\text{NC}_6\text{H}_5(\rightarrow \text{O})$$

$$\xrightarrow{\text{hydrolysis}} \text{C}_8\text{H}_5\text{N(H)}\text{—C—COCHO}$$

Indole and oxalyl chloride react in dry ether to give 2-indolylglyoxalyl chloride, m.p. 138–139°,[374] which on hydrolysis gives 2-indolylglyoxylic acid, m.p. 224–225°. The latter compound yields indole-2-carboxylic acid on heating with alkali. 2-Methylindole and oxalyl chloride gave a tarry product which on hydrolysis gave 2-methyl-3-indolylglyoxylic acid, m.p. 190°.

Indole reacts with diazo esters yielding 3-substituted and 1,3-disubstituted derivatives.[375]

[374] Giua, *Gazz. chim. ital.*, **54,** 593 (1924).

[375] Jackson and Manske, *Can. J. Research*, **13,** 170 (1935); *Chem. Abstr.*, **30,** 455 (1936); *Can. J. Research*, **B 14,** 1 (1936); *Chem. Abstr.*, **30,** 2565 (1936). Manske, U. S. patent 2,079,416; *Chem. Abstr.*, **31,** 4343 (1937).

CHAPTER II

Carbazole

Introduction

Carbazole, dibenzopyrrole, 9-azafluorene (*R.I.* 1675), is a colorless substance which crystallizes from alcohol, benzene, toluene, and glacial acetic acid in beautiful flakes. The preparation of carbazole in a pure state is not easy and as a result the melting-point values reported in the literature

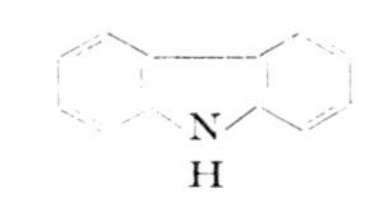

are quite often low, the value 235° being encountered frequently. In fact it is quite evident that a decisive value for the melting point of carbazole has not been established.[1] The following are among the more probable values which have been reported: 245°[2]; 246°[3]; 247°[4]; 245.6°[5]; 244.8°.[6] It is interesting to note that although carbazole has been shown to exhibit little or no fluorescence, the seemingly pure sample of Aristov[3] fluoresced brilliantly.

Carbazole was discovered in 1872[7] in the crude anthracene fraction of coal tar.[8] Different workers have at times applied various systems of numberings to the formulas of carbazole and of tetrahydrocarbazole.[9] The system of numbering employed here is the one used in the current American and British literature, by *Chemical Abstracts*, and given in the "Ring

[1] Campbell and Barclay, *Chem. Revs.*, **40**, 360 (1947).

[2] Tucker, *J. Chem. Soc.*, **1926**, 546.

[3] Aristov, *J. Chem. Ind.* (*Moscow*), **5**, 721 (1938); *Chem. Abstr.*, **23**, 138 (1929).

[4] Kirby, *J. Soc. Chem. Ind.*, **40**, 274 T (1921).

[5] Zelinsky, Titz, and Gaverdowskaja, *Ber.*, **59**, 2590 (1926).

[6] Senseman and Nelson, *Ind. Eng. Chem.*, **15**, 382 (1923).

[7] Graebe and Glaser, *Ber.*, **5**, 12 (1872); *Ann.*, **163**, 343 (1872).

[8] Compare Ziedler, *Ann.*, **191**, 296 (1878).

[9] See pages 3 and 4 of Cohn's monograph, *Die Carbazolgruppe*. Thieme, Leipzig, 1919.

Index."[10] Unfortunately another system of numbering in which the nitrogen atom is numbered 5 was given authoritative support in 1925[11] although it was subsequently replaced by the system used herein. Indiscriminate use of the two systems, sometimes without designating which has been used, has led at times to some confusion.

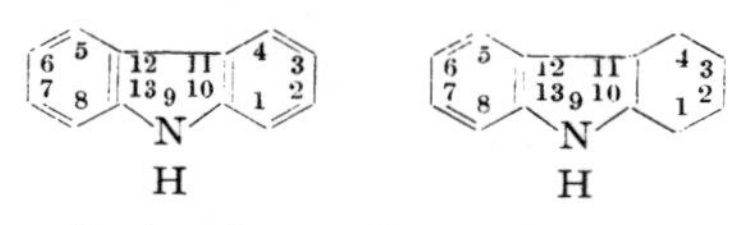

Carbazole Tetrahydrocarbazole

Carbazole may be identified as its picrate, bright red needles, m.p. 185°,[12] and as its trinitrobenzene derivative, red needles, m.p. 166°,[12] or styphnate, m.p. 178.5–179.5°.[13] The addition compound of carbazole with *m*-dinitrobenzene has been prepared[14] as has thecomplex with 1,3,5-trinitronaphthalene.[15] Other complexes of carbazole with aromatic nitro compounds have been studied.[16] Many substituted carbazoles may be similarly characterized as complexes with polynitro compounds.[17] Complexes of carbazoles with quinones[18] have been studied as has carbazole perchlorate[19] as well as an addition compound of carbazole and tetrachlorophthalic anhydride.[20]

As an indole derivative, carbazole gives the pine splinter test (formation of a red color when a pine splinter soaked in hydrochloric acid is held in the vapor of an alcoholic solution of the carbazole). Many color tests have been developed for the detection of carbazole.[21] The most familiar of

[10] Patterson and Capell, *The Ring Index*, Reinhold, New York, 1940, p. 229, No. 1675.

[11] Patterson, *J. Am. Chem. Soc.*, **47**, 556 (1925).

[12] Graebe and Glaser, *Ber.*, **5**, 12 (1872); *Ann.*, **163**, 343 (1872).

[13] Hoo, Ma, and Sah, *Chem. Abstr.*, **28**, 3692 (1934); *Science Repts. Natl. Tsing Hua Univ.*, **2**, 191–199 (1933).

[14] Karve and Sudborough, *J. Indian Inst. Sci.*, **4**, 159-176 (1921); *Chem. Abstr.*, **16**, 65 (1922).

[15] Sudborough, Picton, and Karve, *J. Indian Inst. Sci.*, **4**, 43 (1921); *Chem. Abstr.*, **16**, 560 (1922).

[16] Kremann and Strzelbe, *Monatsh.*, **42**, 167 (1921). Bornstein, Schliewiensky, and Szczesny-Heyl, *Ber.*, **B 59**, 2812 (1926). Orchin and Woolfolk, *J. Am. Chem. Soc.*, **68**, 1727 (1946). Kofler, *Z. physik. Chem.*, **A 190**, 287 (1942). Ciusa and Vecchiotti, *Atti accad. Lincei*, [II] **21**, 161 (1921); *Gazz. chim. ital.*, **43**, II, 91 (1913).

[17] Kent, *J. Chem. Soc.*, **1935**, 976. Kent and McNeil, *ibid.*, **1938**, 8. Sudborough, *ibid.*, **109**, 1339 (1916).

[18] Weitz and Schmidt, *J. prakt. Chem.*, **158**, 211 (1941).

[19] Hofmann, Metzler, and Lecher, *Ber.*, **43**, 178 (1910).

[20] Pfeiffer, *ibid.*, **B 55**, 413 (1922).

[21] See page 43 of Cohn's monograph, *Die Carbazolgruppe*. Thieme, Leipzig, 1919.

these is the development of a bluish-green color when a trace of carbazole is dissolved in concentrated sulfuric acid and a drop of nitric acid is then added. The formation of a stable deep blue color with xanthydrol forms the basis for the colorimetric determination of carbazole.[22]

Preparation of Carbazole

Graebe-Ullmann Synthesis

The method of Graebe and Ullmann[23] is of wide application in the synthesis of carbazoles. *o*-Aminodiphenylamine is treated with nitrous acid giving 1-phenyl-1,2,3-benzotriazole, which loses nitrogen on heating to give a quantitative yield of carbazole. The reaction has been utilized[24] for the preparation of a number of carbazoles. Carbazole is also obtained in the distillation of 1-phenyl-1,2,3-benzotriazole-5-carboxylic acid with lime.[25] 7-Bromo-1-phenyl-1,2,3-benzotriazole-5-carboxylic acid on heating with lime gives carbazole rather than the expected 1-bromocarbazole. Similarly 1-phenyl-1,2,3-benzotriazole-5-sulfonic acid on heating gives carbazole and not carbazole-3-sulfonic acid.[26]

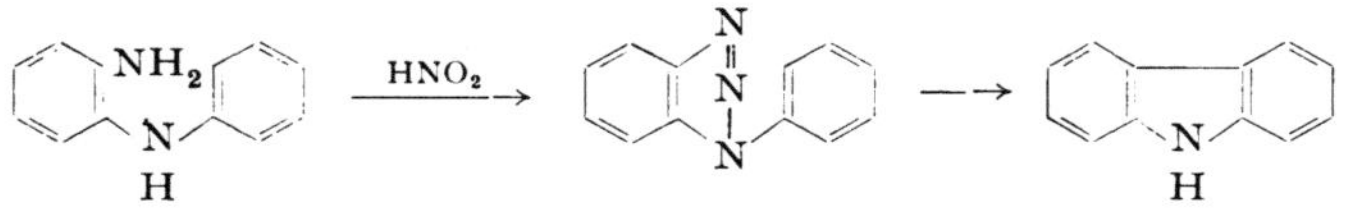

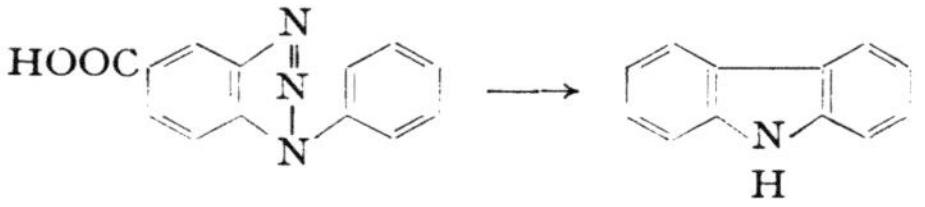

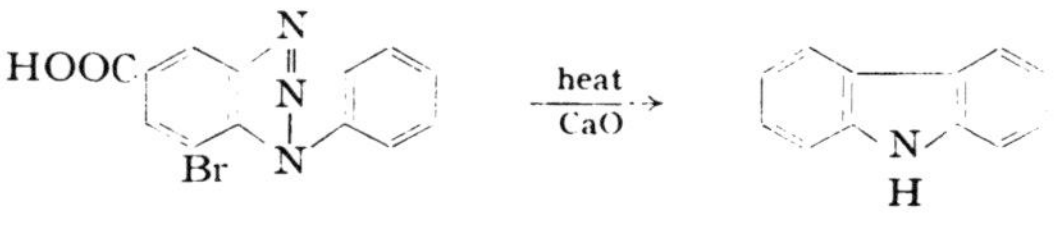

[22] Arreguine, *Rev. univ. nacl. Cordoba (Arg.)*, **31,** 1706 (1944). *Chem. Abstr.*, **39,** 3222 (1945).

[23] Graebe and Ullmann, *Ann.*, **291,** 16 (1896).

[24] Ullmann, *ibid.*, **332,** 82 (1904). Preston, Tucker, and Cameron, *J. Chem. Soc.*, **1942,** 500. Bremer, *Ann.*, **514,** 279, (1934). Borsche and Feise, *Ber.*, **40,** 378 (1907). Kehnmann, Onlevay, and Regis, *ibid.*, **46,** 3712 (1913). Hughes, Mannsell, and Wright, *J. Proc. Roy. Soc. N. S. Wales*, **71,** 428 (1938). Campbell and MacLean, *J. Chem. Soc.*, **1942,** 504. Coker, Plant, and Turner, *ibid.*, **1951,** 110. Tomlinson, *ibid.*, **1951,** 809. Clifton and Plant, *ibid.*, **1951,** 461.

[25] Ullmann, *Ber.*, **31,** 1697 (1898); *Ann.*, **332,** 84 (1904). Deletra and Ullmann, *J. Chem. Soc.*, **86,** i, 270 (1904); *Arch. sci. phys. nat.*, [V] **17,** 78–92 (1904).

[26] Schwalbe and Wolff, *Ber.*, **44,** 237 (1911).

A synthesis of *N*-alkylcarbazoles paralleling the Graebe-Ullmann synthesis has been effected[27] but must take place through a different mechanism since the formation of a benzotriazole as an intermediate cannot be postulated. *N*-Ethylcarbazole, m.p. 67–68°, and *N*-methylcarbazole, m.p. 88°, have both been prepared by this method.

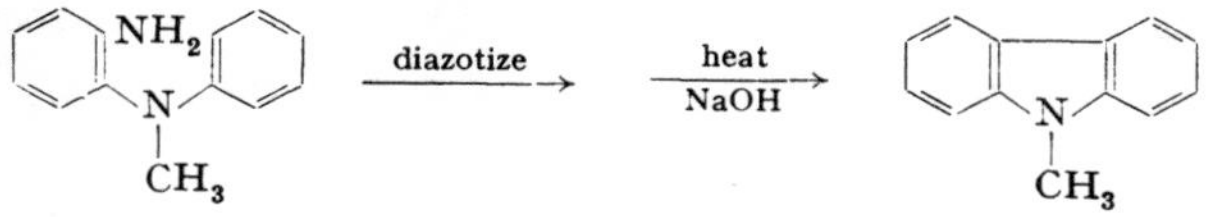

Method of Borsche

Tetrahydrocarbazole is obtained when cyclohexanone phenylhydrazone is heated with dilute sulfuric acid,[28] the preparation constituting an application of the Fischer indole synthesis. Homologs of tetrahydrocarbazole can be prepared in similar fashion.[29] Reagents other than sulfuric acid have been employed frequently.[30,31] Glacial acetic acid, especially, gives a purer, cleaner product.

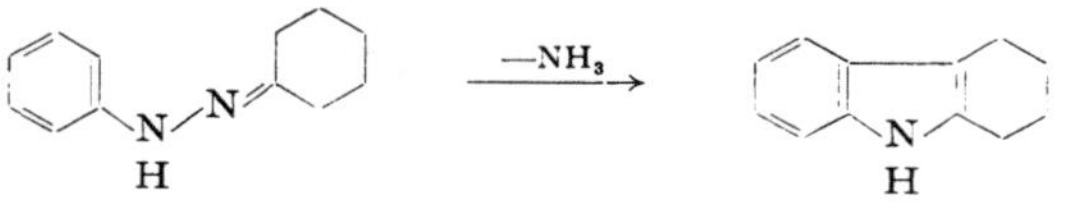

The preparation of carbazoles through this reaction depends on subsequent dehydrogenation of the tetrahydrocarbazoles. Borsche[32] used lead oxide for this purpose as did other workers.[33] The dehydrogenation has also been effected through the agency of sulfur and quinoline,[34] by a catalyst

[27] Burton and Gibson, *J. Chem. Soc.*, **125**, 2501 (1924).

[28] Baeyer, *Ann.*, **278**, 105 (1894). Baeyer and Tutein, *Ber.*, **22**, 2178 (1889). Borsche, Witte, and Bothe, *Ann.*, **359**, 52 (1908). Drechsel, *J. prakt. Chem.*, [2] **38**, 69 (1888). Godchot, *Compt. rend.*, **176**, 448 (1923). Rogers and Corson, *Organic Syntheses*, **30**, 90 (1950). Rogers and Corson, *J. Am. Chem. Soc.*, **69**, 2910 (1947).

[29] Plancher, Cecchetti, and Ghigi, *Gazz. chim. ital.*, **59**, 334 (1929). Ghigi, *ibid.*, **60**, 194 (1930).

[30] Perkin and Plant, *J. Chem. Soc.*, **119**, 1825 (1921). Elmo, Perkin, and Robinson, *ibid.*, **125**, 1751 (1924).

[31] Grammaticakis, *Compt. rend.*, **204**, 502 (1937); **209**, 317 (1939); **210**, 569 (1940).

[32] Borsche, *Ann.*, **359**, 52 (1908).

[33] Perkin and Plant, *J. Chem. Soc.*, **119**, 1825 (1921). v. Braun and Haensel, *Ber.*, **B 59**, 1999 (1926). Britisch patent 337,821; *Chem. Abstr.*, **25**, 2302 (1931). French patent 698,148; *Chem. Abstr.*, **25**, 3012 (1931). German patent 544,621; *Chem. Abstr.*, **26**, 3522 (1932). U. S. patent 1,834,015; *Chem. Abstr.*, **26**, 1000 (1932).

[34] Oakeshott and Plant, *J. Chem. Soc.*, **1926**, 1210. Briscoe and Plant, *ibid.*, **1928**, 1990. Moggridge and Plant, *ibid.*, **1937**, 1125.

composed of oxides of copper, chromium, and barium[35] as well as by naphthalene in the presence of a nickel catalyst,[36] by mercurous acetate,[33, 37] and by palladous chloride.[38] Chloranil[39] has been found to be superior to these other reagents for this purpose and a number of substituted carbazoles as well as carbazole itself have been obtained in good yields through its use. This dehydrogenation has also been accomplished through the agency of cinnamic acid and palladium black[40, 41] while phenol as well as cyclic ketones and a nickel catalyst have been similarly enployed.[42]

In a similar reaction cyclohexanone cyanohydrin and phenylhydrazine react to give a hydrazine nitrile which is cyclized by hydrochloric acid to give tetrahydrocarbazole.[43]

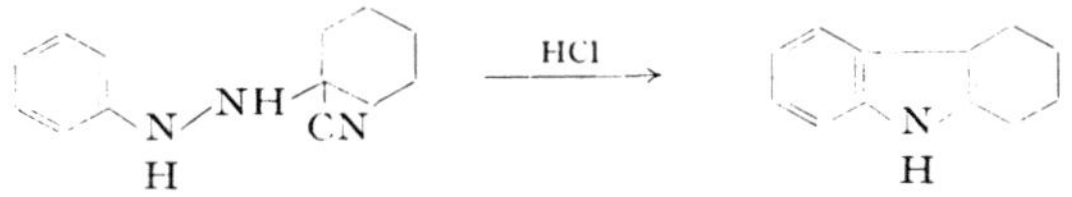

When 2-alkylcyclohexanone phenylhydrazones are employed both 1-substituted tetrahydrocarbazoles and 11-substituted tetrahydrocarbazolenines are obtained. Thus both 1-methyltetrahydrocarbazole (I), m.p. 72°, and 11-methyltetrahydrocarbazolenine (II), m.p. 68°, were obtained through the action of sulfuric acid, zinc chloride, or a Grignard reagent on 2-methylcyclohexanone phenylhydrazone.[44, 45] II has also been prepared by the

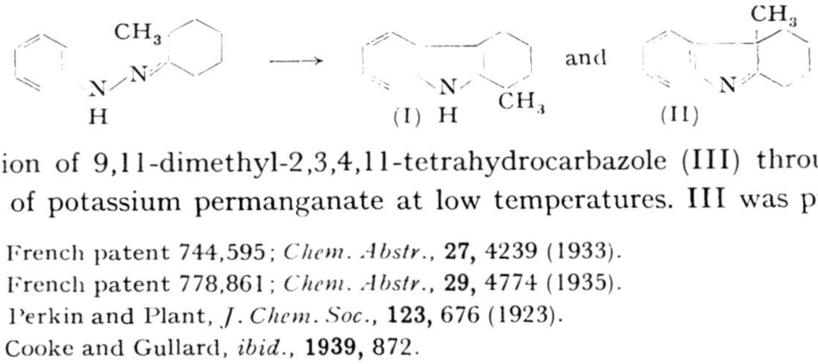

oxidation of 9,11-dimethyl-2,3,4,11-tetrahydrocarbazole (III) through the action of potassium permanganate at low temperatures. III was prepared

[35] French patent 744,595; *Chem. Abstr.*, **27**, 4239 (1933).

[36] French patent 778,861; *Chem. Abstr.*, **29**, 4774 (1935).

[37] Perkin and Plant, *J. Chem. Soc.*, **123**, 676 (1923).

[38] Cooke and Gullard, *ibid.*, **1939**, 872.

[39] Barclay and Campbell, *ibid.*, **1945**, 530.

[40] Hoshino and Takuira, *Bull. Chem. Soc. Japan*, **11**, 218 (1936). Babcock and Pausacker, *J. Chem. Soc.*, **1951**, 1373.

[41] Horning, Horning and Walker, *J. Am. Chem. Scc.*, **70**, 3935 (1948).

[42] British patent 436,110; *Chem. Abstr.*, **30**, 1391 (1936).

[43] Bucherer and Brandt, *J. prakt. Chem.*, **140**, 129 (1934).

[44] Grammaticakis, *Comt. rend.*, **210**, 569 (1940).

[45] Plancher and Testoni, *Atti accad. Lincei*, [5] **9**, 218 (1900). Bauer, Pausacker, and Schubert, *J. Chem. Soc.*, **1949**, 1381. Pausacker and Schubert, *ibid.*, **1949**, 1384; *Nature*, **163**, 289 (1949). Pausacker, *J. Chem. Soc.*, **1950**, 621. Pausacker and Schubert, *ibid.*, **1950**, 1814. Barnes, Pausacker, and Badcock, *ibid.*, **1951**, 730.

from tetrahydrocarbazole through the action of methyl iodide at 120–140°.[46] Other 11-substituted tetrahydrocarbazolenines have been prepared similarly

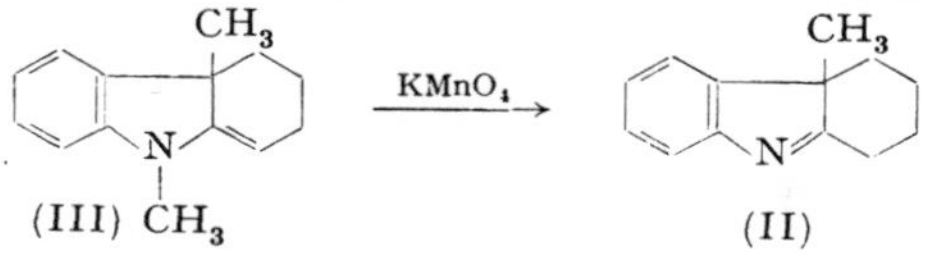

(III) → $KMnO_4$ → (II)

from the corresponding 2-substituted cyclohexanone phenylhydrazones.[47]

The Borsche synthesis, like the Graebe-Ullmann synthesis, serves for the preparation of carbazole derivatives and also for the determination of structure. As in the Fischer indole synthesis ambiguity arises as to the structure when *m*-substituted phenylhydrazones are used. Two products (the 5- and 7-derivatives) are possible theoretically and in actual practice both are obtained frequently. Both carbazole-5-carboxylic acid and carbazole-7-carboxylic acid were obtained from *m*-hydrazinobenzoic acid.[48] The product[49] obtained from cyclohexanone *m*-nitrophenylhydrazone in the Fischer synthesis has been shown to be a mixture of 5- and 7-nitrotetrahydrocarbazoles.[50] The two compounds have been separated by chromatographic methods.[51]

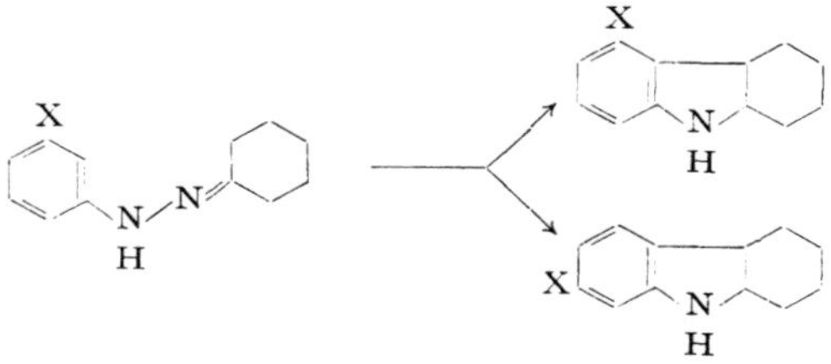

The treatment of α-tetralone phenylhydrazone with either sulfuric acid or zinc chloride yields 1,2-benzo-3,4-dihydrocarbazole (VI), m.p. 60°.[52] Dehydrogenation of this substance yields 1,2-benzocarbazole (VII).

[46] Zanetti and Levi, *J. Chem. Soc.*, **68**, 54 (1895); *Gazz. chim. ital.*, **24**, ii, 111–118 (1899).

[47] Lions, *J. Proc. Roy. Soc. N. S. Wales*, **71**, 192–208 (1938); *Chem. Abstr.*, **32**, 5843 (1938). Harrance and Lions, *J. Proc. Roy. Soc. N. S. Wales*, **73**, 14–21 (1939); *Chem. Abstr.*, **33**, 8196 (1939). Plancher and Carrasco, *Atti accad. Lincei*, [v], **13**, 632–636 (1904); *J. Chem. Soc.*, **86**, 777 (1904). Plancher, Testoni, and Olivari, *Chem. Abstr.*, **15**, 3103 (1921); *Giorn. chim. ind. ed. appl.*, **2**, 458 (1920). Cecchetti and Ghigi, *Gazz. chim. ital.*, **60**, 185 (1930).

[48] Plant and Collar, *J. Chem. Soc.*, **1926**, 808. Plant and Moggridge, *ibid.*, **1937**, 1125. Plant and Wilson, *ibid.*, **1939**, 237.

[49] Plant, *ibid.*, **1936**, 899.

[50] Borsche, Witte, and Bothe, *Ann.*, **359**, 52 (1908).

[51] Barclay and Campbell, *J. Chem. Soc.*, **1945**, 530.

[52] Ghigi, *Gazz. chim. ital.*, **60**, 194 (1930); *Chem. Abstr.*, **24**, 3797 (1930).

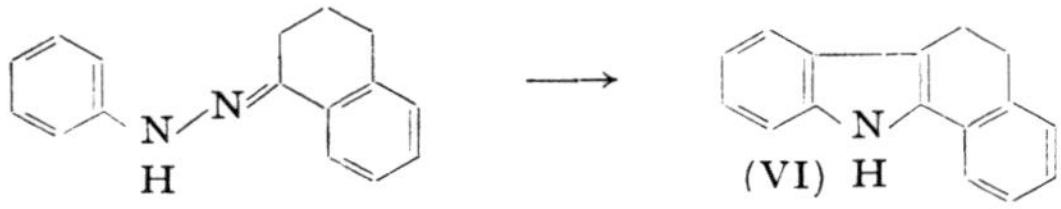

(VI) H

Treatment of β-tetralone phenylhydrazone with sulfuric acid yields 1,2-dihydro-3,4-benzocarbazole (V), m.p. 100°.[53] It has also been stated[53]

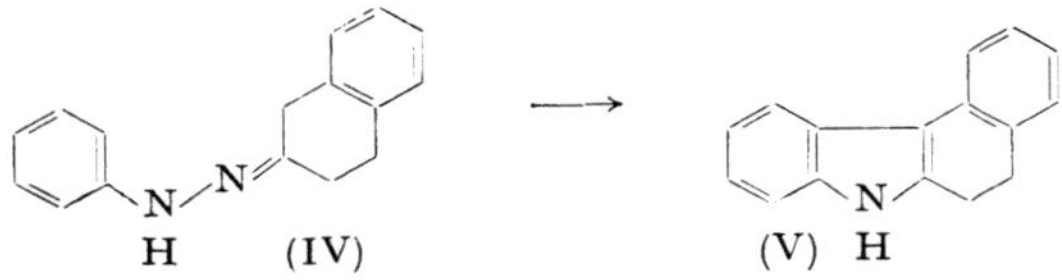

H (IV) (V) H

that treatment of β-tetralone phenylhydrazone with zinc chloride gives 1,2-benzocarbazole (VII), m.p. 228°. Just how this compound VII could have

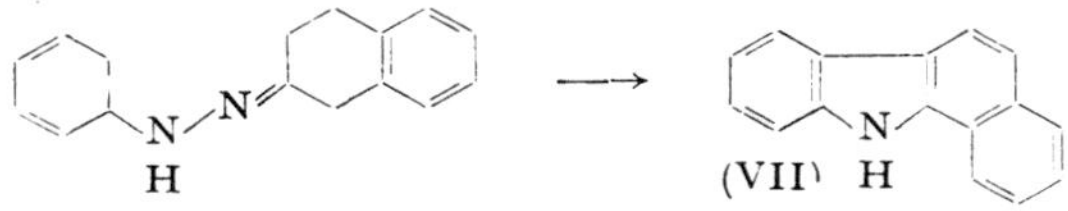

H (VII) H

been formed (or why dehydrogenation should take place here) is obscure as ring closure of IV should certainly yield either V or VIII and not VII as claimed by Ghigi.

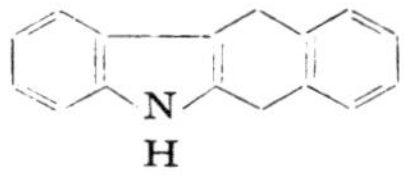

(VIII)

Cyclization of the monophenylhydrazone of cyclohexane-1,2-dione yields 1-keto-1,2,3,4-tetrahydrocarbazole, m.p. 169–170°[54]. The same procedure has also been utilized for the preparation of derivatives of 1-keto-1,2,3,4-tetrahydrocarbazole.[55] Clemmensen reduction of 1-ketotetrahydrocarbazole yields tetrahydrocarbazole. Cyclization of the monophenylhydrazone of 1,3-cyclohexanedione similarly yields 4-keto-1,2,3,4-tetrahydrocarbazole.[56]

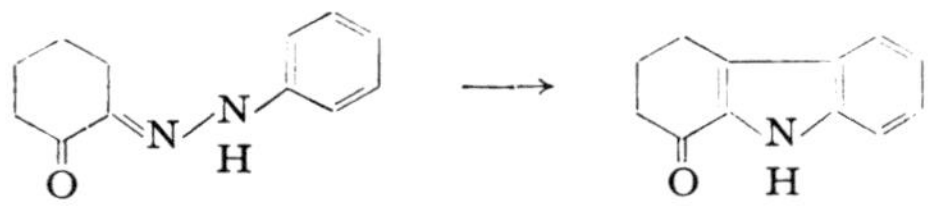

[53] Ghigi, *Gazz. chim. ital.*, **61,** 43 (1931).

[54] Lions, *J. Proc. Roy. Soc. N. S. Wales*, **66,** 516–525 (1923). Sen and Ghosh, *Quart. J. Indian Chem. Soc.*, **4,** 477 (1927); *Chem. Abstr.*, **22,** 1145 (1928).

[55] Kent, *J. Chem. Soc.*, **1935,** 976. Kent and McNeil, *ibid.*, **1938,** 8. Mears, Oakeshott and Plant, *ibid.*, **1934,** 272. Anderson and Campbell, *ibid.*, **1950,** 2855.

[56] Clems and Felton, *ibid.*, **1951,** 700.

Many phenols react in the keto form when heated with phenylhydrazine to give carbazole derivatives. This procedure was employed to prepare 3,4-benzocarbazole[57] from 3-hydroxy-2-naphthoic acid. 1,2,3,4-Dibenzocarbazole[58] has been prepared in similar fashion by heating 9-hydroxyphenanthrene with phenylhydrazine at 200°. 3,4-Benzocarbazole,[59] m.p. 134–135°, has been prepared from β-naphthol and 1,2-benzocarbazole, m.p. 225.5°, obtained similarly from α-naphthol. This method of synthesizing carbazoles has found further application in the synthesis of other carbazole derivatives.[60] The 1,2-benzocarbazole described above had previously been prepared by heating benzophenotriazine with copper powder.[61]

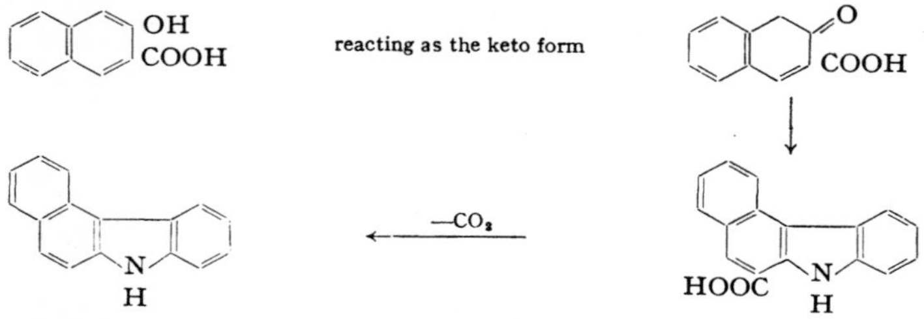

A modification of this procedure due to Bucherer[62] treats either the phenol or the corresponding aryl amine with sodium bisulfite and phenylhydrazine (the amine is converted into the phenol through the Bucherer reaction). The mechanism suggested by Bucherer is as follows:

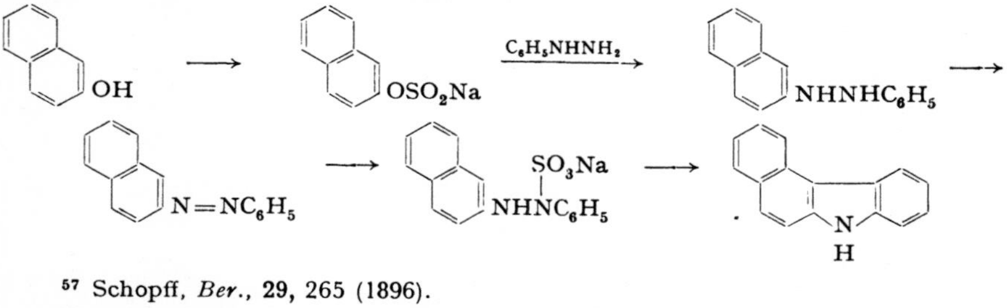

[57] Schopff, *Ber.*, **29**, 265 (1896).

[58] Japp and Findlay, *J. Chem. Soc.*, **71**, 1115 (1897).

[59] Japp and Maitland, *ibid.*, **83**, 269 (1903).

[60] Royer, *Ann. chim. phys.*, **I**, 395 (1946).

[61] Kym, *Ber.*, **23**, 2465 (1890).

[62] Bucherer and Seyde, *J. prakt. Chem.*, [2] **77**, 403 (1908). Bucherer and Schmidt, *ibid.*, [2] **79**, 369 (1909). Bucherer and Sonnenburg, *ibid.*, [2] **81**, 1 (1910).

Friedländer,[63] on the other hand, has suggested that in the Bucherer synthesis of carbazoles the phenol reacts in the keto form first forming a bisulfite addition compound containing the $-CH_2-C(OH)(SO_3Na)$ group. In support of this more probable mechanism Friedländer submits the fact observed

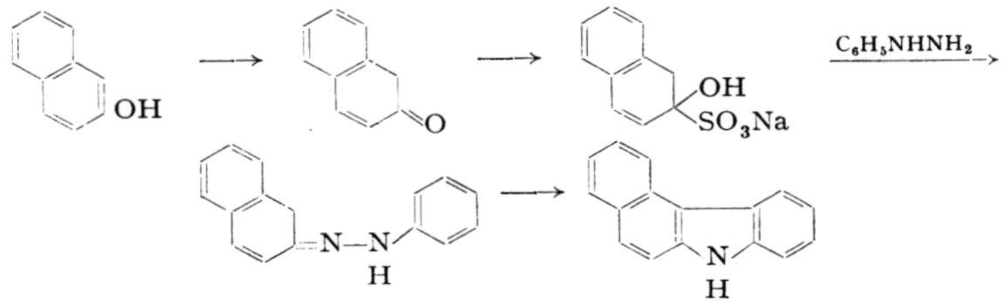

OH → O → OH SO$_3$Na $\xrightarrow{C_6H_5NHNH_2}$

=N—N— H → N H

by Bucherer and Seyde[62] that when unsymmetrical methylphenylhydrazine is used in place of phenylhydrazine the corresponding 9-methyl-3,4-benzocarbazole is obtained. This is impossible with the mechanism suggested by Bucherer.[64]

Other Methods of Synthesis of Carbazoles

Treatment of diphenylhydroxylamine with a mixture of acetic and sulfuric acids gives a mixture of carbazole and *p*-hydroxydiphenylamine.[65] Passage of a mixture of benzene vapor and ammonia through a heated furnace resulted in the formation of aniline while at the same time carbazole and benzonitrile were formed through further condensation.[66] Carbazole has also been prepared through the distillation of aniline with lime[67] as well as by passing aniline vapor through a heated porcelain tube.[68] Better yields resulted when diphenylamine was substituted for aniline in this procedure. Carbazole has also been prepared from 2,2′-diaminodiphenyl[69] by autoclaving with sulfuric or hydrochloric acid at 200° for fifteen hours. This

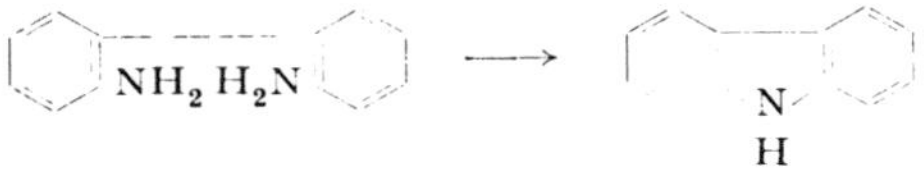

NH_2 H_2N → N H

[63] Friedländer, *Ber.*, **B 54**, 620 (1921).

[64] For other applications of the Bucherer reaction see: German patent 533,470; *Chem. Abstr.*, **26**, 479 (1932). French patent 713,500; *Chem. Abstr.*, **26**, 1620 (1932). French patent 716,158; *Chem. Abstr.*, 26, 1943 (1932).

[65] Wieland and Muller, *Ber.*, 46, 3304 (1913).

[66] Meyer and Tanzen, *ibid.*, 46, 3183 (1913).

[67] Braun and Grieff, *ibid.*, 5, 276 (1872).

[68] Graebe, *ibid.*, 5, 976 (1872); *Ann.*, **167**, 125 (1873); **174**, 177 (1874).

[69] Tauber, *Ber.*, **24**, 197 (1891).

procedure has been used for the synthesis of a number of substituted carbazoles.[70] Carbazole is formed also when the vapor of *o*-aminodiphenyl is passed over heated catalysts.[71] A substituted carbazole[72] has been prepared from a derivative of 2-amino-2′-hydroxydiphenyl by distillation with zinc dust.

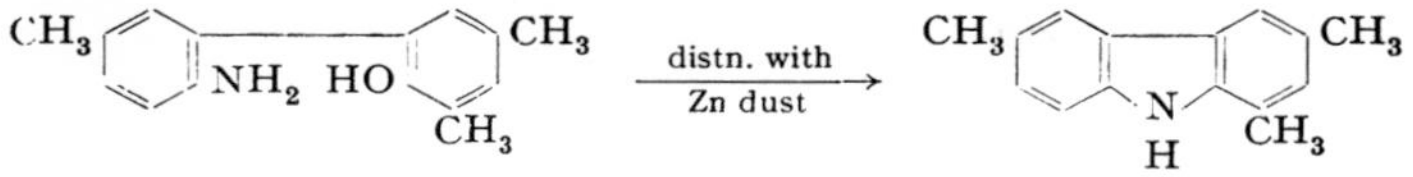

distn. with Zn dust

Ring closure has been effected in 2,2′-dihalogen derivatives of diphenylamine through zinc dust distillation.[72] Carbazole has also been obtained by

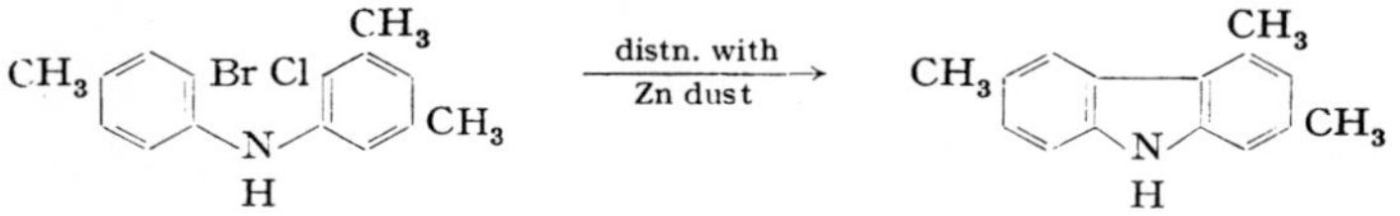

distn. with Zn dust

heating thiodiphenylamine with metallic copper.[73] The same reaction has been used for the synthesis of 1,2-benzocarbazole[74] and of 3,4,5,6-dibenzocarbazole.[75] Carbazole has also been prepared by heating diazotized 2,2′-diaminophenyl with a solution of potassium sulfide,[76] with copper powder,[77]

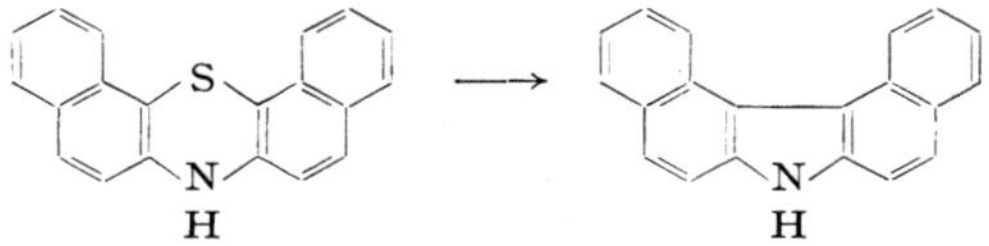

or with cuprous bromide.[78,79] Carbazole is obtained also when 6,6′-diamino-

[70] Tauber, *ibid.*, **23,** 3266 (1890); **25,** 128 (1892). Tauber and Loewenber, *ibid.*, **24,** 1033 (1891). Zelinsky, Titz, and Gaverdovskaja, *ibid.*, **B 59,** 2590 (1926). German patent 542,422; *Chem. Abstr.*, **26,** 2469 (1932). Roosmalen, *Rec. trav. chim.*, **53,** 359 (1934). Shin-ichi Sako, *Bull. Chem. Soc. Japan*, **9,** 55 (1934); **11,** 144 (1936); *Chem. Abstr.*, **30,** 5984 (1936). Niementowski, *Ber.*, **35,** 3325 (1901). King and King, *J. Chem. Soc.*, **1945,** 824. Porai-Koshits and Salyamon, *J. Gen. Chem. (U. S. S. R.)*, **14,** 1019 (1944); *Chem. Abstr.*, **39,** 4599 (1945). Carlin and Forshey, *J. Am. Chem. Soc.*, **72,** 793 (1950).

[71] Blank, *Ber.*, **24,** 306 (1891); U. S. patent 2,456,378; *Chem. Abstr.*, **43,** 1808 (1949). U. S. patent 2,479,211; *Chem. Abstr.*, **43,** 9086 (1949).

[72] Fries, Baker, and Wallbaum, *Ann.*, **509,** 73 (1934).

[73] Goske, *Ber.*, **20,** 532 (1877).

[74] Kym, *ibid.*, **23,** 2465 (1890).

[75] Ris, *ibid.*, **19,** 2242 (1886).

[76] Tauber, *ibid.*, **26,** 1703 (1903).

[77] Niementowsky, *ibid.*, **34,** 3331 (1901).

[78] Dobbie, Fox, and Gauge, *J. Chem. Soc.*, **99,** 1618 (1911).

[79] Compare Kerschbaum, *Ber.*, **28,** 2803 (1895); Nietzki and Goll, *ibid.*, **18,** 3259 (1885); Vesely, *ibid.*, **38,** 136 (1905).

diphenyl-2,2′-dicarboxylic acid is heated with either barium oxide or barium hydroxide.[80] 2,3,6,7-tetranitrocarbazole[81] has been prepared by heating

COOH COOH

NH$_2$ H$_2$N → N H

O$_2$N NO$_2$ O$_2$N NO$_2$ H$_3$CO OCH$_3$ → O$_2$N NO$_2$ O$_2$N NO$_2$ N H

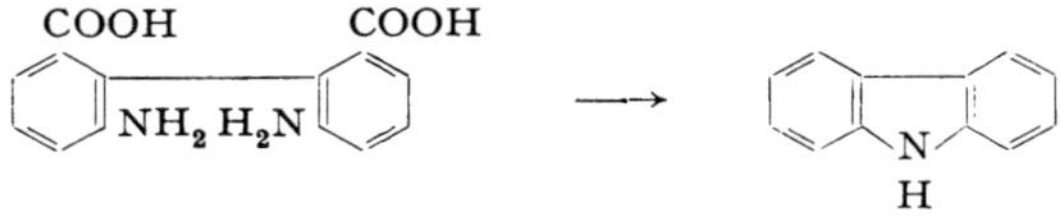

2,2′-dimethoxy-4,5,4′,5′-tetranitrodiphenyl with alcoholic ammonia. 2,3,6,7-Dibenzocarbazole, m.p. 159°, has been obtained in similar fashion from the corresponding dihydroxybinaphthyl by heating with ammonia at temperatures above 200°.[82]

Carbazole was obtained in the catalytic pyrolysis of *o*-xenylamine.[83] 3-Methylcarbazole[84] has been prepared by heating 1-*p*-toluidinocyclopentane-1-carboxylic acid with potassium hydroxide and sodium ethoxide for thirty minutes at 350°. The same compound was obtained by dehydrogenation of 6-methyltetrahydrocarbazole with sulfur and quinoline.

Tetrahydrocarbazole can be prepared through the interaction of 2-chlorocyclohexanone and aniline.[85] The use of substituted aniline derivatives leads to the formation of derivatives of tetrahydrocarbazole.

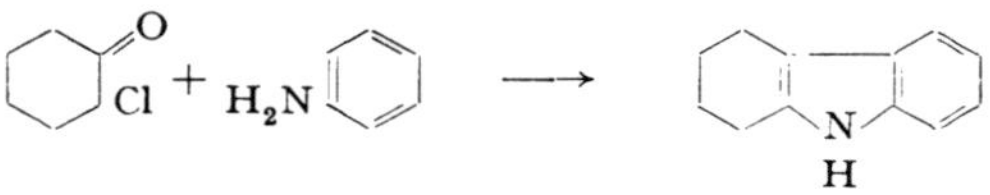

O Cl + H$_2$N → N H

A recent synthesis of carbazole[86] from *o*-azidobiphenyl through the agency of heat or ultraviolet light seems to be generally applicable to the synthesis of substituted carbazole derivatives as well.

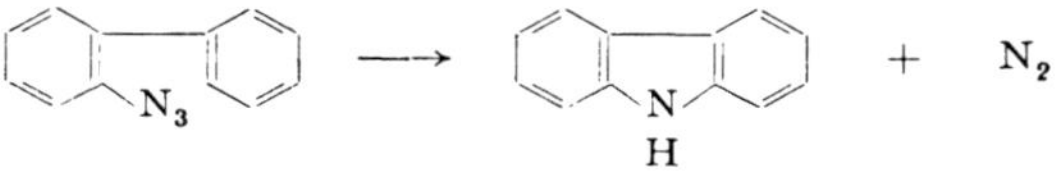

N$_3$ → N H + N$_2$

[80] Schmidt and Kampf, *ibid.*, **36**, 3747 (1903).

[81] Borsche and Scholten, *ibid.*, **50**, 606 (1917). Radunitz, *ibid.*, **B 60**, 738 (1927).

[82] German patent 624,563; *Chem. Abstr.*, **30**, 4873 (1936).

[83] Morgan and Wells, *J. Soc. Chem. Ind.*, **57**, 358 T (1938).

[84] Oakeshott and Plant, *J. Chem. Soc.*, **1926**, 1210.

[85] German patent 374,098; *Chem. Abstr.*, **18**, 2175. Campbell and McCall, *J. Chem. Soc.*, **1950**, 2870.

[86] Smith and Brown, *J. Am. Chem. Soc.*, **73**, 2435, 2438 (1951). Smith and Bayer, *ibid.*, **73**, 2626 (1951).

Another recent synthesis of carbazole was accomplished through heating 2-nitrodiphenyl with ferrous oxalate at 205–215° for one-half hour.[87]

NO_2 ⟶ N H

Nitro Derivatives of Carbazole

Substituted derivatives of carbazole can be prepared by nitration, halogenation, and sulfonation but in many instances the product is a mixture which is difficult to separate into its components.

The 3- and 6-positions in carbazole are the most reactive ones with the 1- and 8-positions being somewhat less active. This was shown by the work Lindeman[88,89,90] who found that treatment of a solution of carbazole in glacial acetic acid with sodium nitrite followed by nitric acid yielded principally 3-nitro-9-nitrosocarbazole which on heating with KOH or glacial acetic acid gave 3-nitrocarbazole, m.p. 215.5–216.5°,[92] along with a small quantity of 1-nitrocarbazole, m.p. 187°. Exhaustive nitration gives 1,3,6,8-tetranitrocarbazole[91]. The following scheme[92] of reactions has been suggested for the preparation of 3-nitrocarbazole:

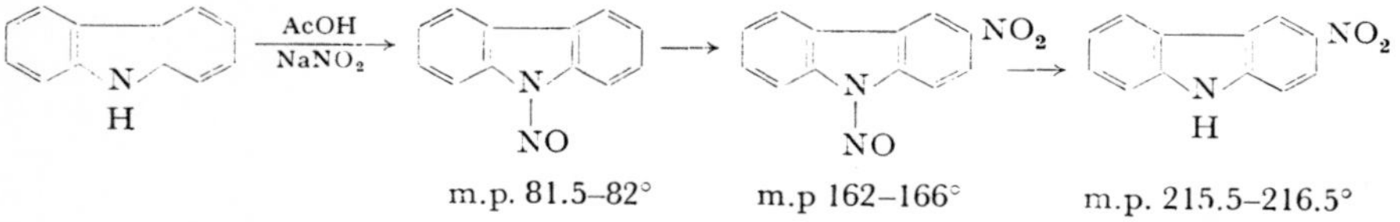

N H —AcOH / $NaNO_2$⟶ N NO (m.p. 81.5–82°) ⟶ NO_2 N NO (m.p 162–166°) ⟶ NO_2 N H (m.p. 215.5–216.5°)

1-Nitrocarbazole, m.p. 187°, was first prepared by Votocek.[93] It was shown to be 1-nitrocarbazole through reduction to 1-aminocarbazole[94] which

[87] Watermann and Vivian, *J. Org. Chem.*, **14,** 289 (1949).

[88] Lindemann, *Ber.*, **57,** 555 (1924).

[89] Ruff and Stein, *ibid.*, **34,** 1668 (1901).

[90] Tucker and Stevens, *J. Chem. Soc.*, **123,** 2140 (1923). Kehrmann and Zweifel, *Helv. Chem. Acta*, **11,** 1213 (1938). Fürst and Bosse, *Chem. Ber.*, **84,** 83 (1951).

[91] Ziersch, *Ber.*, **42,** 3800 (1909). Compare Graebe and Alderskron, *Ann.*, **202,** 26 (1880); Escales, *Ber.*, **37,** 3596 (1904); Ciamician and Silber, *Gazz. chim. ital.*, **12,** 277 (1882).

[92] Eikman, Lukashevich, and Silaeva, *Chem. Abstr.*, **33,** 7297 (1939); *Org. Chem. Ind.* (*U. S. S. R.*), **6,** 93 (1939). Compare Weiland and Susser, *Ann.*, **392,** 172 (1912); and Blom, *J. prakt. Chem.*, **94,** 77 (1916).

[93] Votocek, *J. Chem. Soc.*, **72,** 439 (1897); *Rozpravy Ceské akad.*, **1896,** 5, Kl. ii, Nr. 22.

[94] Lindemann and Werther, *Ber.*, **57,** 1316 (1924). Lindemann and Wessel, *ibid.*, **58,** 1221 (1925).

had been synthesized by means of the Graebe-Ullmann reaction. 1-Nitrocarbazole was prepared also by nitrating 3,6-carbazoledicarboxylic acid and subsequently decarboxylating.[95] The compound has been prepared similarly by removal of the sulfonic acid groups from 1-nitrocarbazole-3,6,8-tri-

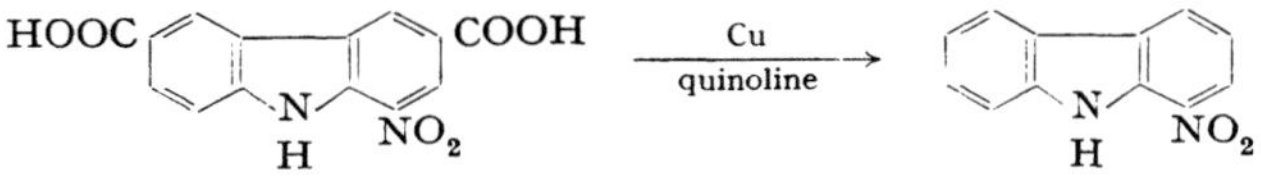

HOOC COOH N H NO_2 —Cu, quinoline→ N H NO_2

sulfonic acid.[96] A product, m.p. 164°, was obtained by nitrating carbazole in acetic acid and mistaken for 1-nitrocarbazole[97] but subsequently shown to be a molecular compound[98] of 1-nitrocarbazole and 3-nitrocarbazole.

Ring closure of cyclohexanone *m*-nitrophenylhydrazone yielded what was at first thought to be a single compound[99] but was subsequently shown to be a mixture[100] of 5-nitrotetrahydrocarbazole and 7-nitrotetrahydrocarbazole. The isomers were separated by chromatographic methods[101] and their structures established by reduction to the corresponding amino compounds and comparison with 5-aminotetrahydrocarbazole prepared from cyclohexanone-2′-chloro-5′-nitrophenylhydrazone.

NO_2 Cl N H N → NO_2 Cl N H

(1) reduction
(2) dehalogenation

N H

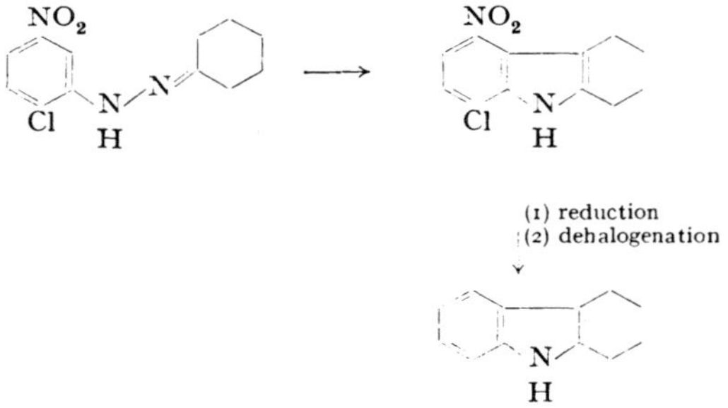

The preparation of 3-nitro-9-ethylcarbazole from 2,4-dinitro-*N*-ethyldiphenylamine through reduction and subsequent application of the Graebe-Ullmann synthesis has been reported.[102] Later studies established that in the reduction of 2,4-dinitro-*N*-ethyldiphenylamine it is the nitro group in

[95] Tucker, Preston, and Cameron, *J. Chem. Soc.*, **1942,** 500.
[96] German patent 507,797; *Chem. Abstr.*, **25,** 716 (1931).
[97] Ziersch, *Ber.*, **42,** 3798 (1909). Whitner, *J. Am. Chem. Soc.*, **46,** 2326 (1924).
[98] Morgan and Mitchell, *J. Chem. Soc.*, **1931,** 3283.
[99] Borsche, Witte, and Bothe, *Ann.*, **359,** 52 (1908).
[100] Plant, *J. Chem. Soc.*, **1936,** 899.
[101] Barclay and Campbell, *ibid.*, **1945,** 530.
[102] Deletra and Ullmann, *ibid.*, **66,** I, 270 (1904); *Arch. sci. nat. phys. Geneve*, **17,** 78 (1904).

position 4[103] which is reduced and not that in position 2. It follows that 3-nitro-9-ethylcarbazole was not prepared in the first study.

3,6-Dinitrocarbazole[104] has been prepared by the action of nitric acid on a solution of *N*-nitrosocarbazole in benzene and by the action of nitric acid on carbazole[105] and on 3-nitrocarbazole.[106] Studies of the action of N_2O_3 on a benzene solution of carbazole have shown that 9-nitroso-, 3-nitroso-, 3-nitro-, 9-nitroso-3-nitro-, 3-nitro-6-nitroso-, and 3,6-dinitrocarbazole are formed successively with the last named being the final and principal product of the reaction.[107]

The nitration of carbazole to 1,3,6,8-tetranitrocarbazole has been effected through the agency of ethyl nitrate.[108] Nitration of 9-*p*-toluenesulfonylcarbazole is reported to take place in position 1 (1-nitroderivative, m.p. 134°).[109] Long heating of this derivative with concentrated hydrochloric acid gave 1-nitrocarbazole. On the other hand, bromination of 9-*p*-toluenesulfonylcarbazole gave the 3-bromo derivative and iodination with iodine monochloride gave the 3,6-diiodo derivative.

The mononitration of tetrahydrocarbazole and its *N*-alkyl derivatives with concentrated nitric and sulfuric acids gives the 6-nitro derivatives[110] while the *N*-acyl derivatives nitrate in the 7-position.[111] Under suitable conditions the *N*-acyltetrahydrocarbazoles react in a quite different manner, the nitric acid adding at the double bond of the reduced ring.[112] Ring fission

N — COC_6H_5 ⟶ N — COC_6H_5 (NO_2, OH)

[103] Stevens and Tucker, *J. Chem. Soc.*, **123,** 2140 (1923). Storrie and Tucker, *ibid.*, **1931,** 2255.

[104] German patent 128,853; *J. Chem. Soc.*, **62, I,** 495 (1902).

[105] U. S. patent 2,392,067; *Chem. Abstr.*, **40,** 1964 (1946).

[106] Eikhman, Lukashevich, and Silaeva, *Chem. Abstr.*, **33,** 7297 (1939); *Org. Chem. Ind.* (U. S. S. R.); **6,** 93–95 (1939).

[107] Il'inskii, Maksarov, and Elagin, *Chem. Abstr.*, **22,** 3888 (1928); *J. Chem. Ind.* (*Moscow*), **5,** 469 (1928).

[108] Raudnitz, *Ber.*, **B 60,** 738 (1927).

[109] Menon, Menon, and Peacock, *J. Chem. Soc.*, **1942,** 509; Compare, however, Kulka and Manske, *J. Org. Chem.*, **17,** 1501 (1952).

[110] Perkin and Plant, *J. Chem. Soc.*, **119,** 1825 (1921). Perkin and Plant, *ibid.*, **123,** 676 (1923). Plant and Rosser, *ibid.*, **1928,** 2454. Plant and Rutherford, *ibid.*, **1929,** 1970. Manjunath and Plant, *ibid.*, **1926,** 2260.

[111] Plant, *ibid.*, **1936,** 899. Moggridge and Plant, *ibid.*, **1937,** 1125.

[112] Perkin and Plant, *ibid.*, **123,** 676 (1923).

occurs on treatment with alkali, δ-(*o*-benzoylaminobenzoyl)valeric acid being formed.

NO_2 OH N COC_6H_5 → OH OH N COC_6H_5 ↓ CO CO N COC_6H_5 ← $CO(CH_2)_4COOH$ $NHCOC_6H_5$

N-Acetyltetrahydrocarbazole reacts differently, 9-acetyl-10,11-dihydroxyhexahydrocarbazole (I) resulting.[113, 114] Treatment of this compound with hot alkali gives an orange compound (III)[115] while treatment with acetic acid yields the oxindole derivative (IV).[116] Compound I has also been prepared by the action of osmium tetroxide of *N*-acetyltetrahydrocarbazole.[114a] 11-Hydroxytetrahydrocarbazolenine (II) is also obtained

C CO N $COCH_3$ (IV) ←H^+— OH N OH $COCH_3$ (I) Colorless m.p. 204° —OH^-→ OH N (II) Colorless m.p. 159° —OH^- or H^+→ CO N C H (III) Orange m.p. 77–79°

through catalytic oxidation of tetrahydrocarbazole in ethyl acetate over platinum followed by subsequent gentle reduction.[116]

Compound III is reduced by $LiAlH_4$ to V which in the presence of acids gives, through a Wagner-Meerwein shift, tetrahydrocarbazole.[116] The *N*-

CO N C H (III) —$LiAlH_4$→ CHOH N C H (V) → N H

[113] Plant and Tomlinson, *ibid.*, **123,** 688 (1923); *ibid.*, **1950,** 2127.

[114] Plant, Robinson, and Tomlinson, *Nature*, **165,** 928 (1950). Plant and Robinson, *ibid.*, **165,** 36 (1950). Moore and Plant, *J. Chem. Soc.*, **1951,** 3475.

[114a] Ockenden and Schofield, *Nature*, **168,** 603 (1951); *J. Chem. Soc.*, **1953,** 612.

[115] Witkop, *J. Am. Chem. Soc.*, **72,** 614, 1428 (1950). Witkop and Goodwin, *ibid.*, **75,** 3371 (1953).

[116] Patrick and Witkop, *ibid.*, **72,** 633 (1950). Witkop and Patrick, *ibid.*, **73,** 713, 1558, 2188, 2196, 2641 (1951); **75,** 2400 (1953).

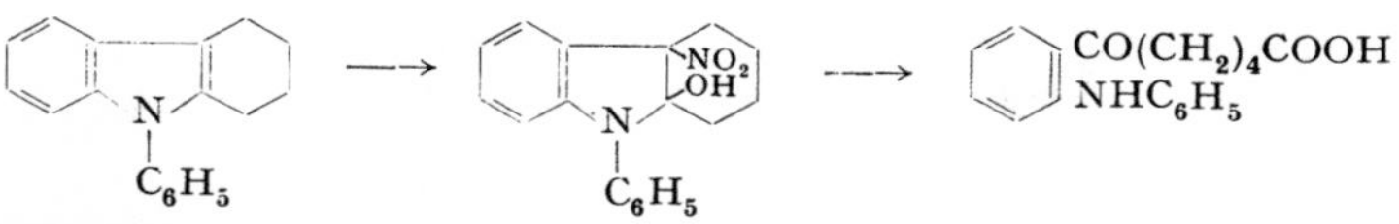

C_6H_5 → C_6H_5 (NO$_2$, OH) → $CO(CH_2)_4COOH$, NHC_6H_5

phenyl derivative behaves like the *N*-benzoyl derivative.[117] On treatment of a glacial acetic acid solution of 9-acetyltetrahydrocarbazole with fuming nitric acid both addition and substitution take place giving 6-nitro-10,11-dihydroxy-9-acetylhexahydrocarbazole (VI), m.p. 238° (dec.).[118] The same compound was obtained from 6-nitro-9-acetyltetrahydrocarbazole through the agency of nitric acid in glacial acetic acid.

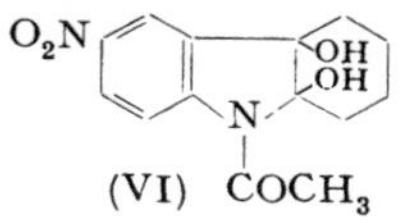

O_2N OH OH N (VI) $COCH_3$

Compound VI was converted into products described[118] as 6-nitro-11-hydroxytetrahydrocarbazolenine, yellow, m.p. 192°, and 5′-nitro-spiro-[cyclopentane-1,2′-ψ-indoxyl], m.p. 241°. It seems evident, however, from the work of Witkop[115,116] that this characterization is in error and that the yellow compound, m.p. 192°, is the 5′-nitro-spiro-(cyclopentane-1,2′-ψ-indoxyl) (5′-nitro derivative of III) while the compound, m.p. 241°, is 5′-nitro-spiro(cyclopentane-1,3′-ψ-oxindole). Compound VI was also obtained from 10,11-dihydroxy-9-acetylhexahydrocarbazole through the action of fuming nitric acid.[118] The nitration in position 6 seems to be in conflict with the finding of Plant[111] to the effect that acyl derivatives of tetrahydrocarbazole nitrate in position 7 rather than in position 6. However, Gurney and Plant[119] found that 9-acetylhexahydrocarbazole nitrates in position 6. It seems probable that in the preparation of VI from 9-acetyltetrahydrocarbazole addition at the 10,11 double bond takes place giving a hexahydrocarbazole derivative which then substitutes in position 6.

Aminocarbazoles

3-Aminocarbazole, m.p. 254°, is readily obtained through the reduction of 3-nitrocarbazole.[120] The compound has also been obtained through the

[117] Perkin and Linnell, *J. Chem. Soc.*, **125**, 2451 (1924).

[118] Massey and Plant, *ibid.*, **1931**, 2218.

[119] Gurney and Plant, *ibid.*, **1927**, 1314.

[120] Kehrmann and Zeifel, *Helv. Chim. Acta*, **11**, 1213 (1928). Whitner, *J. Am. Chem. Soc.*, **46**, 2326 (1924). Novelli, *Chem. Abstr.*, **34**, 6621 (1940); *Ann. asoc. quim. Argentina*, **28**, 87–90 (1940). Mazzara and Leonardi, *Gazz. chim. ital.*, **21**, II, 380 (1891). Ruff and Stein, *Ber.*, **34**, 1679 (1901). Ziersch, *ibid.*, **43**, 3798 (1909). Eikhman, Lukashevich, and Silaeva, *Chem. Abstr.*, **33**, 7297 (1937); *Org. Chem. Ind. (U. S. S. R.)*, **6**, 93–95 (1939). Anderson and Campbell, *J. Chem. Soc.*, **1950**, 2904.

reduction of 9-nitroso-3-nitrocarbazole with sodium sulfide,[121] of 3-nitroso-carbazole with ammonium sulfide,[122] and through the application of the Graebe-Ullmann reaction.[123]

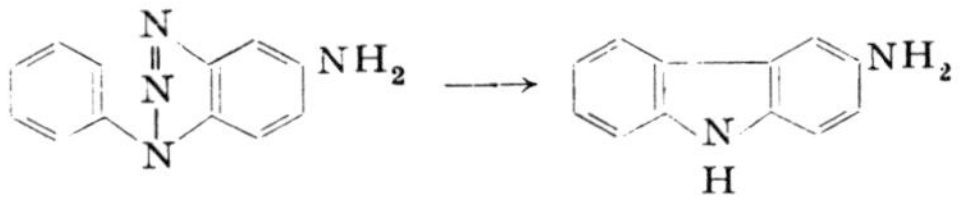

1-Aminocarbazole has been prepared through the reduction of 1-nitro-carbazole,[124] by synthesis from 1-phenyl-7-amino-1,2,3-benzotriazole and from 1-phenyl-5-carboxy-7-amino-1,2,3-benzotriazole.[124] The compound was also prepared by reducing 3,6-dibromo-1-nitrocarbazole with hydriodic acid.[125] The product described by Whitner[120] as being 1-aminocarbazole was

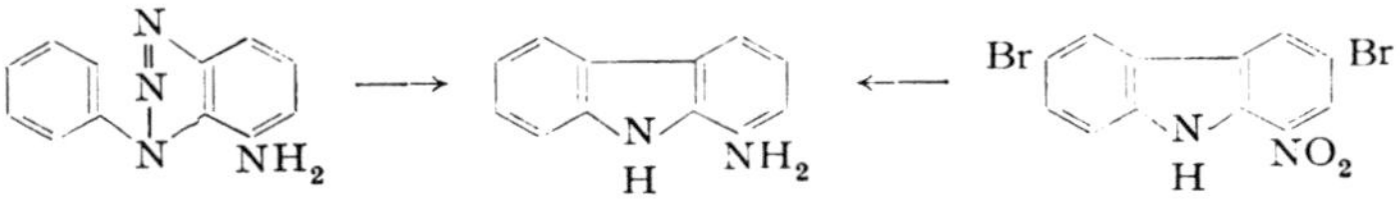

undoubtedly impure 3-aminocarbazole[126] since the nitrocarbazole that he reduced has been shown to be a complex[98] of 1-nitrocarbazole and 3-nitro-carbazole. 1-Aminocarbazole has also been prepared by heating 1-hydroxy-carbazole with ammonium bisulfite solution,[127] through the removal of the sulfonic acid groups from 1-amino-3,6,8-carbazoletrisulfonic acid[128] and through the reduction of 1-amino-3,6-dichlorocarbazole and of 1-amino-3,6-diiodocarbazole with hydrazine hydrate.[129]

2-Aminocarbazole has been prepared through catalytic reduction of 2-nitrocarbazole[120] as well as through reduction of 2-amino-3,6-dichloro-carbazole with hydrazine hydrate.[129] A preparation of 2-aminocarbazole through heating of 2,2′-diaminodiphenyl[130] with lime has been reported.

[121] Schwalbe and Wolff, *Ber.*, **44**, 234 (1911).

[122] German patent 134,983; *Chem. Zentr.*, **1902**, II, 1165; *Friedl.*, **6**, 61 (1904).

[123] Ullmann, *Ber.*, **31**, 1697 (1898); *Ann.*, **332**, 97 (1904).

[124] Lindemann and Werther, *Ber.*, **57**, 1316 (1924). Lindemann and Wessel, *ibid.*, **58**, 1221 (1925). Compare Mottier, *Helv. Chim. Acta*, **17**, 1130 (1934).

[125] Campbell and MacLean, *J. Chem. Soc.*, **1942**, 504.

[126] Campbell and Barclay, *Chem. Revs.*, **40**, 370 (1947).

[127] British patent 316,962; *Chem. Abstr.*, **24**, 1868 (1930). French patent 654,074; *Chem. Abstr.*, **23**, 3715 (1929). U. S. patent 1,878,168; *Chem. Abstr.*, **27**, 313 (1933).

[128] German patent 504,797; *Chem. Abstr.*, **25**, 716 (1931).

[129] British patent 340,625; *Chem. Abstr.*, **25**, 5892 (1931). German patent 522,960; *Chem. Abstr.*, **25**, 3668 (1931).

[130] Blank, *Ber.*, **24**, 306 (1891).

Subsequently doubt has been expressed[131] as to the validity of this latter synthesis.

3-Hydroxycarbazole, m.p. 260–261°, has been prepared by heating diazotized 3-aminocarbazole with water.[132] The diazotized solution from 3-amino-9-methylcarbazole on treatment with Na_2HAsO_3 gives 9-methylcarbazole-3-arsonic acid.[133] The Sandmeyer reaction has been employed in a number of other cases for transforming aminocarbazoles to other carbazole derivative.[134]

9-Aminocarbazole was obtained[135,136] through the reduction of 9-nitrosocarbazole with zinc and acetic acid. The compound is a hydrazine derivative and as such forms hydrazones[137] with aldehydes and ketones.

2,7-Diaminocarbazole has been prepared by heating 2,2′-diaminobenzidine with sulfuric acid at 180–190° for ten hours.[69,70] 3,6-Diaminocarbazole has been prepared similarly from 3,3′-diaminobenzidine[69,70] and by reduction of 3,6-dinitrocarbazole.[91,92]

Halogen Derivatives of Carbazole

Direct halogenation of carbazole gives the 3- and 3,6-derivatives. The 3-halogen carbazoles are first formed but these react readily with more halogen to give the 3,6-dihalogen derivatives.

3-Chlorocarbazole, m.p. 201.5°, has been prepared through the action of thionyl chloride[138] on a solution of carbazole in chloroform, from 1-phenyl-5-chloro-1,2,3-benzotriazole,[139] and from 3-aminocarbazole through the Sandmeyer reaction.[140] 2-Chlorocarbazole, m.p. 244°, was prepared by

[131] Barclay and Campbell, *Chem. Revs.*, **40**, 370 (1947).

[132] Ruff and Stein, *Ber.*, **34**, 1668 (1901).

[133] Bruton and Gibson, *J. Chem. Soc.*, **1947**, 2386. Sherlin and Berlin, *J. Gen. Chem.* (*U. S. S. R.*), **5**, 938 (1935).

[134] Mizuch, *J. Gen. Chem.* (*U. S. S. R.*), **10**, 844–851 (1940); *Chem. Abstr.*, **35**, 2509 (1941). British patent 303,520; *Chem. Abstr.*, **23**, 4483 (1929). French patent 66,450; *Chem. Abstr.*, **24**, 1391 (1930). British patent 328,933; *Chem. Abstr.*, **24**, 5307 (1930). German patent 504,340; *Chem. Abstr.*, **24**, 5309 (1930). German patent 510,436; *Chem. Abstr.*, **25**, 967 (1931). French patent 688,603; *Chem. Abstr.*, **25**, 1099 (1931). U. S. patent 1,807,682; *Chem. Abstr.*, **25**, 4412 (1931). British patent 347,193; *Chem. Abstr.*, **26**, 2874 (1932). U. S. patent 1.973,012; *Chem. Abstr.*, **28**, 7035 (1934).

[135] Wieland and Susser, *Ann.*, **392**, 172 (1912).

[136] Blom, *J. prakt. Chem.*, **94**, 77 (1916).

[137] Gluzmann, *Chem. Abstr.*, **34**, 1314 (1940); **38**, 743 (1944); **41**, 959 (1947).

[138] Mazzara and Lamberti-Zanardi, *Gazz. chim. ital.*, **26**, II, 238 (1896).

[139] Ullmann, *Ann.*, **332**, 93 (1904).

[140] Tucker, *J. Chem. Soc.*, **125**, 1144 (1924).

the Graebe-Ullmann synthesis.[141] 3,6-Dichlorocarbazole, m.p. 202–203°, has been prepared by the action of sulfuryl chloride on a solution of carbazole in chloroform[142] and through the action of thionyl chloride on a nitrobenzene solution of carbazole.[143] The substance has also been prepared through the direct chlorination of carbazole in carbon disulfide.[144] The chlorination of carbazole to tetrachlorocarbazole, m.p. 213°, has been effected through the agency of methyl-*N*,*N*-dichlorocarbamate.[145] A tetrachlorocarbazole, m.p. 223–224°, and an octachlorocarbazole, m.p. 276–277°, were prepared[146] through the action of chlorine on a solution of carbazole in carbon tetrachloride. A trichlorocarbazole, m.p. 180°, and a hexachlorocarbazole, m.p. 225° (dec.), prepared through direct chlorination of carbazole have been described but their structures were not determined.[147] Continued action of chlorine on a solution of carbazole in acetic acid gave hexachlorocarbazole while octachlorocarbazole was prepared by the action of antimony pentachloride on hexachlorocarbazole.[147,148]

Ring closure in the *m*-chlorophenylhydrazone of cyclohexanone gives both 5- and 7-chlorotetrahydrocarbazoles.[149] Dehydrogenation with sulfur and quinoline gave 4-chloro- and 2-chlorocarbazole, respectively. All of the monochlorocarbazoles and all of the monobromocarbazoles have been prepared[150] from the corresponding chloro- or bromotetrahydrocarbazoles through dehydrogenation with chloranil.

9-Methylcarbazole has been converted into 3-chloro-9-methylcarbazole through the agency of sulfuryl chloride.[151] 9-Ethylcarbazole behaves similarly.

Bromination of carbazole yields 3-bromo-, 3,6-dibromo-, 1,3,6-tribromo-, and 1,3,6,8-tetrabromocarbazole depending on the quantity of bromine used. 3-Bromocarbazole, m.p. 199°, was prepared through the

[141] Ullmann, *Ber.*, **31**, 1697 (1898); *Ann.*, **332**, 97 (1904).

[142] Mazzara and Lamberti-Zanardi, *Gazz. chim. ital.*, **26**, II, 240 (1896). Buu-Hoi and Royer, *J. Org. Chem.*, **16**, 1198 (1951).

[143] German patent 294,016; *J. Chem. Soc.*, **112**, (1) 53 (1917).

[144] British patent 340,625; *Chem. Abstr.*, **25**, 4892 (1931).

[145] Saulnier, *Ann. chim. phys.*, **17**, 353 (1942). Bougault and Chabrier, *Compt. rend.*, **213**, 400 (1941).

[146] Zal'kind and Konarenko, *J. Applied Chem.* (*U. S. S. R.*), **12**, 1134–1136 (1939); *Chem. Abstr.*, **34**, 3266 (1940).

[147] Graebe and Knecht, *Ann.*, **202**, 27 (1880).

[148] See also Merz and Weith, *Ber.*, **16**, 2875 (1883).

[149] Moggridge and Plant, *J. Chem. Soc.*, **1937**, 1125.

[150] Barclay and Campbell, *ibid.*, **1945**, 530.

[151] Buu-Hoi and Royer, *Rec. trav. chim.*, **66**, 533 (1947).

action of a mixture of potassium bromide and potassium bromate in glacial acetic acid on carbazole[152] as well as from 3-aminocarbazole through the Sandmeyer reaction.[153] The compound has also been prepared by brominating carbazole with *N*-bromosuccinimide.[154] 3-Bromocarbazole, 3,6-dibromocarbazole, and 1,3,6-tribromocarbazole have been prepared through the action of bromide-bromate mixture on carbazole.[155] Carbazole-3-sulfonic acid on bromination gave 1,3,8-tribromocarbazole-3-sulfonic acid which was converted to 1,3,8-tribromocarbazole, m.p. 184°. The latter compound was further brominated to 1,3,6,8-tetrabromocarbazole, m.p. 230.5–231.5°.[156] 3-Bromocarbazole has been prepared also by brominating 9-acetylcarbazole[157] or 9-benzoylcarbazole[158] and removing the acyl group.

The bromination of 9-methylcarbazole to 3-bromo-9-methylcarbazole and to 3,6-dibromo-9-methylcarbazole has been accomplished through the agency of *N*-bromosuccinimide, of bromide-bromate mixture, and through the direct action of bromine.[159]

3-Iodocarbazole has been prepared through the action of potassium iodide and potassium iodate on carbazole in glacial acetic acid[160] as well as from 3-aminocarbazole through the Sandmeyer reaction.[161] 3,6-Diiodocarbazole has been prepared through iodination of carbazole with iodine and mercuric oxide.[160,162] 3-Iodo-9-acetylcarbazole, 3-iodo-9-*p*-toluenesulfonylcarbazole, 3,6-diiodo-9-*p*-toluenesulfonylcarbazole as well as the 3,6-diodo derivatives of both 9-methyl- and 9-ethyl carbazole have been prepared.[160,163]

2,7-Diiodocarbazole, m.p. 265–266°,[164] was prepared from 4,4′-diiodo-2,2′-diaminodiphenyl through diazotization and treatment with potassium

[152] Vaubel, *Z. angew. Chem.*, **14,** 784 (1901).

[153] Tucker, *J. Chem. Soc.*, **125,** 1144 (1924).

[154] Schmidt and Karrer, *Helv. Chim. Acta*, **29,** 573 (1946).

[155] Mizuch and Sanchenko, *J. Gen. Chem.* (*U. S. S. R.*), **10,** 852–854 (1940); *Chem. Abstr.*, **35,** 2509 (1941).

[156] Lindemann and Muhlhaus, *Ber.*, **B 58,** 2371 (1925).

[157] Ciamician and Silber, *Gazz. chim. ital.*, **12,** 276 (1882).

[158] Mazzara and Leonardi, *ibid.*, **22,** II, 570 (1892).

[159] Buu-Hoi, *Rec. trav. chim.*, **66,** 533 (1947). Cosgrove and Waters, *J. Chem. Soc.*, **1949,** 907.

[160] Tucker, *J. Chem. Soc.*, **1926,** 546.

[161] Tucker, *ibid.*, **125,** 1144 (1924).

[162] Classen, German patent 81,929; *Friedl.*, **4,** 1096 (1899).

[163] Gilman and Kirby, *J. Org. Chem.*, **I,** 146 (1936). Gilman and Spatz, *J. Am. Chem. Soc.*, **63,** 1553 (1941).

[164] Ponte, *Chem. Abstr.*, **28,** 7255 (1934); *Giorn. farm. chim.*, **83,** 185–912 (1934); *Chem. Abstr.*, **29,** 2946 (1934); *Giorn. farm. chim.*, **83,** 347–350 (1934).

iodide. Along with the 2,4,2′,4′-tetraiodobiphenyl there was obtained some 2,7-diiodocarbazole. The same substance was prepared similarly from 2-nitro-2′-amino-4,4′-diiodobiphenyl.

Bromination of the *N*-substituted tetrahydrocarbazoles leads to the formation of unstable 10,11-dibromo derivatives which are easily hydrolysed to give the corresponding dihydroxy derivative.[165]

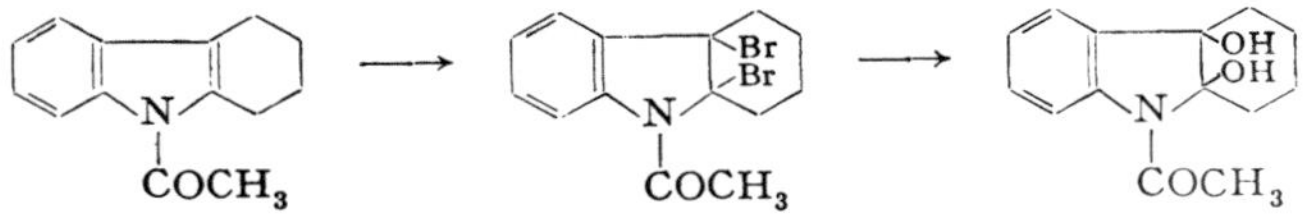

$COCH_3$ → Br, Br, $COCH_3$ → OH, OH, $COCH_3$

The benzoyl derivative differs from the acetyl in that while an intermediate dibromo derivative is obtained only one of the bromine atoms reacts with water, the other combines with a neighboring hydrogen atom to yield hydrogen bromide. The product is 9-benzoyl-1-hydroxy-1,2,3,4-tetrahydrocarbazole,[113–116] which is converted by alcoholic potassium hydroxide to 1-hydroxy-1,2,3,4-tetrahydrocarbazole[113, 114, 166] which reacts with acetic anhydride to give a compound, $C_{24}H_{22}N_2$ (formerly designated 2,3-dihydrocarbazole). The same compound was also prepared from 11-

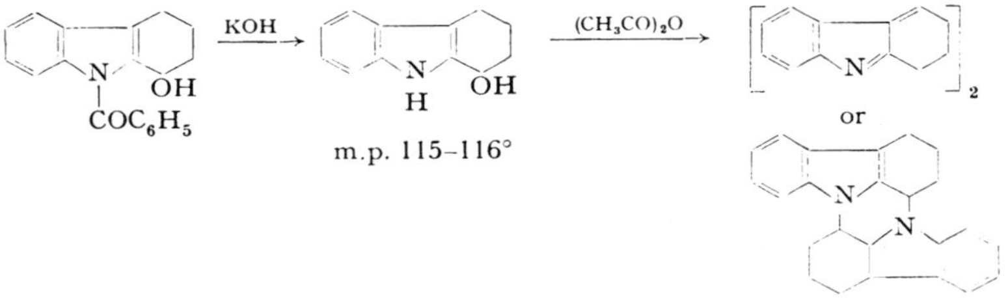

N, OH, COC_6H_5 —KOH→ N, H, OH, m.p. 115–116° —$(CH_3CO)_2O$→ [N]$_2$ or N, N

hydroxytetrahydrocarbazolenine through the agency of acetic anhydride.[113, 114, 115, 167, 116]

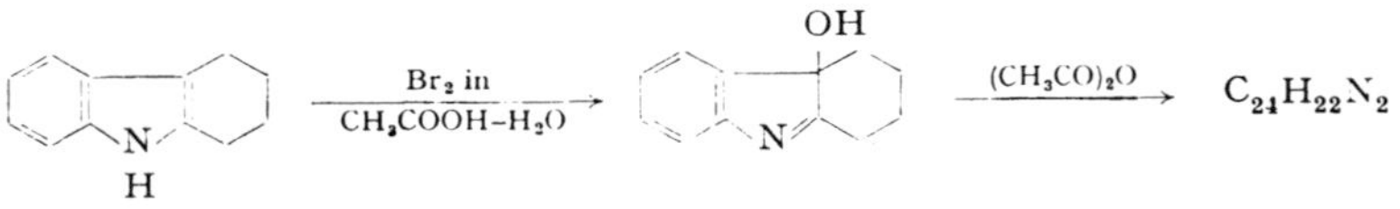

N, H —Br_2 in CH_3COOH–H_2O→ OH, N —$(CH_3CO)_2O$→ $C_{24}H_{22}N_2$

Catalytic oxidation of 1,2,3,4-tetrahydrocarbazole gives 11-hydroperoxytetrahydrocarbazolenine.[115, 166] The latter compound on treatment with alkali yields spiro(cyclopentane-1,2′-pseudo-indoxyl), m.p. 79°; reduction yields 11-hydroxytetrahydrocarbazolenine, m.p. 159°, and the com-

[165] Plant and Tomlinson, *J. Chem. Soc.*, **1931,** 3324.

[166] Beer, McGrath, and Robertson, *Nature,* **164,** 362 (1949); *J. Chem. Soc.*, **1950,** 2118, 3283.

[167] Plant and Tomlinson, *J. Chem. Soc.*, **1933,** 298. Compare also Patrick and Witkop, *J. Am. Chem. Soc.*, **72,** 633 (1950).

pound $C_{24}H_{22}N_2$ while treatment with acid leads to the formation of 1-aza-8,9-benzcyclonona-2,7-dione or the lactam of δ-*o*-aminobenzoylvaleric acid.

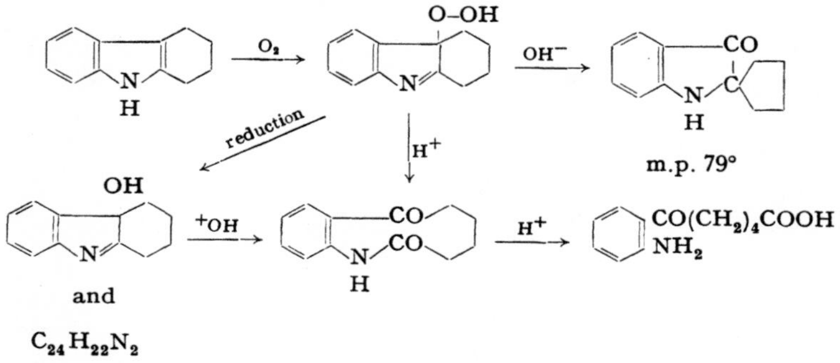

O–OH
O_2
OH^-
CO
N
H
N
N
C
H
reduction
H^+
m.p. 79°
OH
^+OH
CO
CO
H^+
$CO(CH_2)_4COOH$
NH_2
N
N
H
and
$C_{24}H_{22}N_2$

Carbazolesulfonic Acids

The chemistry of the carbazolesulfonic acids presents a picture which at certain points is not clear. The members of the series are for the most part high melting or nonmelting solids and as a consequence it is frequently difficult if not impossible to judge whether the compounds described by different workers are identical or different ones. It appears that the monosulfonic acids obtained through the action of sulfuric acid or of chlorosulfonic acid[168] on carbazole or on the 9-alkylcarbazoles[169] are the carbazole-3-sulfonic acids. Unfortunately 1-phenyl-1,2,3-benzotriazole-5-sulfonic acid gives not carbazole-3-sulfonic acid but carbazole itself. However, a preparation of carbazole-3-sulfonic acid from 3-aminocarbazole has been

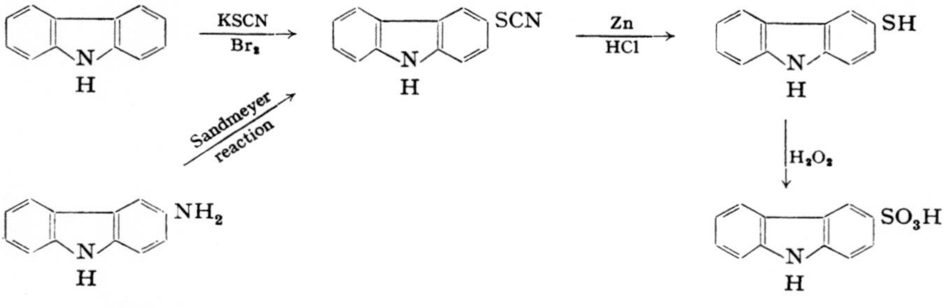

KSCN
Br_2
SCN
Zn
HCl
SH
N
H
N
H
N
H
Sandmeyer reaction
H_2O_2
NH_2
SO_3H
N
H
N
H

[168] Borodkin and Mal'kova, *Zhur. Priklad. Khim.*, **21,** 1032 (1948); *Chem. Abstr.*, **43,** 6205 (1949). Borodkin, *J. Applied Chem.* (*U. S. S. R.*), **23,** 803 (1950); *Chem. Abstr.*, **46,** 8089 (1952).

[169] German patent 260,898; *Chem. Abstr.*, **7,** 3199 (1913). British patent 9960; *Chem. Abstr.*, **8,** 438 (1914). German patent 275,795; *Chem. Abstr.*, **9,** 385 (1915). U. S. patent 1,128,369; *Chem. Abstr.*, **9,** 1122 (1915).

effected[170] yielding a product which the author states is identical with commercial carbazolemonosulfonic acid. Carbazole is only slightly attacked by cold sulfuric acid.[171] On warming there is obtained besides the monosulfonic acid a mixture of di- and trisulfonic acids from which it is difficult to separate a definite homogeneous compound, although from this mixture supposedly pure disulfonic acid samples have been prepared. The identity of the products obtained by various workers has not been established. The properties of the sulfonamides prepared by the several workers are not in agreement and it is difficult if not impossible to reach any definite conclusions from the data of these workers. Cohn[172] states that one of the known carbazoledisulfonic acids must be the 3,6-derivative since it does not react with *p*-nitrosophenol to give an indophenolsulfonic acid (the *p*-position to the imido group being occupied) but his statement throws little light on the question since he is not clear as to which of the several disulfonic acid preparations he refers.

The preparation of 1,3,6-carbazoletrisulfonic acid has been described.[173,174] Several references[175] are to be found in the literature to the nitration of 1,3,6-carbazoletrisulfonic acid to yield 1-nitro-3,6,8-carbazoletrisulfonic acid.

A carbazoletetrasulfonic acid has been[176] prepared through the action of fuming sulfuric acid on carbazole. It is claimed[177] that this compound is 2,3,6,8-carbazoletetrasulfonic acid and it is further stated that fusion with potassium hydroxide gives 2,8-dihydroxycarbazole-3,6-disulfonic acid which on heating with acid gives 2,8-dihydroxycarbazole. The entrance of a sulfonic

[170] Mizuch, *J. Gen. Chem.* (*U. S. S. R.*), **10**, 844–851 (1940); *Chem. Abstr.*, **35**, 2509 (1941).

[171] Schultz and Hauenstein, *J. prakt. Chem.*, [2], **76**, 335 (1907). Graebe and Glaser, *Ann.*, **163**, 347 (1872). Bechhold, *Ber.*, **23**, 2144 (1890). Schwalbe and Wolff, *ibid.*, **44**, 237 (1911). Compare, however, Borodkin, *Zhur. Priklad. Khim.*, **24**, 1202-7 (1951); *Chem. Abstr.*, **46**, 7092 (1952).

[172] *Die Carbazolgruppe*. Thieme, Leipzig, 1919, p. 107.

[173] German patent 258,298; *Chem. Abstr.*, **7**, 2836 (1913). German patent 275,975; *Chem. Abstr.*, **8**, 3503 (1914).

[174] Borodkin and Mal'kova, *Zhur. Priklad. Khim.*, **21**, 849 (1948); *Chem. Abstr.*, **43**, 6205 (1949).

[175] French patent 654,129; *Chem. Abstr.*, **23**, 3715 (1929). British patent 316,962; *Chem. Abstr.*, **24**, 1868 (1930). British patent 320,641; *Chem. Abstr.*, **24**, 2614 (1930). German patent 507,797; *Chem. Abstr.*, **25**, 716 (1931).

[176] German patent 224,952; *Chem. Zentr.*, **1910**, II, 700; *Friedl.*, **10**, 146 (1913).

[177] U. S. patent 1,981,301; *Chem. Abstr.*, **29**, 618 (1935). French patent 754,224; *Chem. Abstr.*, **28**, 1050 (1934). German patents 615,578 and 615,579, *Chem. Abstr.*, **29**, 8006 (1935). British patent 417,794; *Chem. Abstr.*, **29**, 1438 (1935).

acid group in position 2 (or 7) during the sulfonation of carbazole appears highly improbable and one is tempted to conclude that the carbazoletetrasulfonic acid must be the 1,3,6,8-derivative and not the 2,3,6,8-tetrasulfonic acid as claimed. Certainly the 1,3,6,8-structure is in better accord with the behavior of carbazole on nitration and halogenation and with the usual rules of orientation.

Carbazole-2,7-disulfonic acid has been synthesized from 2,2′-diaminodiphenyl-4,4′-disulfonic acid through heating to 150° in the presence of ammonium sulfate or ammonium chloride.[178] The carbazole 2,7-disulfonic acid has been converted to 2-hydroxycarbazole-7-sulfonic acid by heating to 260–280° with alkali and conversion of the resulting salt to the free acid.[179] 2-Aminocarbazole-7-sulfonic acid is obtained when 2,2′,4′-triaminodiphenyl-4-sulfonic acid is heated with dilute acid at 150–179°. At 180–200° 2-aminocarbazole is obtained.[180] 2-Hydroxycarbazole reacts with chlorosulfonic acid to give 2-hydroxycarbazole-3-sulfonic acid.[181]

Reduction of Carbazole

It might be expected that the hydrogenation of carbazole would yield, first, 1,4-dihydrocarbazole and second, 1,2,3,4-tetrahydrocarbazole. The preparation of 1,4-dihydrocarbazole by reduction of carbazole with sodium

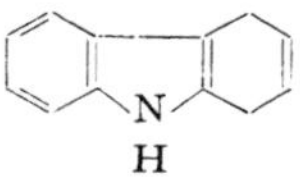

1,4-Dihydrocarbazole

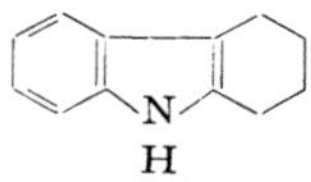

1,2,3,4-Tetrahydrocarbazole

and boiling amyl alcohol has been reported[182] but has subsequently been shown[183-185] to be a mixture containing at least 50% unreduced carbazole.

1,2,3,4-Tetrahydrocarbazole, m.p. 116°, can be prepared either by reduction of carbazole with sodium and alcohol[186] or from cyclohexanone

[178] British patent 358,056; *Chem. Abstr.*, **26**, 5971 (1932). German patent 538,450; *Chem. Abstr.*, **26**, 1623, (1932). U. S. patent 1,867,864; *Chem. Abstr.*, **26**, 5105 (1932).

[179] U. S. patent 1,879,425; *Chem. Abstr.*, **27**, 427 (1933).

[180] U. S. patent 1,867,864; *Chem. Abstr.*, **26**, 5105 (1932).

[181] German patent 645,881; *Chem. Abstr.*, **31**, 6478 (1937). British patent 458,268; *Chem. Abstr.*, **31**, 3069 (1937).

[182] Schmidt and Schall, *Ber.*, **40**, 3225 (1907).

[183] Barclay, Campbell, and Gow, *J. Chem. Soc.*, **1946**, 997.

[184] Compare, however, Sanna, *Gazz. chim. ital.*, **80**, 572 (1950).

[185] Compare, however, Witkop, *J. Am. Chem. Soc.*, **72**, 614, 1428 (1950); Patrick and Witkop, *ibid.*, **72**, 633 (1950).

[186] Zanetti, *Ber.*, **26**, 2006 (1893).

phenylhydrazone.[187] This compound is converted into indole-α-carboxylic acid through fusion with sodium hydroxide. 1,2,3,4,10,11-Hexahydrocarbazole is obtained when carbazole is reduced with phosphorus and hydriodic acid[188] and more readily by reduction of tetrahydrocarbazole with tin and hydrochloric acid[188] or by electrolytic reduction.[189]

Reduction of *N*-methylcarbazole and *N*-ethylcarbazole with a nickel catalyst and hydrogen at 210°[190] and 25 atmospheres pressure gave a mixture of the tetrahydro- and octahydro-*N*-alkylcarbazoles along with some unchanged starting material. The pyrrole double bonds in both of these derivatives remain intact during catalytic reduction but are attacked

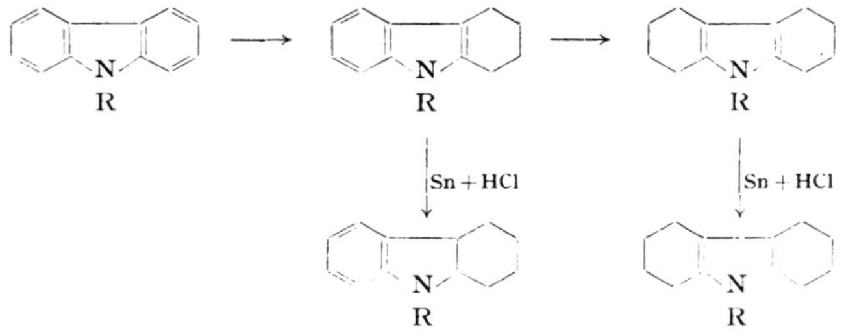

N R → N R → N R

Sn + HCl ↓ N R   Sn + HCl ↓ N R

by tin and hydrochloric acid to give the hexahydro and decahydro derivatives, respectively. The latter compound is extremely resistant to further reduction. The position of the double bonds in the octahydro derivatives was shown[191] by the reduction of 6,9-dimethyl- and 3,9-dimethyl-1,2,3,4-tetrahydrocarbazoles to yield the same octahydrocarbazole. The position of the

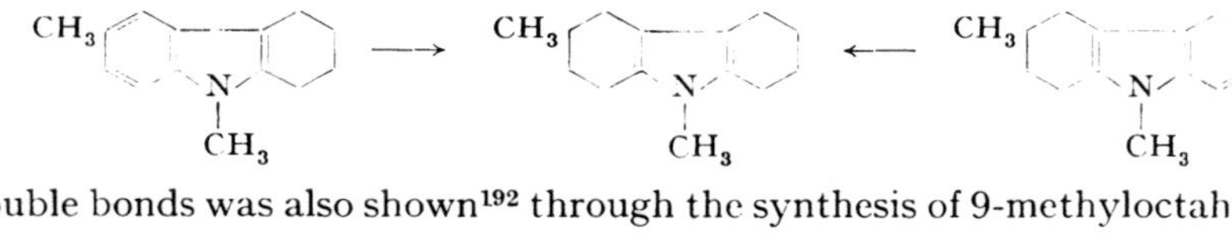

CH_3 N CH_3 → CH_3 N CH_3 ← CH_3 N CH_3

double bonds was also shown[192] through the synthesis of 9-methyloctahydrocarbazole from 2,2′-dicyclohexanone and methylamine.

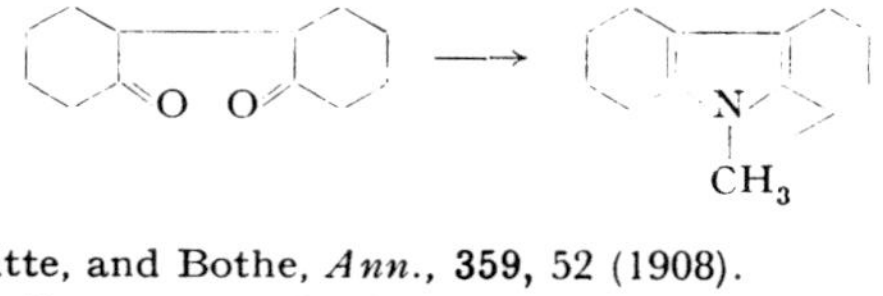

O O → N CH_3

[187] Borsche, Witte, and Bothe, *Ann.*, **359**, 52 (1908).

[188] Graebe and Glaser, *Ber.*, **5**, 12 (1872). Schmidt and Sigwart, *ibid.*, **45**, 1779 (1912).

[189] Perkin and Plant, *J. Chem. Soc.*, **125**, 1503 (1924). Carrasco, *Gazz. chim. ital.*, **38**, 301 (1908); *Chem. Abstr.*, **2**, 3229 (1908).

[190] v. Braun and Ritter, *Ber.*, **55**, 3792 (1922).

[191] v. Braun and Schorning, *ibid.*, **58**, 2156 (1925).

[192] Plant, *J. Chem. Soc.* **1930**, 1595.

A second octahydrocarbazole, m.p. 102°, was prepared from cyclohexylidine azine.[193, 194] The compound was at first thought to be 1,2,3,4,5,6,7,8-octahydrocarbazole[193] but it appears more probable that the compound is

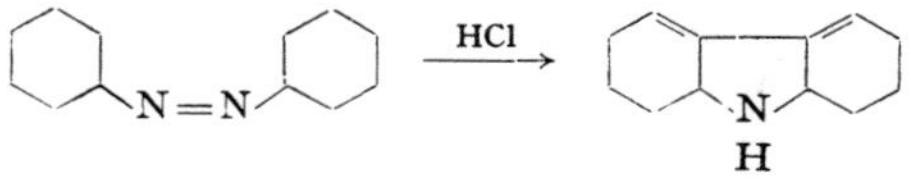

in reality 1,2,3,6,7,8,10,13-octahydrocarbazole.[194, 195] Reduction of the octahydrocarbazole, m.p. 102°, with tin and hydrochloric acid gave a decahydrocarbazole, m.p. 75°,[193] and electrolytic reduction gave a dodecahydrocarbazole. The decahydrocarbazole was stable to electrolytic reduction and therefore could not be an intermediate in the reduction to dodecahydrocarbazole which is perhaps formed as shown in the scheme:

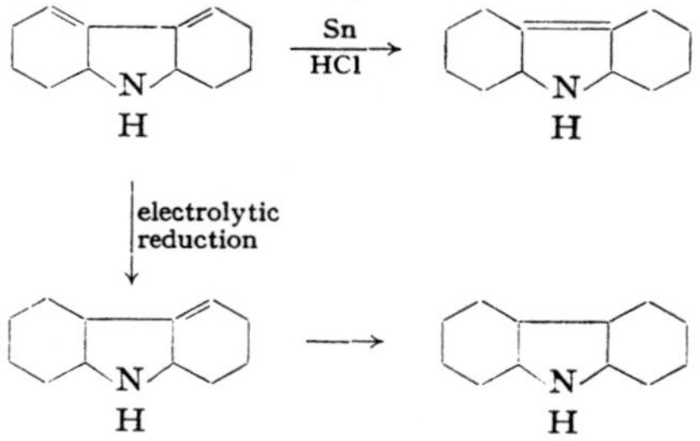

Not isolated

It has been reported[196] that catalytic hydrogenation of carbazole in the presence of finely divided nickel at 200–220° for twelve to eighteen hours gave mainly a product, m.p. about 95°, thought to be 2,3-diethylindole but it seems more likely that this product was a mixture of hydrogenated carbazoles such as was obtained by Adkins and Coonradt.[197] These workers obtained 1,2,3,4-tetrahydrocarbazole, 1,2,3,4,10,11-hexahydrocarbazole, and dodecahydrocarbazole, in proportions which varied with the catalyst and other experimental conditions, through hydrogenation of carbazole with Raney nickel or copper chromite catalysts. The hydrogenation of the 9-alkylcarbazoles yielded the corresponding tetrahydro and dodecahydro derivatives but did not yield the hexahydro derivatives. Hexahydrocarbazole

[193] Perkin and Plant, *ibid.*, **125**, 1503 (1924).
[194] Benary, *Ber.*, **67**, 208 (1934).
[195] v. Braun and Schorning, *ibid.*, **58**, 2156 (1925).
[196] Padoa and Chiaves, *Gazz. chim. ital.*, **38**, 236 (1908).
[197] Adkins and Coonradt, *J. Am. Chem. Soc.*, **63**, 1563 (1941). Compare Shah, Tilac, and Venkataraman, *Proc. Indian Acad. Sci.*, **A 28,** 142–150 (1948).

exists in two forms, *cis* and *trans*.[56, 184, 198] The hydrogenation of tetrahydro-

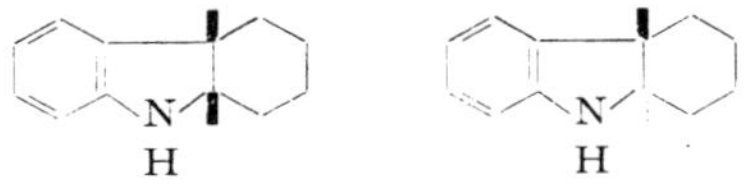

Cis *Trans*

carbazole gives preponderately the more stable *cis* form, only one to two per cent of the *trans* modification being found in the reaction product.[198]

Oxidation of Carbazole

The oxidation of carbazole has been effected by means of sodium dichromate[199] in a mixture of sulfuric and acetic acids and by potassium permanganate[200] in acetone. Using the latter reagent three products were isolated: (A), m.p. 220°; (B), m.p. 265°, and (C), amorphous, m.p. about 175°. A was shown to be 9,9′-dicarbazyl[201] since on bromination it gave the same 3,6,3′,6′-tetrabromodicarbazyl obtained by the action of iodine on the potassium salt of 3,6-dibromocarbazole. The possibility that either B or C

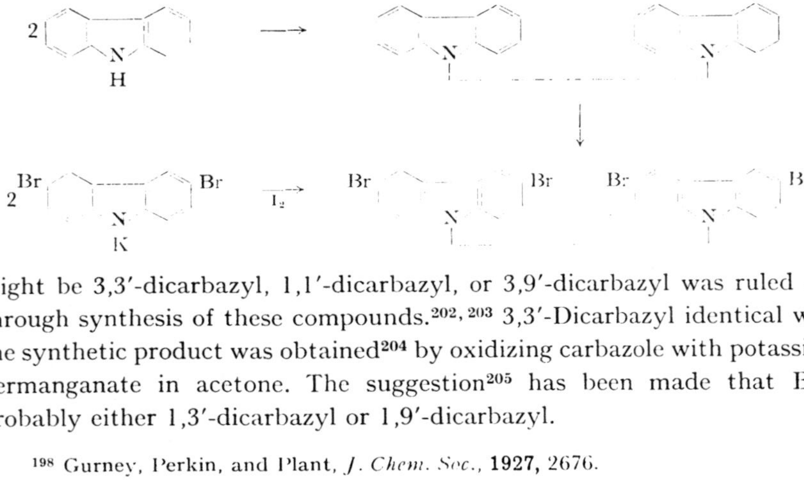

might be 3,3′-dicarbazyl, 1,1′-dicarbazyl, or 3,9′-dicarbazyl was ruled out through synthesis of these compounds.[202, 203] 3,3′-Dicarbazyl identical with the synthetic product was obtained[204] by oxidizing carbazole with potassium permanganate in acetone. The suggestion[205] has been made that B is probably either 1,3′-dicarbazyl or 1,9′-dicarbazyl.

[198] Gurney, Perkin, and Plant, *J. Chem. Soc.*, **1927**, 2676.
[199] Wieland, *Ber.*, **46**, 3296 (1913).
[200] Perkin and Tucker, *J. Chem. Soc.*, **119**, 216 (1921).
[201] Tucker and McLintock, *ibid.*, **1927**, 1214.
[202] Tucker, *ibid.*, **1926**, 3033.
[203] Tucker and Macrae, *ibid.*, **1933**, 1520. Tucker and Nelmes, *ibid.*, **1933**, 1523. Dunlop, Macrae, and Tucker, *ibid.*, **1934**, 1672.
[204] Maitland and Tucker, *ibid.*, **1927**, 1388.
[205] Campbell and Barclay, *Chem. Revs.*, **40**, 368 (1947).

The oxidation of carbazole by silver oxide was studied by Branch and co-workers.[206] Three oxidation products were obtained which did not agree in properties with those described above. Two of Branch's products were thought to exhibit the properties of free radicals containing divalent nitrogen. However, since Branch's compounds were not obtained in pure state and consequently were not well characterized it is difficult to evaluate the merits of his claims. It would appear that further study of these products is in order.

Carbazyl Aldehydes and Ketones

Carbazyl aldehydes are readily prepared through the action of *N*-methylformanilide on 9-alkylcarbazoles and 3,9-dialkylcarbazoles, the formyl group being introduced in position 3 in the first case and in position 6 in the second case.[207]

Carbazole and acyl chlorides yield with aluminum chloride in carbon disulfide the 3,6-diacyl derivatives with but little or no monoacyl derivative being obtained, even when the quantity of acyl halide used is limited. 3,6-Diacetylcarbazole and 3,6-dibenzoylcarbazole were prepared in this fashion.[208] 3,6-Dibenzoylcarbazole has also been prepared through the use of benzoic anhydride and aluminum chloride as well as by heating carbazole-3,6-diphthaloylic acid with copper bronze in quinoline.[209] 3-Benzoylcarbazole has been synthesized through the application of both the Graebe-Ullmann[210] and the Borsche[211] syntheses.

C_6H_5CO N N N ⟶ C_6H_5CO N H

two steps

C_6H_5CO N N H

[206] Branch and Smith, *J. Am. Chem. Soc.*, **42**, 2405 (1920). Branch and Hall, *ibid.*, **46**, 438 (1924).

[207] Buu-Hoi, *ibid.*, **73**, 98 (1951); *J. Org. Chem.*, **16**, 1327 (1951).

[208] Buu-Hoi and Royer, *J. Org. Chem.*, **15**, 123 (1950). Buu-Hoi and Royer, *Rec. trav. chim.*, **66**, 533 (1947). Plant, Rogers, and Williams, *J. Chem. Soc.*, **1935**, 741. Plant and Tomlinson, *ibid.*, **1932**, 2188.

[209] Mitchell and Plant, *J. Chem. Soc.*, **1936**, 1295.

[210] Hunter and Darling, *J. Am. Chem. Soc.*, **53**, 4183 (1931).

The Fries rearrangement has provided the means for the preparation of a number of 3-acyl- and 3,6-diacylcarbazoles. Thus 3-benzoylcarbazole, m.p. 206°, was prepared by heating 9-benzoylcarbazole with aluminum chloride at 120°.[208] The 3-acetyl derivative, m.p. 167°, was prepared in similar manner.[211, 212] The structure of the 3-acetyl derivative was shown by reduction (Clemmensen) to the 3-ethyl derivative, m.p. 142°. When the Fries[213,214] rearrangement of 9-acetylcarbazole is effected through heating with $AlCl_3$ in nitrobenzene some 1-acetylcarbazole is obtained along with 3-acetylcarbazole.[215] The Fries reaction has also been used[216] for the preparation of a number of 3-acyl- and 3,6-diacylcarbazoles in which the acyl groups are stearyl, palmityl, lauryl, and myristyl groups.

The 9-alkylcarbazoles react with acyl halides yielding either the mono or the diacyl derivatives depending on the experimental conditions.[208, 217] The 9-acyl derivatives, on the other hand, give 2,9-derivatives.[218] 9-Acetylcarbazole with acetyl chloride and aluminum chloride gives 2,9-diacetylcarbazole.[219] 9-Benzoylcarbazole on treatment with acetyl bromide and aluminum chloride gave 2-acetyl-9-benzoylcarbazole, m.p. 153°.[220,221] In similar fashion 9-benzoylcarbazole, benzoyl chloride, and aluminum chloride gave 2,9-dibenzoylcarbazole, m.p. 140–142°. In some experiments 3,6-dibenzoylcarbazole was obtained, rearrangement having preceded substitution. Treatment of 9-acetylcarbazole with chloracetyl chloride and aluminum chloride gives 2-chloroacetyl-9-acetylcarbazole.[222, 223]

[211] Plant and Williams, *J. Chem. Soc.*, **1934,** 1142.
[212] Ruberg and Small, *J. Am. Chem. Soc.*, **63,** 736 (1941).
[213] Plant, Rogers, and Williams, *J. Chem. Soc.*, **1935,** 741.
[214] Meitzner, *J. Am. Chem. Soc.*, **57,** 2327 (1935).
[215] Compare Berlin, *J. Gen. Chem. (U. S. S. R.)*, **14,** 1096 (1944); *Chem. Abstr.*, **40,** 4054 (1946). Plant and Williams, *J. Chem. Soc.*, **1934,** 1142.
[216] Ford, *Iowa State Coll. J. Sci.*, **12,** 121 (1937); *Chem. Abstr.*, **32,** 4943 (1938). Ralston and Christiansen, *Ind. Eng. Chem.*, **29,** 194 (1937). U. S. patent 2,101,559; *Chem. Abstr.*, **32,** 1019 (1938). (In these papers and the abstracts thereof the system of nomenclature used is that in which *N* is 5. The notation has been transposed in this text to the Ring Index system, *i.e.*, *N* is 9.)
[217] Plant, Rogers, and Williams, *J. Chem. Soc.*, **1935,** 741.
[218] German patent 555,312; *Chem. Abstr.*, **26,** 5104 (1932).
[219] Borsche and Feise, *Ber.*, **40,** 378 (1907). Berlin, *J. Gen. Chem.*, *(U. S. S. R.)*, **14,** 1096 (1944); *Chem. Abstr.*, **40,** 4054 (1946).
[220] Manske and Kulka, *Can. J. Research*, **B 28,** 443–452 (1950).
[221] Plant and Williams, *J. Chem. Soc.*, **1934,** 1142.
[222] Plant, Rogers, and Williams, *ibid.*, **1935,** 741.
[223] Sherlin and Berlin, *J. Gen. Chem. (U. S. S. R.)*, **7,** 2275 (1937).

The hydrolysis of the 2,9-diacyl derivatives yields the 2-acyl derivatives of carbazole.[219, 224]

The degree of reduction of the carbazole affects the course of the Friedel-Crafts reaction. Treatment of 9-acetylhexahydrocarbazole with acetyl chloride gives 6,9-diacetylhexahydrocarbazole[225] while 7,9-diacetyltetrahydrocarbazole is obtained through the acetylation of 9-acetyltetrahydrocarbazole.[226] 6-Substituted-9-acetyltetrahydrocarbazoles behave similarly.[227]

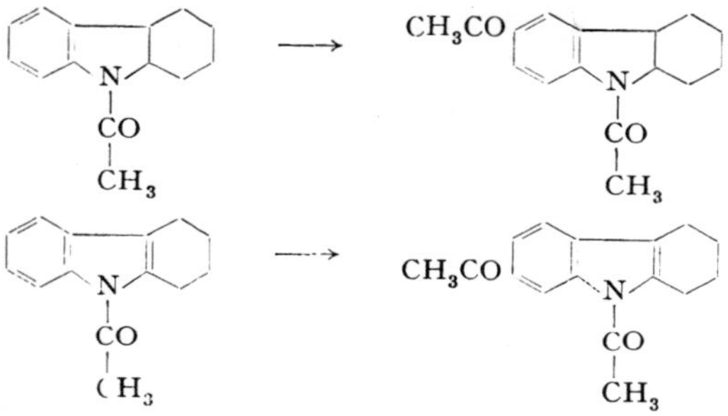

Carbazole and phthalic anhydride condense in the presence of aluminium chloride to give carbazole-9-phthaloylic acid (I), softening at 150°, m.p. 190°.[228] Under slightly different experimental conditions carbazole and phthalic anhydride react in the presence of aluminum chloride to give carbazole-3,6-diphthaloylic acid (II), m.p. 300–301°.[229] The latter is easily converted by sulfuric acid to 2,3,6,7-diphthaloylcarbazole (III). The 9-alkyl

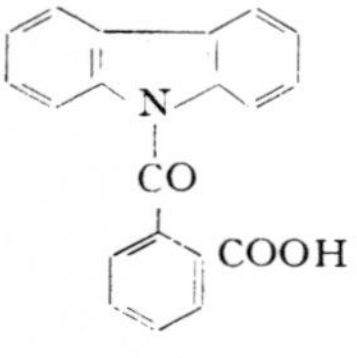

(I)

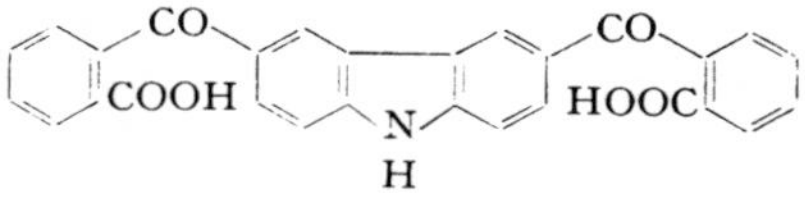

(II)

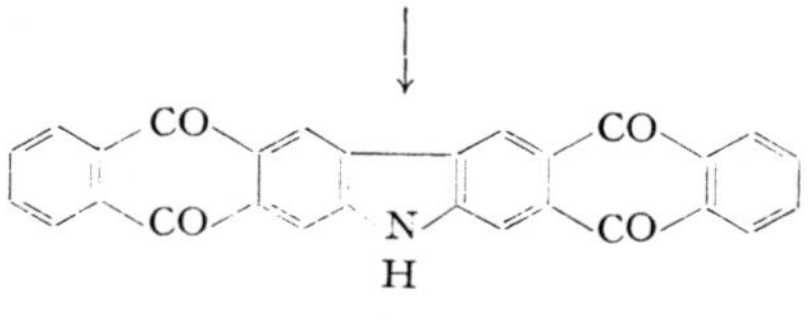

(III)

[224] Plant, Rogers, and Williams, *J. Chem. Soc.*, **1935**, 741.
[225] Mitchell and Plant, *ibid.*, **1936**, 1295.
[226] Plant and Rogers, *ibid.*, **1936**, 40.
[227] Plant and Powell, *ibid.*, **1947**, 937.
[228] Stummer, *Monatsh.*, **28**, 411 (1907).
[229] Scholl and Neovius, *Ber.*, **44**, 1249 (1911). Mitchell and Plant, *J. Chem. Soc.*, **1936**, 1295.

analogs of II and III have been prepared similarly from the corresponding 9-alkylcarbazoles.[230] 9-Ethyl-2,3,6,7-diphthaloylcarbazole is the vat dye Hydron Yellow.[231] Phthalyl chloride reacts with 9-ethylcarbazole in both the symmetrical and the unsymmetrical forms.[232] From the action of the former there results 9-ethylcarbazole-3-phthaloylic acid and 9-ethylcarbazole-3,6-diphthaloylic acid. The unsymmetrical form reacts to give two phthalides formulated as follows:

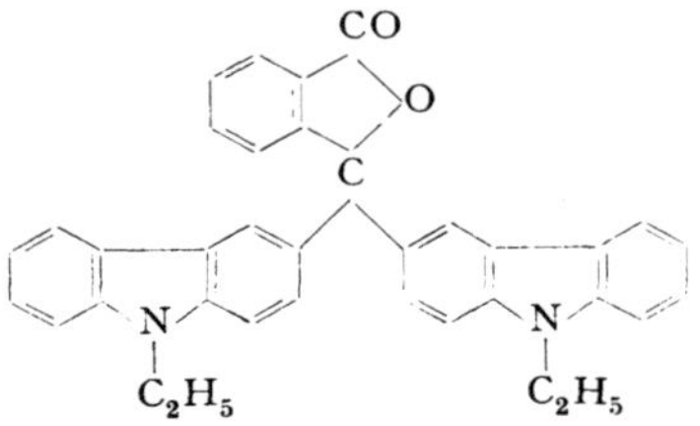

CO O C N N C_2H_5 C_2H_5

Bis-9-ethylcarbazolephthalide

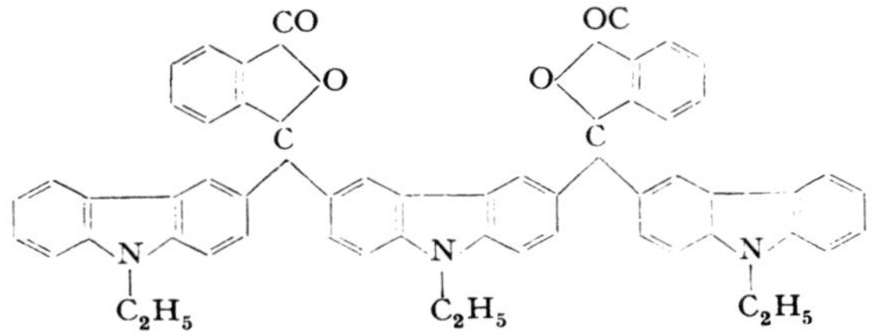

CO O C OC O C N N N C_2H_5 C_2H_5 C_2H_5

Tris-9-ethylcarbazolediphthalide

1,2-Phthaloyl carbazole, m.p. 255°, was prepared from 1-chloroanthraquinone and azimidobenzene.[233]

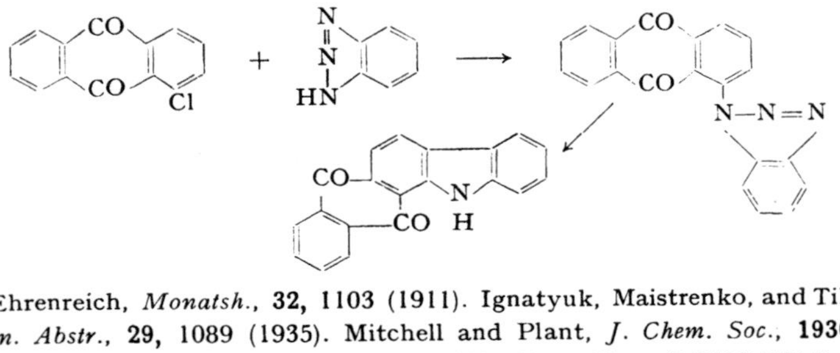

CO CO Cl + N N HN ⟶ CO CO N—N=N

CO N CO H

[230] Ehrenreich, *Monatsh.*, **32,** 1103 (1911). Ignatyuk, Maistrenko, and Tikhonov, *Chem. Abstr.*, **29,** 1089 (1935). Mitchell and Plant, *J. Chem. Soc.*, **1936,** 1295. British patent 22,836; French patent 463,508; *Chem. Abstr.*, **8,** 2489 (1914). German patent 275,670; *Chem. Abstr.*, **9,** 965 (1915). Russian patent 35,189; *Chem. Abstr.*, **30,** 3445 (1936). Russian patent 40,975; *Chem. Abstr.*, **30,** 7128 (1936).

[231] German patent 261,495; *Chem. Abstr.*, **7,** 3547 (1913). British patent 28,874; *Chem. Abstr.*, **7,** 1617 (1913).

[232] Copisarow and Weitzman, *J. Chem. Soc.*, **107,** 878 (1915).

[233] Ullmann and Illgen, *Ber.*, **47,** 383 (1914).

Carbazole-3,6-bis-(γ-ketobutyric acid), m.p. 292–293°, has been prepared[234, 235] through the interaction of carbazole and succinic anhydride in the presence of aluminum chloride. Reduction (Clemmensen) of this compound gave carbazole-3,6-bis(γ-butyric acid), m.p. 197–198°.

Carbazole Carboxylic Acids

The carbonation of potassium carbazole yields carbazole-1-carboxylic acid,[236] m.p. 275–276°, the structure of which is shown by its preparation through dehydrogenation of 1,2,3,4-tetrahydrocarbazole-8-carboxylic acid.[237]

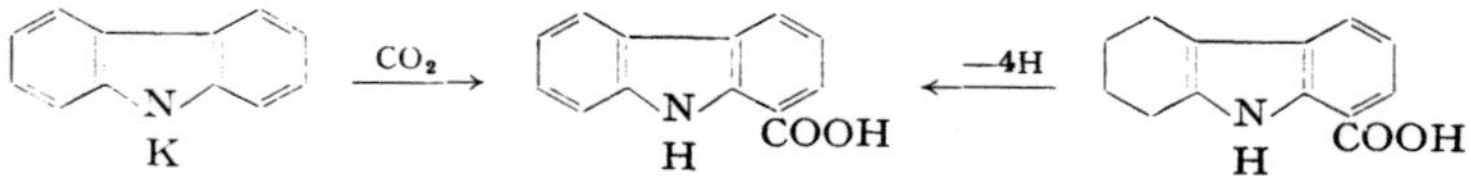

The acid has also been prepared through carbonation of carbazyl-9-magnesium iodide.[238] This reaction has been studied by Gilman[239] who favors mechanism (A) because of the fact that a purple color test (with triaryl-bismuth dihalides) was obtained after heating the MgI compound and before carbonation. The N-MgI compounds do not give this color test while the C-MgI compounds do give the test. Carbazole-1-carboxylic acid has also been prepared from the corresponding 1-aminocarbazole by diazotization and conversion of the diazonium salt to the nitrile and then to the acid by the usual methods.[240] The compound has also been obtained in poor yield

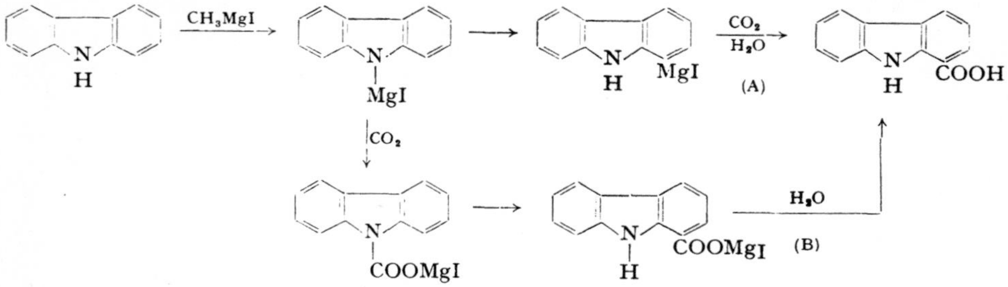

[234] Mitchell and Plant, *J. Chem. Soc.*, **1936,** 1295.

[235] Rejnowski and Suszko, *Arch. Chem. Pharm.*, **3,** 135–140 (1937); *Chem. Abstr.*, **32,** 2939 (1938).

[236] Ciamician and Silber, *Gazz. chim. ital.*, **12,** 272 (1882).

[237] Briscoe and Plant, *J. Chem. Soc.*, **1928,** 1990.

[238] Oddo, *Gazz. chim. ital.*, **41,** I, 255 (1911).

[239] Gilman and Yablunsky, *J. Am. Chem. Soc.*, **63,** 839 (1941).

[240] British patent 303,520; *Chem. Abstr.*, **23,** 4483 (1929). French patent 666,450; *Chem. Abstr.*, **24,** 1391 (1930).

through treatment of carbazole with *n*-butyl lithium and subsequent carbonation.[241]

Carbazole-2-carboxylic acid, m.p. 319°, has been prepared from 2-aminocarbazole,[240] by fusion of 2-benzoylcarbazole[242] or 2,9-diacetylcarbazole[243] with potassium hydroxide and from methyl 1,2,3,4-tetrahydrocarbazole-7-carboxylate[244] through dehydrogenation and saponification. 9-Ethylcarbazole-2-carboxylic acid has been prepared through fusion of 9-ethyl-2-acetylcarbazole with potassium hydroxide.[245]

Carbazole-3-carboxylic acid, m.p. 276–278°, has been prepared through the fusion of 3-acetylcarbazole with potassium hydroxide[246] and through dehydrogenation and subsequent saponification of the methyl ester of the corresponding tetrahydrocarbazolecarboxylic acid.[242] 9-Ethylcarbazole-3-carboxylic acid has been prepared through mercuration of 9-ethylcarbazole[241] and subsequent conversion (in several steps) of the product into the acid. It should be noted that mercuration takes place in position 3 whereas treatment with butyl lithium resulted in metallation in position 1. The Houben reaction has been employed for the synthesis of carbazole-3-carboxylic acid and of carbazole-3,6-dicarboxylic acid.[247]

CN; $\xrightarrow{Cl_3CCN}$ C(=NHHCl)—CCl_3 ⟶ $COCCl_3$; KOH; KOH; N=N—N, CN ⟶ CN ⟶ COOH; N H

The preparation of a carbazoledicarboxylic acid through the interaction of carbon dioxide and potassium carbazole has been reported[248] but the structure of the product was not given (presumably it is the 1,8-derivative). Carbazole-3,6-dicarboxylic acid has been prepared from 3,6-dibenzoyl-

[241] Gilman and Kirby, *J. Org. Chem.*, **1**, 146 (1936).

[242] Plant, Rogers, and Williams, *J. Chem. Soc.*, **1935**, 741.

[243] Borsche and Fiese, *Ber.*, **40**, 378 (1907).

[244] Moggridge and Plant, *J. Chem. Soc.*, **1937**, 1125.

[245] Burton and Lehrman, *J. Am. Chem. Soc.*, **62**, 527 (1940).

[246] Plant and Williams, *J. Chem. Soc.*, **1934**, 1142.

[247] Dunlop and Tucker, *ibid.*, **1939**, 1945. Preston, Tucker, and Cameron, *ibid.*, **1942**, 500.

[248] German patent 263,150; *Chem. Abstr.*, **7**, 3668 (1913).

carbazole[249] through fusion with potassium hydroxide. 9-Ethylcarbazole-3,6-dicarboxylic acid was prepared from 9-ethyl-3,6-dibromocarbazole by metallation and subsequent carbonation.[250] Metallation of *N*-phenylcarbazole by *n*-butyllithium[251] takes place in the ortho position of the phenyl group rather than in the carbazole nucleus.

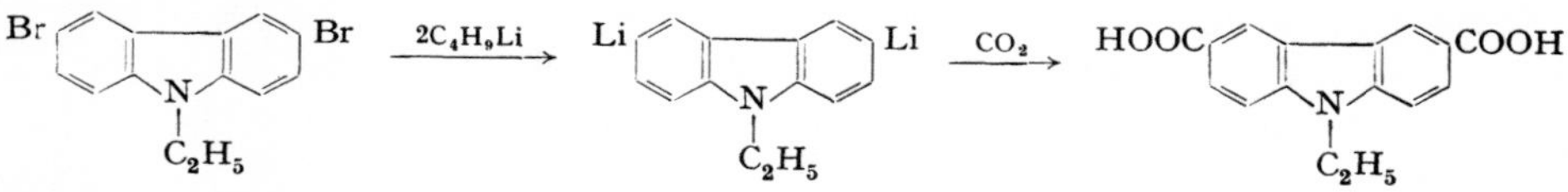

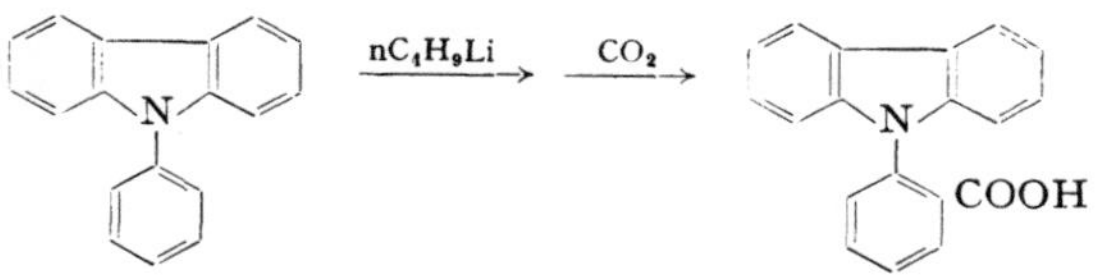

The Willgerodt reaction has been employed for the preparation of carbazole-1-acetic acid, carbazole-2-acetic acid, and carbazole-3-acetic acid from the corresponding acetylcarbazoles.[220, 252]

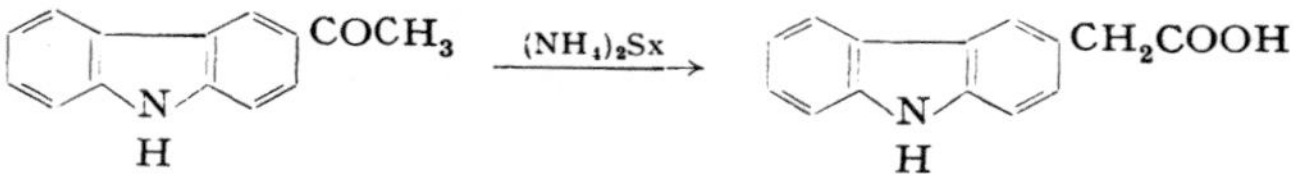

Tetrahydrocarbazole-6-carboxylic acid, m.p. 282°, has been prepared through treatment of cyclohexanone *p*-carboxyphenylhydrazone with sulfuric acid.[253] Cyclohexanone *o*-carboxyphenylhydrazone gave tetrahydrocarbazole-8-carboxylic acid, m.p. 203°, while cyclohexanone *m*-carboxyphenylhydrazone gave a mixture of the tetrahydrocarbazole-5- and 7-carboxylic acids. A tetrahydrocarbazolecarboxylic acid has been prepared[254] from the phenylhydrazone of 3-ketohexahydrobenzoic acid but it was not established whether this was the 2- or the 4-carboxylic acid. Tetrahydrocarbazole-3-carboxylic acid has been prepared from the phenylhydrazone of 4-ketohexahydrobenzoic acid.[255] 1,2,3,4-Tetrahydrocarbazole-6-propionic

[249] Plant and Mitchell, *J. Chem. Soc.*, **1936,** 1295.
[250] Spatz and Gilman, *Proc. Iowa Acad. Sci.*, **47,** 262 (1940). Gilman and Spatz, *J. Am. Chem. Soc.*, **63,** 1553 (1941). Spatz, *Iowa State Coll. J. Sci.*, **17,** 129 (1942).
[251] Gilman, Stuckwisch, and Kendall, *J. Am. Chem. Soc.*, **63,** 1758 (1941). Gilman and Stuckwisch, *ibid.*, **65,** 1729 (1943).
[252] Gilman and Avakian, *ibid.*, **68,** 2104 (1946).
[253] Collar and Plant, *J. Chem. Soc.*, **1926,** 808.
[254] Baeyer and Tutein, *Ber.*, **22,** 2184 (1884).
[255] Perkin, *J. Chem. Soc.*, **85,** 419 (1904).

acid has been prepared from the hydrazone obtained from cyclohexanone and 3-(4-hydrazinophenyl)-propionic acid.[256]

Hydroxycarbazoles are converted into hydroxycarbazolecarboxylic acids through the Kolbe reaction. Thus 3-hydroxycarbazole is converted into 3-hydroxycarbazole-2-carboxylic acid through the action of carbon dioxide and alkali.[257] 2-Hydroxycarbazole-3-carboxylic acid, m.p. 273–274°,

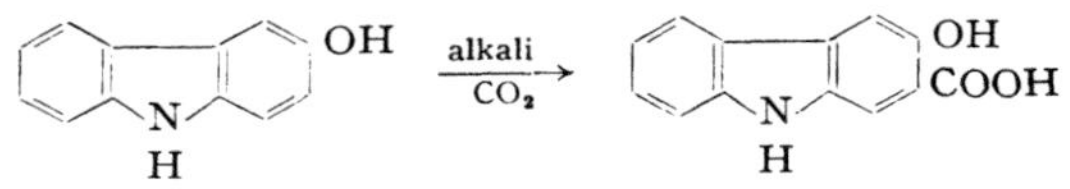

OH $\xrightarrow[CO_2]{alkali}$ OH COOH

and 2-hydroxycarbazole-1-carboxylic acid, m.p. 271–272°, were prepared similarly from 2-hydroxycarbazole. The reaction has also been employed for preparing other hydroxycarboxylic acids of the carbazole series.[258]

Ethyl 1-keto-1,2,3,4-tetrahydrocarbazole-2-carboxylate, m.p. 145–146°, has been prepared as shown in the following scheme:[259]

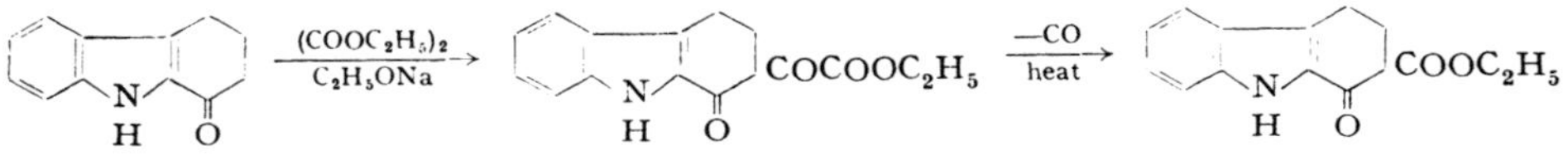

$\xrightarrow[C_2H_5ONa]{(COOC_2H_5)_2}$ $COCOOC_2H_5$ $\xrightarrow[heat]{-CO}$ $COOC_2H_5$

N-Acyl Carbazoles

The 9-acylcarbazoles have been prepared through the action of acyl halides and anhydrides on carbazole,[260,261,263] of acyl halides on potassium carbazole, and of acyl halides on carbazolemagnesium halides. 9-Acetylcarbazole, m.p. 60–70°, has been prepared through the treatment of carbazole with acetic anhydride.[260,261,262] The compound was also

[256] Manske and Kulka, *Can. J. Research,* **B 25,** 376 (1947).
[257] U. S. patent 2,144,704; *Chem. Abstr.*, **33,** 3166, (1939).
[258] U. S. patent 1,819,127; *Chem. Abstr.*, **25,** 5678 (1931). German patent 512,234; *Chem. Abstr.*, **25,** 966 (1931). British patent 303,901; *Chem. Abstr.*, **23,** 4579 (1929). French patent 667,220; *Chem. Abstr.*, **24,** 1121 (1930). German patent 554,645; *Chem. Abstr.*, **26,** 5970 (1932). German patent 566,452; *Chem. Abstr.*, **27,** 1001 (1933). British patent 451,123; *Chem. Abstr.*, **31,** 270 (1937).
[259] Elks, Elliott, and Hems, *J. Chem. Soc.*, **1944,** 624.
[260] Graebe and Glaser, *Ann.*, **163,** 343 (1872).
[261] Boeseken, *Rec. trav. chim.*, **31,** 350 (1912).
[262] Compare Berlin, *J. Gen. Chem. (U. S. S. R.)*, **14,** 438–446 (1944); *Chem. Abstr.*, **36,** 4606 (1945).
[263] Copisarow, *J. Chem. Soc.*, **113,** 816 (1918).

prepared from 2-acetylamino-2′-aminodiphenyl according to the scheme:[264]

NH H_2N / CO / CH_3 $\xrightarrow[HNO_2]{HCl}$ NH N_2 / CO Cl / CH_3 ⟶ N / CO / CH_3

Carbazole reacts with benzoyl chloride at 160–170° to give 9-benzoylcarbazole, m.p. 98.5°.[265] The same compound is obtained when carbazole is heated with benzoic anhydride.

Treatment of carbazolemagnesium halides with benzoyl chloride and with acetyl chloride results in formation of the corresponding 9-acylcarbazoles.[262–272] With ethyl chlorocarbonate carbazolemagnesium iodide gives the ethyl ester of carbazole-9-carboxylic acid, m.p. 77.5°.

Nearly quantitative yields of 9-acyl and 9-alkyl derivatives of carbazole may be obtained when a solution of carbazole in acetone or alcohol is treated with acylating or alkylating agents in the presence of alkali.[273, 274, 275] It was at first reported[276] that carbazole and phosgene did not react but subsequently the acid chloride of carbazole-9-carboxylic acid, m.p. 104°, was prepared through interaction of these two substances in pyridine.[277] 9-(*p*-Toluenesulfonyl)carbazole has been prepared through the interaction of potassium carbazole and *p*-toluenesulfonyl chloride.[278]

N-Alkylcarbazoles

The 9-alkyl derivatives of carbazole are usually prepared by treating

[264] Shin-ichi Sako, *Chem. Abstr.*, **26**, 3246 (1932); *Mem. Coll. Eng. Kyushu Imp. Univ.*, **6**, 263–306 (1932).

[265] Mazzara, *Ber.*, **24**, 278 (1891). Raison, *J. Chem. Soc.*, **1949**, 3319.

[266] Oddo, *Gazz. chim. ital.*, **41**, I, 255 (1911); *Mem. accad. Lincei*, [V] **14**, 510 (1923); *Chem. Abstr.*, **19**, 2492 (1925).

[267] Sanna and Chessa, *Gazz. chim. ital.*, **58**, 121 (1928).

[268] Robinson and Tomlinson, *J. Chem. Soc.*, **1934**, 1524.

[269] Perkin and Plant, *ibid.*, **12**, 3676 (1923).

[270] Plant and Rosser, *ibid.*, **1928**, 2454.

[271] Plant and Rutherford, *ibid.*, **1929**, 1970.

[272] Manjunath and Plant, *ibid.*, **1926**, 2260.

[273] Stevens and Tucker, *ibid.*, **123**, 2140 (1923).

[274] Dunlop and Tucker, *ibid.*, **1939**, 1945.

[275] v. Braun, *Ber.*, **43**, 2879 (1910).

[276] Paschkowezky, *ibid.*, **24**, 2905 (1891).

[277] Ruigh, U. S. patent 2,089,985; *Chem. Abstr.*, **31**, 7069 (1937).

[278] German patent 224,951 (1910). U. S. patent 996,092 (1910).

potassium carbazole with alkyl halides or alkyl sulfates.[279-282] 9-Methylcarbazole, m.p. 87°, and 9-ethylcarbazole, m.p. 67–68°, have been prepared in this manner. Potassium carbazole and ethyl acetate react readily to give ethyl carbazole-9-acetate.[283] 9-(2-Diethylaminoethyl)carbazole has been prepared through the action of diethylaminoethyl chloride and sodium amalgam on carbazole[284] as well as through the action of diethyl amine on 9-(2-chloroethyl)carbazole.[285] 9-(2-Chloroethyl)carbazole has been prepared through the interaction of carbazole and β-chloroethyl *p*-toluenesulfonate.[286] The compound is readily converted into 9-vinylcarbazole.

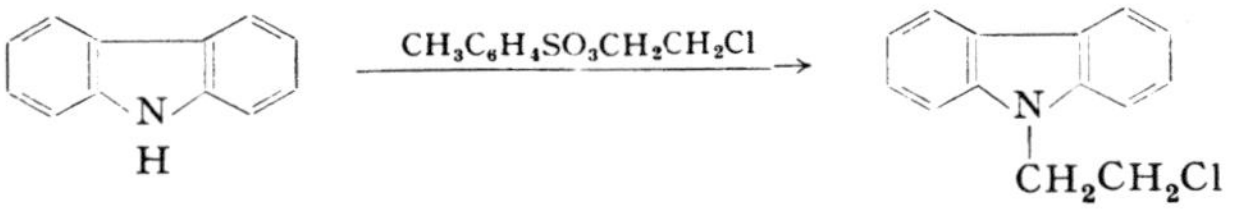

$$\text{Carbazole (N–H)} \xrightarrow{CH_3C_6H_4SO_3CH_2CH_2Cl} \text{9-}(CH_2CH_2Cl)\text{carbazole}$$

m.p. 130–131°

$$\text{9-}(CH{=}CH_2)\text{carbazole}$$

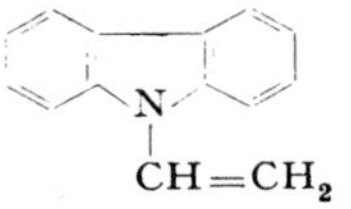

Carbazole reacts readily with acrylo-nitrile to yield 9-cyanoethylcarbazole, m.p. 155.5°.[287] Hexahydrocarbazole similarly yields 9-cyanoethyl-1,2,3,4,10,11-hexahydrocarbazole.[288] 9-Phenylcarbazole, m.p. 91.5°, has been prepared from carbazole through treatment with iodobenzene, K_2CO_3, and copper bronze.[289, 290] The same compound has also been

[279] Graebe and Adlerskon, *Ann.*, **202**, 23 (1880).

[280] Ehrenreich, *Monatsh.*, **32**, 1103 (1911).

[281] Cade, *Chem. Met. Eng.*, **29**, 319 (1923). U. S. patent 1,494,879; *Chem. Abstr.*, **18**, 2173 (1924). U. S. patent 1,662,061; *Chem. Abstr.*, **22**, 1366 (1928). Tikhonov and Inatyuk-Maistrenko, *Org. Chem. Ind.* (*U. S. S. R.*), **8**, 206 (1938); *Chem. Abstr.* **32**, 7915 (1938).

[282] Levy, *Monatsh.*, **33**, 177 (1912).

[283] German patent 255,304; *Chem. Abstr.*, **8**, 1212 (1914). British patent 101,433; *Chem. Abstr.*, **11**, 212 (1917). Seka, *Ber.*, **B 57**, 1527 (1924).

[284] Eisleb, *Ber.*, **B 74**, 1433 (1941). Burckhalter, Stephens, and Hall, *J. Am. Pharm. Assoc.*, **39**, 271 (1950).

[285] Fourneau and de Lestrange, *Bull. soc. chim. France*, **1947**, 827.

[286] Clemo and Perkin, *J. Chem. Soc.*, **125**, 1804 (1924).

[287] Whitmore, Mosher, *et al.*, *J. Am. Chem. Soc.*, **66**, 725 (1944). German patent 641,597; *Chem. Abstr.*, **31**, 5812 (1937).

[288] French patent 47,827; *Chem. Abstr.*, **32**, 4608 (1938). Smith, *J. Am. Chem. Soc.*, **72**, 4313 (1950).

[289] Dunlop and Tucker, *J. Chem. Soc.*, **1939**, 1945.

[290] German patent 224,951. See Hager, *Organic Syntheses*, Collective Vol. I, Wiley, New York, 1932, p. 532.

prepared by the action of *o*-halobenzoic acids on carbazole in nitrobenzene in presence of alcoholic alkali and subsequent splitting out of carbon dioxide by heating.[291] Derivatives of 9-phenylcarbazole have been prepared in similar fashion,[292,293] while 9-anthraquinonylcarbazole, m.p. 252–254°, has been prepared[294] through the use of 1-iodoanthraquinone. 9-[2-(1-Pyrrolidyl)ethyl]carbazole was obtained through the action of 2-(1-pyrrolidyl)ethyl chloride on the sodium salt of carbazole.[295]

Potassium carbazole condenses with nitrobenzene[296] to give 9-*p*-nitrophenylcarbazole. The structure of this product was shown by its reduction and conversion to the *p*-chlorophenyl derivative through the Sandmeyer reaction and by its synthesis from carbazole and *p*-chloronitrobenzene.[297]

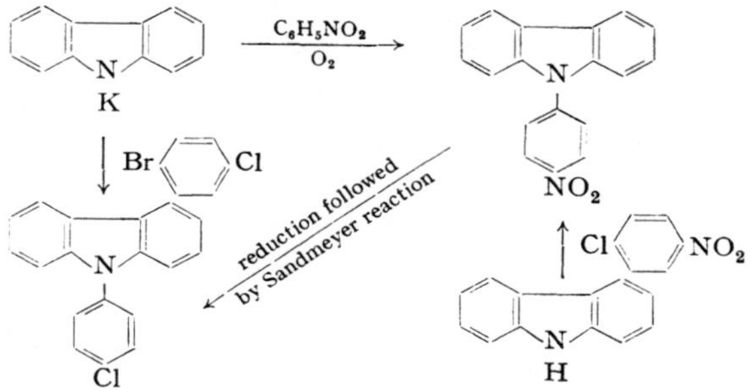

It has been found that 9-*o*-carboxyphenyl-3-nitrocarbazole can be resolved,[298] the lack of symmetry being due to restricted rotation.

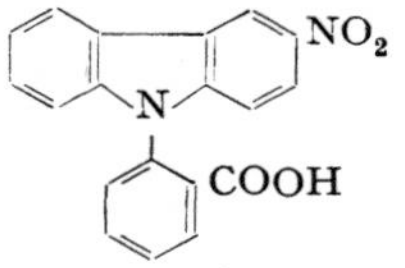

Carbazole and formaldehyde condense in alcoholic solution in the presence of potassium carbonate to give 9-hydroxymethylcarbazole, m.p. 127–128°.[299] The compound is hydrolyzed by boiling with water giving again carbazole and formaldehyde. Treatment of the compound with mineral acids

[291] Eckert, Seidel, and Endler, *J. prakt. Chem.*, **104**, 85 (1922).
[292] Dunlop and Tucker, *J. Chem. Soc.*, **1939**, 1945.
[293] Preston and Tucker, *ibid.*, **1943**, 659.
[294] Lambe, *Ber.*, **40**, 3563 (1907).
[295] Wright, *J. Am. Chem. Soc.*, **71**, 1028 (1949).
[296] de Montmollin and de Montmollin, *Helv. Chim. Acta*, **6**, 94 (1923).
[297] Nelmes and Tucker, *J. Chem. Soc.*, **1933**, 1523.
[298] Patterson and Adams, *J. Am. Chem. Soc.*, **55**, 1069 (1933).
[299] German patent 256,757; *Chem. Zentr.*, **1913**, I, 974; *Friedl.*, **11**, 176 (1915).

converts it into 3,3′-methylenedicarbazole. The latter compound was first prepared through the action of formaldehyde on carbazole at 100°[300] and later through the interaction of dimethylenegluconic acid and carbazole.[301] The compound has been formulated as the 9,9-derivative[302] but strong evidence has been given supporting the 3,3-structure.[303] A compound characterized as 9,9′-methylenedicarbazole has been prepared[304] through the interaction of 9-acetoxymethylcarbazole and carbazole-9-magnesium iodide. In further support of the 3,3-structure for the product obtained by the condensation of carbazole with formaldehyde at 100° is the fact that 9-ethylcarbazole undergoes a similar condensation with formaldehyde.[305]

9-Acetoxymethylcarbazole, m.p. 81–82°, can be prepared through the action of acetic anhydride on a pyridine solution of 9-hydroxymethylcarbazole.[306] The 9-benzoxymethyl derivative, m.p. 99.8–100.8°, was prepared in similar manner through the agency of benzoic anhydride.

Treatment of 9-acetoxymethylcarbazole with alkylmagnesium halides gives the 9-alkylcarbazoles.[304]

$$\text{9-}(CH_2\text{-}O\text{-}COCH_3)\text{carbazole} \xrightarrow{C_2H_5MgBr} \text{9-}(CH_2\text{-}CH_2\text{-}CH_3)\text{carbazole}$$

Carbazole condenses with ethylene oxide in the presence of potassium hydroxide to give (9-β-hydroxymethyl)carbazole[307] which can also be prepared through interaction of carbazole and β-chloroethanol in presence of alkali.[308] Dehydration of the β-hydroxyethylcarbazole gives 9-vinylcarbazole, m.p. 66°. The latter compound was also prepared through the treat-

[300] Pulvermacher and Loeb, *Ber.*, **25**, 2766 (1892).
[301] Votocek and Vesely, *ibid.*, **40**, 410 (1907).
[302] Feldman and Wagner, *J. Org. Chem.*, **7**, 31 (1942).
[303] Dutt, *J. Chem. Soc.*, **125**, 802 (1924).
[304] Milzuch, *J. Gen. Chem.* (*U. S. S. R.*), **16**, 1471–1474 (1926); *Chem. Abstr.*, **41**, 5508 (1947); **43**, 3819 (1949).
[305] German patent 293,587 (1913).
[306] Milzuch and Gel'fer, *J. Appl. Chem.* (*U. S. S. R.*), **19**, 939–944 (1946); *Chem. Abstr.*, **42**, 565 (1948).
[307] Otsuki, Okana, and Takeda, *J. Soc. Chem. Ind. Japan*, **49**, 169–170 (1946); *Chem. Abstr.*, **42**, 6354 (1948). Matsui, *J. Soc. Chem. Ind. Japan*, **45**, 1192 (1942); *Chem. Abstr.*, **42**, 6159 (1948).
[308] Flowers, Miller, and Flowers, *J. Am. Chem. Soc.*, **70**, 3019 (1948).

ment of 9-β-chloroethylcarbazole with a methanol solution of potassium hydroxide.[309]

$$\text{Carbazole (N–H)} \xrightarrow[\text{KOH}]{H_2C\text{—}CH_2\ (\text{O})} \text{9-}CH_2CH_2OH\text{-carbazole} \xrightarrow{H_2O} \text{9-}CH{=}CH_2\text{-carbazole}$$

9-Vinylcarbazole can be prepared also by pyrolysis of acyl derivatives of 9-β-hydroxyethylcarbazole[310] and through the interaction of acetylene and carbazole in the presence of an alkaline catalyst.[311] 9-Vinylcarbazole readily yields a thermoplastic polymer which is marketed as Luvican.[312]

[309] Clemo and Perkin. *J. Chem. Soc.*, **125,** 1804 (1924). Ishii and Hayashi, *J. Soc. Org. Synthet. Chem. Japan,* **7,** 41 (1949); *Chem. Abstr.*, **44,** 3970 (1950). British patent 620,733; *Chem. Abstr.*, **43,** 6669, (1949).

[310] U. S. patent 2,426,465; *Chem. Abstr.*, **42,** 224 (1948). British patent 620,734; *Chem. Abstr.*, **43,** 6669 (1949).

[311] German patent 618,120; *Chem. Abstr.*, **30,** 110 (1936). German patent 642,939; *Chem. Abstr.*, **31,** 5816 (1937). U. S. patent 2,123,733; *Chem. Abstr.*, **32,** 7055 (1938).

[312] G. M. Kline, *Modern Plastics,* **24,** [3] 157–158, 194 (1946); *Chem. Abstr.*, **41,** 883 (1947). German patent 664,231; *Chem. Abstr.*, **33,** 784 (1939). French patent 792,820; *Chem. Abstr.*, **30,** 4178 (1936).

CHAPTER III

Isatin

Early studies on the action of oxidizing agents on indigo led to the discovery[1] of an oxidation product, $C_8H_5NO_2$, to which the name isatin was given. This compound could be obtained by the action of nitric and chromic acids on indigo, and mono- and dihalogenated isatins were formed by the use of halogens for the oxidation.

Evidence for a benzene ring in isatin was obtained in the formation of chloranil by treating chloroisatin with chlorine,[1] in the formation[2] of aniline and chloroaniline by heating isatin and chloroisatin with strong alkali, and in the formation of 5-nitrosalicylic acid by the action of nitrous acid on isatin.[3]

Reduction of isatin through the then unknown dioxindole and oxindole gave indole,[4] and isatin was found to dissolve in alkali to give the salt of an acid, isatic acid (isatinic acid). This brought the quick realization that isatic acid was *o*-aminobenzoylformic acid (I) and that isatin was its lactam (2,3-diketo-2,3-dihydroindole) (II).[5] This suggestion was quickly accepted,

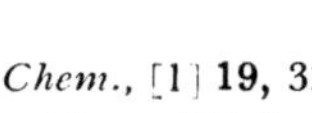

$COCO_2H$
NH_2
(I)

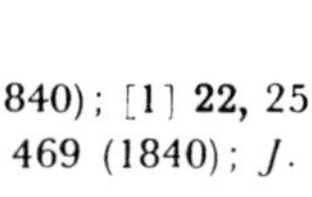

4, 5, 6, 7, 3 CO, 2 CO, 1 N
(II) H

and the relationship of the reduction products to isatin correctly formulated.[6]

This structure for isatin was given definite confirmation by its synthesis

[1] Erdman, *J. prakt. Chem.*, [1] **19,** 321 (1840); [1] **22,** 257 (1841); [1] **24,** 1 (1841). Laurent, *Ann. chim. phys.*, [3] **3,** 372, 469 (1840); *J. prakt. Chem.*, [1] **25,** 434 (1842).

[2] Hofmann, *Ann.*, **53,** 11 (1845).

[3] Hofmann, *ibid.*, **115,** 279 (1860).

[4] Baeyer, *Ber.*, **1,** 17 (1868). Baeyer and Knop, *Ann.*, **140,** 1 (1866).

[5] Kekulé, *Ber.*, **2,** 748 (1869).

[6] Baeyer, *ibid.*, **11,** 582, 1228 (1878).

from *o*-nitrobenzoyl chloride (III)[7] and from *o*-nitrophenylpropiolic acid (IV) by treatment with alkali.[8]

$$C_6H_4(COCl)(NO_2)\ (III) \xrightarrow{KCN} C_6H_4(COCN)(NO_2) \longrightarrow C_6H_4(COCO_2H)(NH_2) \longrightarrow C_6H_4\langle_{NH}^{CO}\rangle CO$$

$$C_6H_4(NO_2)C{\equiv}C{-}CO_2H\ (IV) \xrightarrow[-CO_2]{H_2O} C_6H_4\langle_{NH}^{CO}\rangle CO$$

Preparation of Isatin

Of the many methods which have been developed for the preparation of isatin and its derivatives, the two procedures developed by Sandmeyer are the most general. One of these starts with thiocarbanilide (I), which, with lead carbonate and hydrogen cyanide, is converted into the nitrile anilide. Treatment of the latter material with ammonium sulfide yields the thioamide. Cyclization with sulfuric acid, followed by hydrolysis, produces isatin.[9]

The second of the Sandmeyer procedures[10] involves the reaction of an aniline, chloral hydrate and hydroxylamine to form an isonitrosoacetanilide (II), which is converted to isatin on treatment with concentrated sulfuric

$$C_6H_5NH{-}CSNHC_6H_5\ (I) \xrightarrow[HCN]{PbCO_3} C_6H_5NH{-}C(CN){=}NC_6H_5 \xrightarrow{(NH_4)_2S} C_6H_5NH{-}C(CSNH_2){=}NHC_6H_5 \xrightarrow{H_2SO_4}$$

$$C_6H_4\langle_{NH}^{CO}\rangle C{=}NC_6H_5 \xrightarrow{H_2O} C_6H_4\langle_{NH}^{CO}\rangle CO$$

$$C_6H_5NH_2 \xrightarrow[H_2NOH]{Cl_3CCHO} C_6H_5NH{-}CO{-}CH{=}NOH\ (II) \xrightarrow{H_2SO_4} C_6H_4\langle_{NH}^{CO}\rangle CO$$

acid. The reaction has been employed for the preparation of numerous

[7] Claisen and Shadwell, *ibid.*, **12**, 350 (1879).

[8] Baeyer, *ibid.*, **13**, 2259 (1880). Forrer, *ibid.*, **17**, 976 (1884).

[9] Sandmeyer, *Z. Farben Texilchem.*, **2**, 129 (1903). Bonnefoy and Martinet, *Compt. rend.*, **172**, 220 (1921). Ferber and Schmolke, *J. prakt. Chem.*, [2] **155**, 234 (1940).

[10] Sandmeyer, *Helv. Chim. Acta*, **2**, 234 (1919).

derivatives,[11] and although it has been reported that isonitrosoacetanilides derived from nitroanilines fail to form isatins[12] 7-nitroisatin has been prepared by this reaction.[13]

A similar method of preparation is that involving the reaction of aryl amines with dichloroacetic acid.[14] The intermediate anilinoöxindole (III) is

2 $C_6H_5NH_2$ $\xrightarrow{Cl_2CHCO_2H}$ (III) $\longrightarrow$ (IV) $\longrightarrow$ isatin

(III): indoline ring with CHNHC$_6$H$_5$, CO, NH; (IV): C=NC$_6$H$_5$, CO, NH; isatin: CO, CO, NH

is converted to the isatin by oxidation to the anilide (IV), followed by hydrolysis. An equally valuable method for obtaining many isatins depends on the condensation of anilines with oxomalonic ester.[15] Hydrolysis and decarboxylation of the ester yield dioxindole (V) in the absence of air, or isatin in the presence of atmospheric oxygen.

$C_6H_5NH_2$ + $COCO_2C_2H_5$ / $CO_2C_2H_5$ $\longrightarrow$ C(OH)–$CO_2C_2H_5$, CO, NH (oxindole) $\xrightarrow{H_2O}$ CHOH, CO, NH (V); $\xrightarrow{H_2O,\ O_2}$ CO, CO, NH

[11] v. Braun *et al.*, *Ann.*, **507**, 14 (1933); **451**, 1 (1926). Gullard, Robinson, Scott, and Thornley, *J. Chem. Soc.*, **1929**, 2924. Inagaki, *J. Pharm. Soc. Japan*, **58**, 961 (1938). Martinet and Coisset, *Compt. rend.*, **172**, 1234 (1921). Marvel and Hiers, *Organic Syntheses*, Collective Vol. I, 2nd ed., Wiley, New York, 1941, p. 327. Mayer and Schulze, *Ber.*, **58**, 1465 (1925). Morsch, *Monatsh.*, **55**, 144 (1930). Ressy and Ortodoscu, *Bull. soc. chim.*, **33**, 637, 1297 (1923). Shibata, Okuyama, and Okamura, *J. Soc. Chem. Ind. Japan*, **36**, Suppl. binding 569 (1933); *Chem. Abstr.*, **28**, 474 (1934). Wibaut and Geerling, *Rec. trav. chim.*, **50**, 41 (1931). Bachman and Picha, *J. Am. Chem. Soc.*, **68**, 1599 (1946). Baker, Schaub, Joseph, McEvoy, and Williams, *J. Org. Chem.*, **17**, 149, 157 (1952). Akahoshi, *J. Pharm. Soc. Japan*, **71**, 710 (1951); *Chem. Abstr.*, **46**, 2097 (1952).

[12] Borsche, Weussmann, and Fritzsche, *Ber.*, **57**, 1149 (1924). Rupe and Kersten, *Helv. Chim. Acta*, **9**, 578 (1926).

[13] Buchman, McClosky, and Seneker, *J. Am. Chem. Soc.*, **69**, 380 (1947).

[14] Mayer, *Ber.*, **16**, 2262 (1883). Duisburg, *ibid.*, **18**, 190 (1885). Heller, *Ann.*, 358, 349 (1907). Heller and Aschkenasi, *ibid.*, **375**, 261 (1910). Ostromisslenski, *Ber.*, **40**, 4972 (1907); **41**, 3032 (1908). Paucksch, *ibid.*, **17**, 2800 (1884).

[15] Martinet, *Compt. rend.*, **166**, 851, 998 (1918); *Ann. chim.*, **11**, 85 (1919).Martinet and Vacher, *Bull. soc. chim.*, **31**, 435 (1922). Bonnefoy and Martinet, *Compt. rend.*, **172**, 220 (1921). Guyot and Martinet, *ibid.*, **156**, 1625 (1913). Halberkann, *Ber.*, **54**, 3079 (1921). Hinsberg, *ibid.*, **21**, 117 (1888). Kalb, *ibid.*, **44**, 1455 (1911). Kalb and Berrer, *ibid.*, **57**, 2105 (1924). Langenbeck, Hellrung, and Juttemann, *Ann.*, **499**, 201 (1932); **512**, 276 (1934).

N-Substituted isatins are obtained conveniently from *N*-substituted anilines and oxalyl chloride, cyclizing the anilide with aluminium chloride.[16]

$$C_6H_5NHR + (COCl)_2 \longrightarrow C_6H_5N(R)COCOCl \xrightarrow{AlCl_3} \text{N-R-isatin}$$

Isatin is the product resulting when thioöxanilide (VI) is treated with concentrated sulfuric acid,[17] and is also obtained from *o*-hydroxyamino-mandelonitrile (VII) by heating with hydrochloric acid.[18]

$$C_6H_5NH{-}CO{-}CS{-}NHC_6H_5 \ \text{(VI)} \xrightarrow{H_2SO_4} \text{isatin}$$

$$o\text{-}C_6H_4(CHOHCN)(NHOH) \ \text{(VII)} \xrightarrow{HCl} \text{isatin}$$

Nuclear substituted isatins have been prepared by hydrolysis of *o*-aminobenzoylformic esters (VIII)[19] and by treating the azlactone of *o*-nitro-benzaldehydes (IX) with alkali.[20]

$$(CH_3O)_2C_6H_2(COCO_2C_2H_5)(NH_2) \ \text{(VIII)} \longrightarrow \text{dimethoxyisatin}$$

$$HO(OCH_3)C_6H_2(NO_2){-}CH{=}C{-}CO{-}O{-}C(C_6H_5){=}N \ \text{(IX)} \longrightarrow \text{hydroxy-methoxyisatin}$$

Both isatin and nuclear substituted isatins have been synthesized by treating substituted imide chlorides of oxalic acid with sulfuric acid.[21]

Isatin is obtained commercially by the oxidation of indigo, and a number of substituted isatins have been prepared from the appropriately

[16] Stolle, *J. prakt. Chem.*, **105,** 137 (1922); *Ber.*, **46,** 3915 (1913). D. R. P. 281,046, 341,112.

[17] Reissert, *Ber.*, **37,** 3710 (1904).

[18] Heller, *ibid.*, **40,** 1291 (1907).

[19] Fetsher and Bogert, *J. Org. Chem.*, **4,** 71 (1939).

[20] Burton and Stoves, *J. Chem. Soc.*, **1937,** 402.

[21] Bauer, *Ber.*, **40,** 2650 (1907); D. R. P. 193,633; *Ber.*, **41,** 450 (1908); **42,** 2111 (1909). Ostromisslenski, *ibid.*, **40,** 4972 (1907); **41,** 3032 (1908).

$$\text{C}_6\text{H}_5\text{N}{=}\text{CCl}{-}\text{C(Cl)}{=}\text{NC}_6\text{H}_4 \xrightarrow{H_2SO_4} \text{(indole ring) } \text{C}{=}\text{NC}_6\text{H}_5,\ \text{N(H)}{-}\text{CO} \xrightarrow{H_2O} \text{(isatin: CO, N(H)-CO)}$$

substituted indigo derivatives by oxidation with a large variety of reagents.[22] It is reported that isatin is excreted by rabbits following the ingestion of *o*-nitrophenylglyoxylic acid.[23]

N-Hydroxyisatin and derivatives have been prepared,[24] but attempts to obtain *N*-aminoisatin have not been successful.[25]

Properties of Isatin

Isatin forms red needles melting at 200–201 °C., when crystallized from water, alcohol, or acetic acid. It dissolves in concentrated hydrochloric or sulfuric acid, and in sodium or potassium hydroxide, forming the sodium or potassium salt of isatin. These salts may also be prepared by treating isatin in absolute alcohol solution with the appropriate ethoxide. The silver salt of isatin is produced by the action of silver nitrate on the sodium salt, or by treating isatin in alcohol with silver acetate. With perchloric acid isatin forms a stable perchlorate.[26] Mercurous[27] and mercuric[28] salts have also been prepared.

The question of the structures of the alkali and silver salts has enter-

[22] D. R. P. 229,815. Forrer, *Ber.*, **17,** 976 (1884). Erdmann, *J. prakt. Chem.*, [1], **24,** 1 (1841). Gericke, *ibid.*, [1] **23,** 278 (1890). Hofmann, *Ann.*, **53,** 11 (1845). Kalb and Berrer, *Ber.*, **57,** 2105 (1924). Kalb and Vogel, *ibid.*, **57,** 2117 (1924). Knape, *J. prakt. Chem.*, [2] **43,** 211 (1891). Knoevenagel, *ibid.*, [2] **89,** 46 (1914). Kranzlein, *Ber.*, **70,** 1776 (1937). Laurent, *J. prakt. Chem.*, [1] **25,** 434 (1842). Pummerer, Fiesselmann, and Müller, *Ann.*, **544,** 206 (1940). Sumpter and Amundsen, *J. Am. Chem. Soc.*, **54,** 1917 (1932). Saha, *J. Indian Chem. Soc., Ind. News Ed.*, **13,** 178 (1950); *Chem. Abstr.*, **45,** 8005 (1951).

[23] Böhm, *Z. physiol. Chem.*, **265,** 210 (1940).

[24] Reissert, *Ber.*, **29,** 657 (1896); **41,** 3921 (1908). Reissert and Hessert, *ibid.*, **57,** 964 (1924). Alessandri, *Gazz. chim. ital.*, **57,** 195 (1927). Heller, *Ber.*, **39,** 2341 (1906); **42,** 470 (1910). Heller and Sourlis, *ibid.*, **41,** 373 (1908). Arndt, Eistert, and Partale, *ibid.*, **61,** 1113 (1928); **60,** 1364 (1927). Giovannini, and Portmann, *Helv. Chim. Acta*, **31,** 1381 (1948).

[25] Neber and Keppler, *Ber.*, **57,** 778 (1924). Stolle and Becker, *ibid.*, **57,** 1123 (1924).

[26] Hofmann, Metzler, and Hobold, *Ber.*, **43,** 1080 (1910).

[27] Hantzsch, *ibid.*, **58,** 685 (1925).

[28] Peters, *ibid.*, **40,** 236 (1907).

tained considerable discussion.[29] The most plausible representation for the anion is as a hybrid of the two charged forms (I and II) with a smaller contribution from the quinoid structure (III).

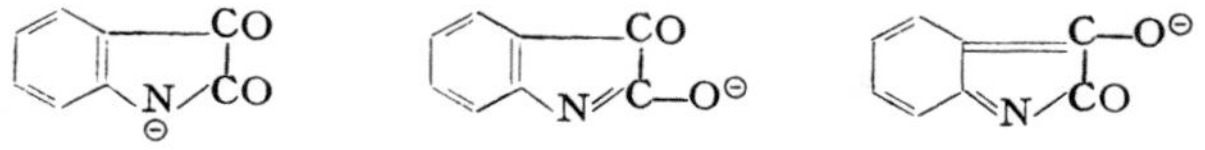

(I) (II) (III)

Tautomerism

The possibility for tautomerism in isatin was early represented in the form of structures IV and V.[30] The situation is not unlike that of acetoacetic

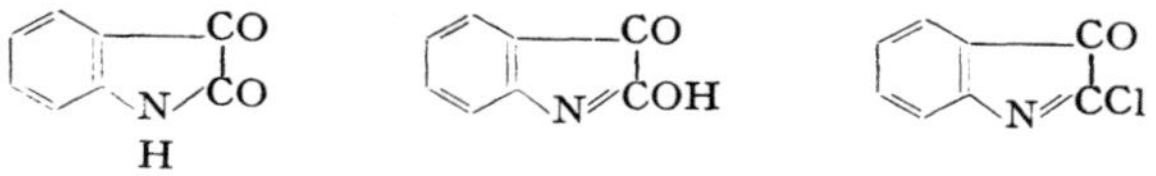

(IV) (V) (VI)

ester, in that *N*-alkyl derivatives are formed with sodium salts, while the use of silver salts produces *O*-alkyl derivatives. Further evidence for the lactim structure (V) is observed in the reaction of isatin with phosphorus pentachloride in benzene to give isatin-α-chloride (VI).[31] Absorption spectra aid little in this problem, since isatin, *N*-alkyl-, and *O*-alkylisatins have very similar absorption curves.[32] The conflicting early reports[33] of absorption spectra are explained by a recent study of the polarographic behavior of isatin and its *N*-alkyl derivatives.[34] This work has shown that solutions of

[29] Claascz, *ibid.*, **50,** 511 (1917). Deussen, Heller, and Nötzel, *ibid.*, **40,** 1300 (1907). Hantzsch, *ibid.*, **54,** 1221, 1257 (1921); **55,** 3180 (1922). Heller, *ibid.*, **40,** 1291 (1907); **49,** 2757 (1916); **50,** 1199 (1917); **54,** 2214 (1921); **55,** 2681 (1922). Schlenk and Thal, *ibid.*, **46,** 2841 (1913).

[30] Baeyer, *ibid.*, **16,** 2188 (1883). Baeyer and Oekonomides, *ibid.*, **15,** 2093 (1882). Compare also Sidgwick, Taylor and Baker, *Organic Chemistry of Nitrogen*, Oxford, London, 1937, p. 505.

[31] Baeyer, *Ber.*, **12,** 456 (1879).

[32] Arbuzov, *Bull. acad. sci. U. R. S. S. Classe sci., chim.*, **1940,** 89; *Chem. Abstr.*, **35,** 2508 (1941). Ault, Hirst, and Morton, *J. Chem. Soc.*, **1935,** 1653. Cox, Goodwin, and Wagstaff, *Proc. Roy. Soc. London,* **A 157,** 399 (1936). Magini and Passerini, *Boll. sci. facolta chim. ind. Bologna*, **9,** 51 (1941); *Chem. Abstr.*, **46,** 350 (1952).

[33] Dabrowski and Marchlewski, *Bull. soc. chim.*, **53,** 946 (1933). Harley and Dobbie, *Trans. Chem. Soc.*, **77,** 640 (1899). Korczynski and Marchlewski, *Ber.*, **35,** 4337 (1902). Menczel, *Z. physik. Chem.*, **125,** 161 (1927). Morton and Rogers, *J. Chem. Soc.*, **127,** 2698 (1925).

[34] Sumpter, Williams, Wilken, and Willoughby, *J. Org. Chem.*, **14,** 713 (1949). Sumpter, Wilken, Williams, Wedemeyer, Boyer, and Hunt, *ibid.*, **16,** 1777 (1951).

isatin contain several molecular or ionic varieties, depending on the *p*H and the freshness of the solutions. These forms (Fig. 1) have been isolated by virtue of their reduction potentials, and their absorption spectra determined.

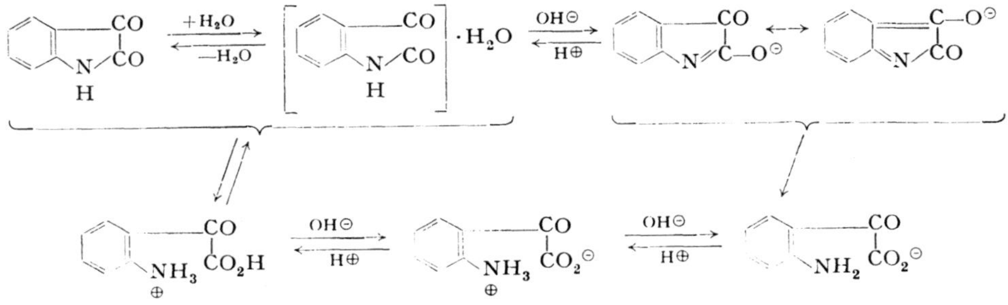

Figure 1

The occurrence of such open ring forms is well known in the case of isatin. When alkaline solutions are heated the ring is opened to yield salts of isatic acid, with loss of color. Acidification of the solutions causes ring closure to occur, with the precipitation of the reformed isatin.

Reactions of Isatin

Chemically, isatin may be characterized as the lactam of *o*-aminobenzoylformic acid. Thus it possesses both amide and keto carbonyl groups, an active hydrogen atom attached to nitrogen (or oxygen), and an aromatic ring which should substitute in the 5 and 7 positions. These functions all appear in the reactions of the molecule.

Oxidation and Reduction

Chromic acid oxidation of isatin in acetic acid solution yields isatoic anhydride (I),[35] prepared also from anthranilic acid (II) and phosgene.[36] The anhydride on hydrolysis and decarboxylation yields anthranilic acid (II). Similar products are formed with substituted isatins,[37] but *N*-acetylisatin is

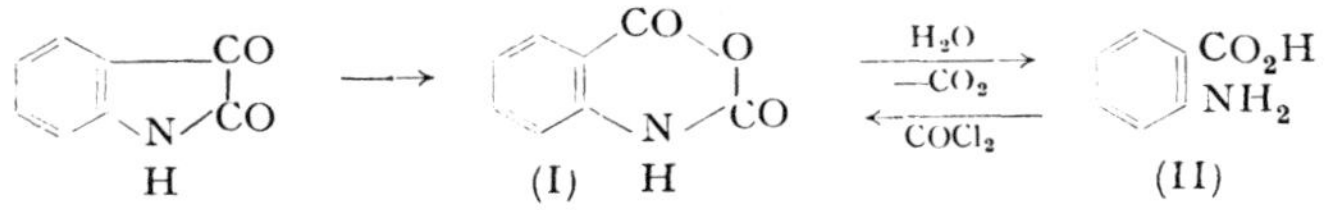

[35] Kolbe, *J. prakt. Chem.*, [2] **30,** 84 (1884).

[36] Erdmann, *Ber.*, **32,** 2162 (1899).

[37] Binz and Heuter, *ibid.*, **48,** 1038 (1915). Majima and Kotake, *ibid.*, **63,** 2237 (1930). Panaotovic, *J. prakt. Chem.*, [2] **33,** 58 (1886). Rupe and Kersten, *Helv. Chim. Acta*, **9,** 578 (1926).

converted directly to *o*-acetaminobenzoic acid.[35] Alkaline hydrogen peroxide also converts isatin into anthranilic acid.[38]

Oxidation with potassium permanganate gives a compound designated as anhydroisatin-α-anthranilide (III),[39] also obtained[40] from indigo in pyridine solution by atmospheric oxidation in the presence of copper salts. The compound has been prepared in other, independent, manners.[41]

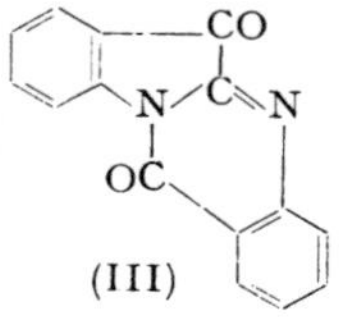

(III)

Reduction of isatin in acid solution gives a bimolecular product, isatide (IV),[42] also prepared by the condensation of isatin with dioxindole in the presence of pyridine.[43] Hydrogen sulfide reduction of isatin gives the sulfur analog, isatin thiopinacol (V).[44]

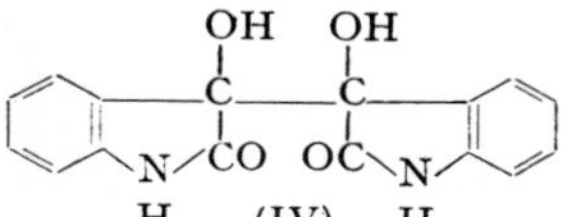

(IV) 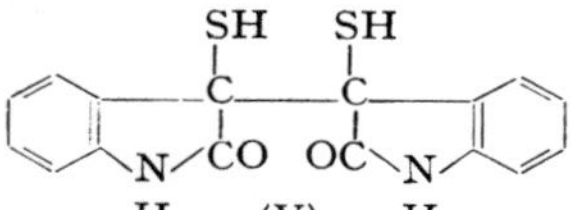 (V)

Further reduction, by the use of sodium amalgam, gives dioxindole

[38] Kalb and Berrer, *Ber.*, **57,** 2105 (1924). Sumpter, *J. Am. Chem. Soc.*, **64,** 1736 (1942). Sumpter und Amundsen, *ibid.*, **54,** 1917 (1932). Sumpter and Jones, *ibid.*, **65,** 1802 (1943).

[39] Friedländer and Roschdestwensky, *Ber.*, **48,** 1841 (1915).

[40] Machmer, *ibid.*, **63,** 1341 (1930).

[41] D. R. P. 287,373, 288,055. Heller and Benade, *Ber.*, **55,** 1006 (1922). Heller and Lauth, *J. prakt. Chem.*, [2] **112,** 331 (1926). Heller and Siller, *ibid.*, [2] **123,** 257 (1929).

[42] Baeyer, *Ber.*, **1,** 17 (1868); **12,** 1309 (1879). Baeyer and Knop, *Ann.*, **140,** 1 (1866). Erdmann, *J. prakt. Chem.*, [1] **24,** 1 (1841). Heller, *Ber.*, **37,** 943 (1904). Laurent, *Ann. chim. phys.*, [3] **3,** 372 (1840); *J. pr. Chem.* [1] **25,** 434 (1842); **47,** 166 (1849); *Ann.*, **72,** 285 (1849). Wahl and Faivret, *Ann. chim.*, [10] **5,** 314 (1926); *Compt. rend.*, **180,** 589 (1925); **181,** 790 (1925); *Ann. chim.* [10] **9,** 277 (1928); *Compt. rend.*, **186,** 378 (1928). Compare also Oliveri-Mandala and Deleo, *Gazz. chim. ital.*, **79,** 337 (1949).

[43] Hansen, *Ann. chim.*, [10] **1,** 94, 126 (1924). Heller, *Ber.*, **37,** 943 (1904). Wahl and Faivret, *Ann. chim.*, [10] **5,** 314 (1926). Walter, *Ber.*, **35,** 1320 (1902).

[44] Laurent, *Ann. chim. phys.*, [3] **3,** 469 (1840). Sander, *Ber.*, **58,** 820 (1925). Wahl and Fericean, *Ann. chim.*, [10] **9,** 277 (1928); *Compt. rend.*, **186,** 378 (1928); **184,** 826 (1927). Wahl and Hansen, *Compt. rend.*, **176,** 1070 (1923); **178,** 214, 393 (1924). Wahl and Lobeck, *ibid.*, **186,** 1313 (1928).

(VI),[45] also prepared by the reduction of isatin with sodium amalgam in alkaline medium, zinc and acetic acid[46] or sodium hydrosulfite.[47] The latter method appears to be superior. Reduction of dioxindole with tin and hydrochloric acid or sodium amalgam and acid gives oxindole (VII).[44] *N*-Alkylisatins behave similarly in these reductions.[48]

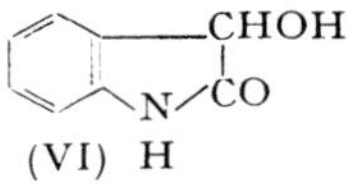

CHOH CO N H (VI) CH_2 CO N H (VII)

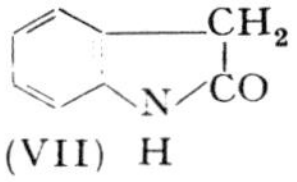

Other reductions of isatin include the formation of oxindole by the Wolff-Kishner reduction[49] and the electrolytic reduction to a rather wide variety of products.[50]

Halogenation and Nitration

Chlorination of isatin gives 5-chloroisatin,[51] and 5,7-dichloroisatin.[52] Bromination behaves similarly.[53] Direct iodination can be effected with

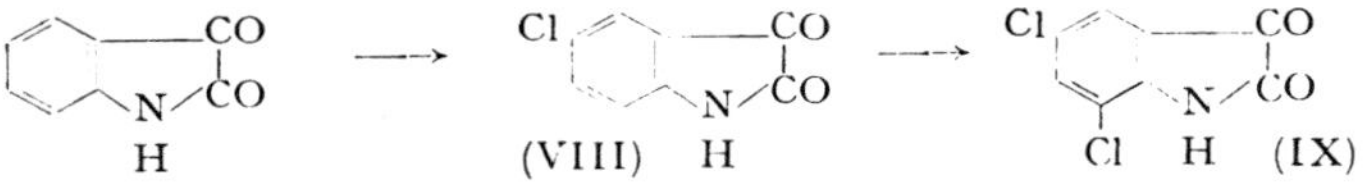

CO CO N H ⟶ Cl CO CO N H (VIII) ⟶ Cl CO CO N Cl H (IX)

[45] Baeyer, *Ber.*, **1,** 17 (1868). Baeyer and Knop, *Ann.*, **140,** 1 (1866).
[46] Heller, *Ber.*, **37,** 943 (1904).
[47] Marshalk, *ibid.*, **45,** 582 (1912); *J. prakt. Chem.*, [2] **88,** 227 (1913). Kalb, *Ber.*, **44,** 1455 (1911).
[48] Colman, *Ann.*, **248,** 121 (1888). Hill and Sumpter, Unpublished dissertation of W. C. Sumpter, Yale University, 1930. Michaelis, *Ber.*, **30,** 2814 (1897). Sumpter, Unpublished work.
[49] Curtius and Land, *J. prakt. Chem.*, [2] **44,** 551 (1890). Curtius and Thun, *ibid.*, **44,** 187 (1890). Borsche and Meyer, *Ber.*, **54,** 2841 (1921). Shapiro, *ibid.*, **62,** 2133 (1929). Siebert, *Chem. Ber.*, **80,** 494 (1947).
[50] Sakurai, *Bull. Chem. Soc. Japan*, **17,** 269 (1942).
[51] Dorsch, *J. prakt. Chem.*, [2] **33,** 49 (1886). Erdmann, *ibid.*, [1] **19,** 321 (1840); [1] **22,** 257 (1841); [1] **24,** 1 (1841). Hofmann, *Ann.*, **53,** 11 (1845). Kambli, *Helv. Chim. Acta*, **24,** 93 (1941). Laurent, *Ann. chim. phys.*, [3] **3,** 372 (1840). Liebermann and Kraus, *Ber.*, **40,** 2492 (1907). Marchlewski, *ibid.*, **29,** 1031 (1896).
[52] Dainala, *Compt. rend.*, **149,** 793, 1385 (1909). Dorsch, *J. prakt. Chem.*, [2] **33,** 49 (1886). Grandmougin, *Compt. rend.*, **174,** 620 (1922). Grandmougin and Seyder, *Ber.*, **47,** 2365 (1914). Heller, *ibid.*, **55,** 2681 (1922). Kalb, *ibid.*, **42,** 3663 (1909). Kambli, *Helv. Chim. Acta*, **24,** 93 (1941).
[53] Baeyer and Oekonomides, *Ber.*, **15,** 2093 (1882). D. R. P. 558,238. Inagaki, *J. Pharm. Soc. Japan*, **58,** 961 (1938). Lindwall, Bandes, and Weinberg, *J. Am. Chem. Soc.*, **53,** 317 (1931).

iodine monochloride to give 5-iodoisatin.[54] The 5,7-diiodoisatin apparently cannot be prepared directly,[55] but has been obtained by indirect methods.[56]

The action of phosphorus pentachloride on isatin in hot benzene solution yields isatin-α-chloride (X),[57] but at room temperature 3,3-dichlorooöxindole

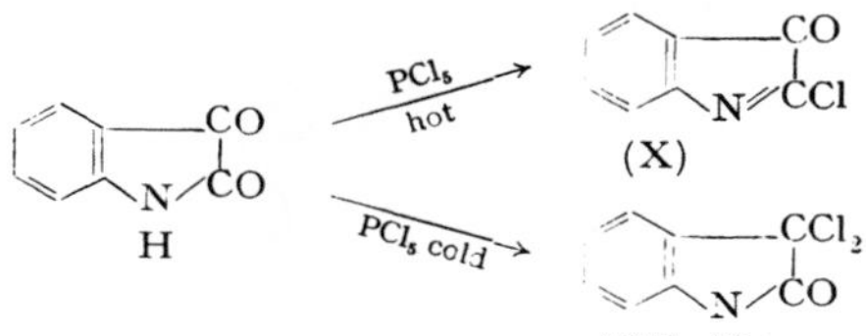

PCl_5 hot → (X)

PCl_5 cold → (XI)

is produced.[58] *N*-Alkyl derivatives give only the 3,3-dichlorooöxindole,[59] under these conditions.

N-Chlorination has been reported to occur by the chlorination of isatin or 5-chloroisatin in aqueous suspension in the presence of potassium iodide, or by treating a hydrochloric acid solution of isatin with sodium hypochlorite. The product was characterized as 1,5-dichloroisatin[60] although the possibility of its being the 2,5-derivatives is not to be overlooked.

Nitration of isatin with potassium nitrate in a concentrated sulfuric acid solution,[61] or with nitric and sulfuric acids,[62] has been shown to yield 5-nitroisatin.[63] The 5,7-dinitro compound has been prepared by both of the above methods.[64]

[54] Borsche, Weussmann, and Fritzsche, *Ber.*, **57**, 1770 (1924). Kalb and Berrer, *ibid.*, **57**, 2105 (1924). Musajo, *Gazz. chim. ital.*, **62**, 566 (1932).

[55] Sumpter and Amundsen, *J. Am. Chem. Soc.*, **54**, 1917 (1932). Sumpter, *ibid.*, **63**, 2027 (1941).

[56] Kalb and Berrer, *Ber.*, **57**, 2105 (1924). Kalb and Vogel, *ibid.*, **57**, 2117 (1924). Sumpter and Amundsen, *loc. cit.*

[57] Baeyer, *Ber.*, **12**, 456 (1879). Kambli, *Helv. Chim. Acta*, **24**, 93 (1941).

[58] Hantzsch, *Ber.*, **54**, 1221, 1257 (1921).

[59] Colman, *Ann.*, **248**, 121 (1888). Fischer and Hess, *Ber.*, **17**, 564, 2195 (1884). Kohn and Klein, *Monatsh.*, **33**, 929 (1912). Kohn and Ostersetzer, *ibid.*, **37**, 25 (1916). Michaelis, *Ber.*, **30**, 2814 (1897). Stolle, *ibid.*, **46**, 3915 (1913). D. R. P. 281,046, 341,112.

[60] D. R. P. 255,772, 255,773, 255,774.

[61] Baeyer, *Ber.*, **12**, 1309 (1879). Liebermann and Kraus, *ibid.*, **40**, 2492 (1907).

[62] Calvery, Noller, and Adams, *J. Am. Chem. Soc.*, **47**, 3059 (1925). Rupe and Kersten, *Helv. Chim. Acta*, **9**, 578 (1926). Rupe and Stocklin, *ibid.*, **7**, 557 (1924).

[63] Sumpter and Jones, *J. Am. Chem. Soc.*, **65**, 1802 (1943). Giovannini and Portmann, *Helv. Chim. Acta*, **31**, 1361 (1948). Taglianetti, *Anais. fac. farm. e odontol. Univ. São Paulo*, **7**, 57 (1949); *Chem. Abstr.*, **45**, 1997 (1951).

[64] Guha and Basu-Mallic, *J. Indian Chem. Soc.*, **11**, 395 (1934). Menon, Perkin, and Robinson, *J. Chem. Soc.*, **1930**, 830. Sumpter, Unpublished work.

Derivatives of isatin behave similarly in these nuclear substitutions. *N*-Alkylisatins furnish 5-substituted compounds.[65] If the 5-position is substituted by an alkyl group, reaction occurs at the 7-position.[66]

Treatment of isatin with fuming sulfuric acid readily furnishes isatin-5-sulfonic acid.[67]

Acylation and Alkylation

The isatin molecule is easily acylated or alkylated at the N atom. *N*-Acetylisatin is formed either by heating isatin with acetic anhydride[68] or by treating the sodium salt of isatin with acetyl chloride.[69] *N*-Benzoylisatin has also been prepared by both of the above techniques.[70] The benzenesulfonyl derivatives[71] as well as other acyl derivatives are known.[72]

The sodium and potassium salts of isatin react with alkyl halides and sulfates to give *N*-alkyl derivatives,[73] or these compounds may be prepared by the direct action of alkyl sulfates on isatin.[74] *N*-Methylisatin (XII) was first obtained from N-methylindole[75] by conversion to 3,3-dibromo-1-methyloxindole (XIII) by alkali hypobromite, followed by hydrolysis to the isatin, showing definitely the position of the alkyl group.

CH, CH, N–CH_3 $\xrightarrow{NaOBr}$ CBr_2, CO, N–CH_3 (XIII) $\xrightarrow{H_2O}$ CO, CO, N–CH_3 (XII)

The silver salt of isatin, on the other hand, reacts with alkyl halides to give *O*-alkyl derivatives.[76] These differ from the isomeric *N*-alkyl derivatives

[65] Borsche, Weussmann, and Fritzsche, *Ber.*, **57**, 1149 (1924).

[66] Buu-Hoi and Guettier, *Bull. soc. chim.*, **1946**, 586.

[67] Martinet and Dornier, *Compt. rend.*, **172**, 330, 1415 (1921). Schlieper and Schlieper, *Ann.*, **120**, 1 (1861).

[68] Suida, *Ber.*, **11**, 585 (1878). Liebermann and Kraus, *ibid.*, **40**, 2492 (1907).

[69] Heller, *ibid.*, **51**, 424 (1918).

[70] Schwartz, *Jahresber.*, **1863**, 557. Schotten, *Ber.*, **24**, 774 (1897). Heller, *ibid.*, **36**, 2763 (1903); **40**, 1291 (1907). Heller and Lauth, *J. prakt. Chem.*, [2] **113**, 225 (1927).

[71] Heller, *Ber.*, **51**, 1270 (1918).

[72] Aeschlimann, *J. Chem. Soc.*, **1926**, 2902. Meyer, *Monatsh*, **26**, 1323 (1905); **28**, 38 (1907).

[73] Baeyer, *Ber.*, **16**, 2188 (1883). Baeyer and Oekonomides, *ibid.*, **17**, 2195 (1884) Fischer and Hess, *Ber.*, **17**, 564, 2195 (1884). Heller, *ibid.*, **40**, 1291 (1907). Pummerer and Steiglitz, *Ber.*, **75**, 1072 (1942).

[74] Borsche and Jacobs, *Ber.*, **47**, 354 (1914). Friedländer and Kielbasinski, *ibid.*, **44**, 3098 (1911). Kohn and Ostersetzer, *Monatsh.*, **34**, 789 (1913); **37**, 25 (1916).

[75] Colman, *Ann.* **248**, 121 (1888).

[76] Baeyer and Oekonomides, *Ber.*, **15**, 2093 (1882). Heller, *ibid.*, **40**, 1291 (1907).

in their ease of hydrolysis. Sodium *N*-methylisatinate is reconverted to the *N*-alkyl isatin on acidification, but dissolving the *O*-methylether in alkali and then acidifying precipitates isatin, rather than a methylisatin.

Treating the sodium salt of isatin with ethyl chloroformate furnishes ethyl isatin-1-carboxylate,[77] and with ethyl chlorooxalate, ethyl isatin-1-glyoxalate is produced.[78] Similarly, ethyl isatin-1-acetate and diethyl isatin-1-malonate have been produced by using ethyl chloroacetate and ethyl chloromalonate, respectively.[79] Cyanoethylation of isatin with acrylonitrile yields 1-cyanoethylisatin.[80]

Isatin will add to the carbonyl group of formaldehyde to give compounds of the type of *N*-hydroxymethylisatin.[81] If this reaction is carried out in the presence of secondary amines[82] a Mannich type condensation occurs to give compounds such as *N*-(diethylaminomethyl)isatin. With phenylisocyanate the imino hydrogen reacts in the usual manner to give ureide type products.[83]

Reactions of the Carbonyl Groups

Grignard Reagents. The reactions of the two carbonyl groups in isatin are typical of such functions. Although the α-carbonyl is part of an amide grouping, and as such does not enter into the usual reactions of a ketone function, there are cases in which isatin behaves as an α-dicarbonyl compound, with both groups taking part in a reaction. For example, with the Grignard reagent isatin normally yields the 3-alkyl(or aryl)-3-hydroxyoxindoles.[84] Thus *N*-methylisatin and phenylmagnesium bromide in equi-

[77] Hantzsch, *ibid.*, **57,** 195 (1924). Heller, *ibid.*, **51,** 424 (1918). Heller and Jacobson' *ibid.*, **54,** 1107 (1921). Heller and Lauth, *J. prakt. Chem.*, [2] **113,** 225 (1927)' Putochin, *Ber.*, **60,** 1636, 2033 (1927).

[78] Heller, *Ber.*, **51,** 424 (1918).

[79] Ainley and Robinson, *J. Chem. Soc.*, **1934,** 1508. Langenbeck, *Ber.*, **60,** 942 (1928). Putochin, *J. Russ. Phys.-Chem. Soc.*, **60,** 1179 (1928); *Chem. Abstr.*, **23,** 2970 (1929).

[80] DiCarlo and Lindwall, *J. Am. Chem. Soc.*, **67,** 199 (1945).

[81] Reissert and Handeler, *Ber.*, **57,** 989 (1924).

[82] Einhorn and Göttler, *ibid.*, **42,** 4850 (1909).

[83] Gumpert, *J. prakt. Chem.*, [2] **32,** 283 (1886). Goldschmidt and Meiseler, *Ber.*, **23,** 278 (1890).

[84] Kohn, *Monatsh.*, **31,** 747 (1910); **32,** 905 (1911). Kohn and Ostersetzer, *ibid.*, **34,** 789, 1714 (1913); **37,** 25 (1916). Myers and Lindwall, *J. Am. Chem. Soc.*, **60,** 2153 (1938). Steinkopf and Hanske, *Ann.*, **541,** 238 (1939). Steinkopf and Hempel, *ibid.*, **495,** 144 (1932). Steinkopf and Wilhelm, *Ber.*, **70,** 2233 (1937). Stolle, Hecht, and Becker, *J. prakt. Chem.*, **135,** 345 (1932). Sumpter, *Trans. Kentucky Acad. Sci.*, **9,** 61 (1941). Sumpter, *J. Am. Chem. Soc.*, **64,** 1736 (1942).

molecular proportions give 1-methyl-3-phenyldioxindole (XIV).[85] The *N*-alkyl- or aryldioxindoles, however, will react further with the Grignard reagent, producing presumably the intermediate glycol (XV).[86] The product

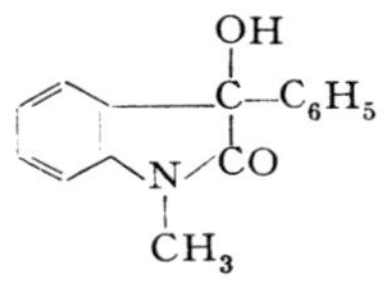

(XIV)

actually isolated on hydrolysis of the Grignard complex with sulfuric acid is a mixture of the isomeric indoxyl (XVI) and oxindole (XVII) derivatives, formed by a Wagner rearrangement.[87,88] Hydrolysis with ammonium

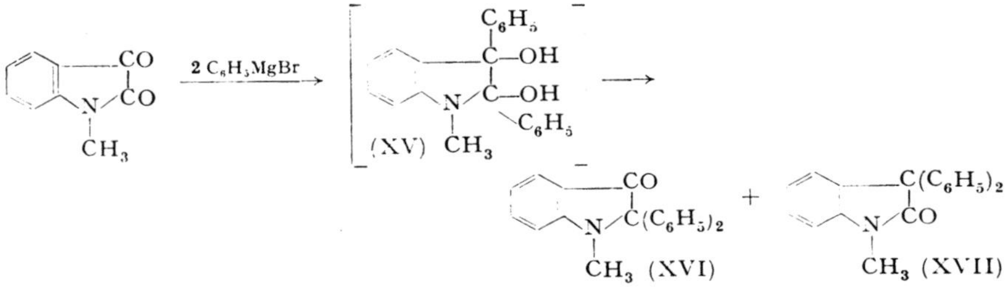

$\xrightarrow{2\ C_6H_5MgBr}$ (XV) $\longrightarrow$ (XVI) + (XVII)

chloride, however, results in essentially pure 3,3-diphenyl-1-methyloxindole (XVII), while hydrolysis with strong hydrochloric acid gives predominantly XVI.[88]

In the Reformatsky reaction *N*-substituted isatins react as expected.[89] However, hydrolysis of the ethyl 3-hydroxy-1-alkyloxindolyl-3-acetate (XVIII) results in ring opening and reclosure to yield the quinoline derivative (XIX).

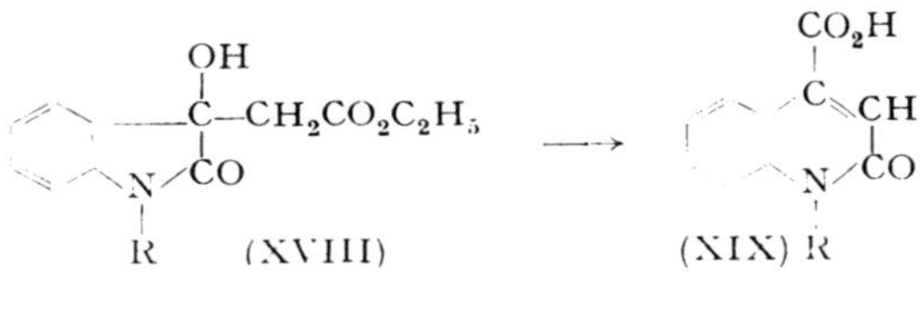

(XVIII) $\longrightarrow$ (XIX)

Amine Derivatives. With carbonyl reagents isatin normally reacts to

[85] Kohn and Ostersetzer, *Monatsh.*, **34,** 1714 (1913).
[86] Myers and Lindwall, *J. Am. Chem. Soc.*, **60,** 644 (1938).
[87] Witkop and Ek, *ibid.*, **73,** 5664 (1951).
[88] Sumpter, Unpublished work.
[89] Myers and Lindwall, *J. Am. Chem. Soc.*, **60,** 2153 (1938).

give the β-derivatives. Thus, hydroxylamine,[90] phenylhydrazine,[91] semicarbazide,[92] and phenylhydroxylamine[93] yield the β-oxime, -phenylhydrazone, -semicarbazone, and -anilide oxide, respectively. In the case of *N*-acetylisatin, both carbonyl groups react to give the dioxime[94] and diphenylhydrazone (osazone).[95]

Since isatin-β-oxime (XX) is identical with nitrosoöxindole (XXI), and can be prepared by treating oxindole with nitrous acid,[96] this common derivative is quite useful for relating substituted oxindoles and isatins. An

C=NOH, CO, N, H (XX) ⇌ CHNO, CO, N, H (XXI) $\xleftarrow{HNO_2}$ CH_2, CO, N, H

CH_2, CO, N, H $\xrightarrow{SeO_2}$ CO, CO, N, H

alternate procedure is to oxidize the oxindole to the isatin with selenium dioxide.[97] Isatin-β-oxime forms insoluble salts with a number of metal ions and has been suggested as an organic precipitant.[98]

Isatin-α-oxime (XXII) has been prepared by the action of nitrous acid on ethyl indoxylic acid (XXIII)[99] and from *O*-methylisatin (XXIV) by reaction

[90] Baeyer and Comstock, *Ber.*, **16**, 1706 (1883). Borsche and Sander, *ibid.*, **47**. 2815 (1914). Brunner, *Monatsh.*, **18**, 533 (1897). Gabriel, *Ber.*, **16**, 518 (1883)' Gabriel and Meyer, *ibid.*, **14**, 2332 (1881). Hantzsch and Barth, *ibid.*, **35**, 220 (1902). Kazak, *Chem. Zentr.*, **1909**, II, 987. Marchlewski, *Ber.*, **29**, 1031 (1896). Zawidski, *ibid.*, **37**, 2300 (1904).

[91] Baeyer, *Ber.*, **16**, 2188 (1883). Fischer, *ibid.*, **17**, 577 (1884). Krause, *ibid.*, **23**, 3602 (1890). Rolla, *Gazz. chim. ital.*, **37**, I, 627 (1907).

[92] Marchlewski, *J. prakt. Chem.*, [2] **60**, 407 (1899).

[93] Allesandri, *Gazz. chim. ital.*, **57**, 195 (1927).

[94] Kazak, *Chem. Zentr.*, **1909**, II, 987. Schunck and Marchlewski, *Ber.*, **28**, 543 (1895); **29**, 203 (1896).

[95] Panaotovic, *J. prakt. Chem.*, [2] **33**, 58 (1886).

[96] Baeyer and Knop, *Ann.* **140**, 1 (1866). Borsche, Weussmann, and Fritzsche, *Ber.*, **57**, 1149 (1924). Gabriel, *ibid.*, **16**, 518 (1883). Hahn and Schulz, *ibid.*, **72**, 1309 (1939). Hahn and Tulus, *ibid.*, **74**, 500 (1941).

[97] Giovannini and Portmann, *Helv. Chim. Acta*, **31**, 1392 (1948).

[98] Hovorka and Sykora, *Collection Czechoslov. Chem. Communs.*, **10**, 83 (1938). *Chem. Abstr.*, **32**, 4460 (1938); *Chem. Listy*, **32**, 241 (1938); *Chem. Abstr.*, **32**, 7914 (1938).

[99] Baeyer, *Ber.*, **15**, 782 (1882); **16**, 2188 (1883).

COC_2H_5 / CCO_2H / N / H (XXIII) → CO / $C{=}NOH$ / N / H (XXII) ← CO / $COCH_3$ / N (XXIV)

with hydroxylamine.[100] The dioxime is also known.[101] Both the α-oxime and its ethyl ether undergo a Beckmann rearrangement on standing in alkaline solution, giving benzoyleneurea (XXV).[102] The β-oxime does not rearrange, but gives *o*-cyanophenylisocyanate (XXVI) with phosphorus pentachloride.[103] Other derivatives behave similarly.[104]

CO / $C{=}NOH$ / N / H → CO / NH / CO / N / H (XXV) ; $C{=}NOH$ / CO / N / H → CN / NCO (XXVI)

For the preparation of the α-phenylhydrazone, either isatin-*O*-methyl ether or isatin-α-chloride may be used with phenylhydrazine.[105] The α-derivative may be converted into the osazone by treatment with phenylhydrazine.[106] Both the α- and β-semicarbazones are known.[107]

With other amines isatin behaves in a similar manner. The β-anilide (XXVII) is readily formed with aniline in alcoholic solution.[108] The α-deriva-

$C{=}NC_6H_5$ / CO / N / H (XXVII)

[100] Heller, *ibid.*, **49**, 2757 (1916).

[101] Reissert and Hessert, *ibid.*, **57**, 964 (1924).

[102] Heller, *ibid.*, **49**, 2757 (1916).

[103] Borsche and Sander, *ibid.*, **47**, 2815 (1914).

[104] Beckmann and Bark, *J. prakt. Chem.*, [2] **105**, 335 (1923).

[105] Albert and Hurtzig, *Ber.*, **52**, 530 (1919). Auwers and Boennecke, *Ann.*, **381**, 303 (1911). Heller, *Ber.*, **43**, 2892 (1911). Reissert and Hoppmann, *ibid.*, **57**, 972 (1924). Schunck and Marchlewski, *ibid.*, **28**, 543 (1895); **29**, 203 (1896).

[106] Heller, *Ber.*, **42**, 470 (1910).

[107] Marchlewski, *J. prakt. Chem.*, [2] **60**, 407 (1899). Holzbecher, *Chem. Listy*, **44**, 126 (1950); *Chem. Abstr.*, **45**, 8005 (1950).

[108] Binz and Heuter, *Ber.*, **48**, 1038 (1915). Engelhardt, *J. prakt. Chem.* [1] **65**, 261 (1855). Knoevenagel, *ibid.*, [2] **89**, 46 (1914). Möhlau and Litter, *ibid.*, [2] **73**, 449 (1906). Ostromisslenski, *Ber.*, **40**, 4972 (1907); **41**, 3032 (1908). Pummerer, *ibid.*, **44**, 338 (1911). Pummerer and Göttler, *ibid.*, **42**, 4269 (1909); **43**, 1376 (1910). Schiff, *Ann.*, **144**, 51 (1867); **210**, 121 (1881); **218**, 192 (1883). Stolle, Bergdoll, Luther, Auerhahn, and Waker, *J. prakt. Chem.*, [1] **28**, 1 (1930). Vorländer, *Ber.*, **40**, 1419 (1907).

tive is an intermediate in the preparation of isatin from thiocarbanilide,[109] and has been prepared from nitrosobenzene and indoxylic acid (XXVIII).[110] That it exists in tautomeric forms[110] has been demonstrated.[111] The α-anilide reacts with phenylhydrazine to give the α-phenylhydrazone,[112] and with

CO, N H, $CHCO_2H$ (XXVIII) $\xrightarrow{C_6H_5NO}$ CO, N H, $C{=}NC_6H_5$ ⇌ CO, N, $CNHC_6H_5$

hydrogen sulfide in acid solution to give α-thioisatin (XXIX).[109] Isatin dianilide can be prepared by the action of aniline on either the α-anilide or *O*-methylisatin.[112]

CO, CS, N H

(XXIX)

A further example of the existence of a dicarbonyl system in isatin is found in the reaction of isatin with *o*-phenylenediamine to form an indoloquinoxaline (XXX).[113]

CO, CO, N H + H_2N, H_2N ⟶ C, N, C, N, N H (XXX)

With ammonia, isatin gives the unstable aldehyde ammonia or the imide, depending on the conditions.[114] Treatment of isatin-β-imide with aniline gives the β-anilide. Isatin-α-imide has also been prepared indirectly.[115]

[109] Sandmeyer, *Z. Farben Textilchem.*, **2,** 129 (1903).

[110] D. R. P. 113,980, 113,981, 277,396.

[111] Callow and Hope, *J. Chem. Soc.*, **1929,** 1911. Compare also Ettinger and Friedländer, *Ber.*, **45,** 2074 (1912), and Rupe and Gugenbühl, *Helv. Chim. Acta*, **8,** 358 (1925); **10,** 926 (1927).

[112] Heller, *Ber.*, **40,** 1291 (1907). Reissert and Hoppmann, *ibid.*, **57,** 972 (1924).

[113] Buraczewski and Marchlewski, *ibid.*, **34,** 4010 (1901). Korczynski and Marchlewski, *ibid.*, **35,** 4334 (1902). Marchlewski, *J. prakt. Chem.*, [2] **60,** 407 (1899). Marchlewski and Radcliffe, *Ber.*, **32,** 1869 (1899); **34,** 1113 (1901). Schunck and Marchlewski, *ibid.*, **28,** 2527 (1895). Gal, *Experientia*, **7,** 261 (1951).

[114] Reissert and Hoppmann, *Ber.*, **57,** 972 (1924). Sommaruga, *Ann.*, **190,** 367 (1877); **194,** 85 (1878); *Ber.*, **11,** 1085 (1878); **12,** 980 (1879); *Monatsh.*, **1,** 575 (1880). Sommaruga and Reichardt, *Ber.*, **10,** 432 (1877). Jacini, *Gazz. chim. ital.*, **71,** 532 (1941). Laurent, *J. prakt. Chem.*, [1] **25,** 456 (1842); **35,** 108 (1845).

[115] Reissert and Hessert, *Ber.*, **57,** 964 (1924).

Isatin, when treated with alcoholic ethylamine, forms the β-ethylimide.[116] An excess of ethylamine is reported to give 1-ethyl-3,3-bis(ethylamino)oxindole (XXXI).

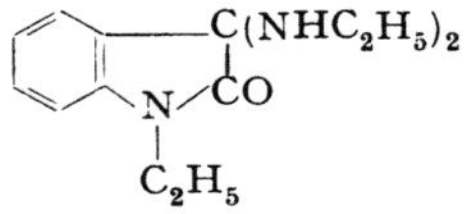

$C(NHC_2H_5)_2$ — N — CO, N–C_2H_5

(XXXI)

The product of the reaction of isatin with β-phenylhydroxylamine was originally considered to be the α-derivative,[117] since the same compound was formed from isatin-α-chloride and β-phenylhydroxylamine. However, this product gave known β-derivatives with phenylhydrazine and on reduction gave 3-anilinoöxindole.[118] The α-anilide oxide (XXXII) has been prepared from o-nitrophenylacetylene and nitrobenzene.[118]

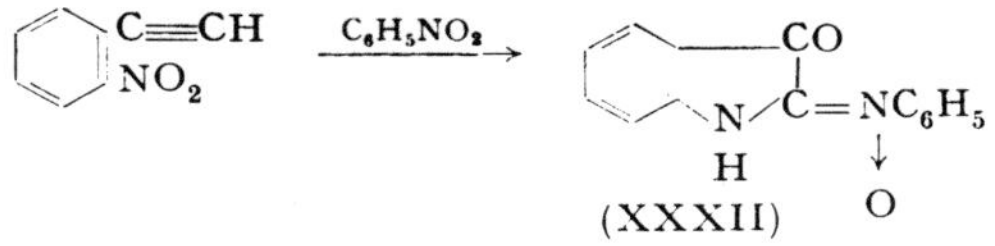

C≡CH, NO_2 $\xrightarrow{C_6H_5NO_2}$ CO, N(H)–C=NC_6H_5 → O

(XXXII)

Miscellaneous Reagents. As a normal ketonic function, the β-carbonyl group of isatin forms a cyanohydrin,[119] and a bisulfite addition product,[120] adds mercaptans,[121] and with activated rings, such as phenols, alkylanilines, toluene,[122] and even benzene in the presence of aluminum chloride,[123] isatin

[116] Haslinger, *ibid.*, **40**, 3598 (1907); **41**, 1444 (1908).

[117] Rupe and Apotheker, *Helv. Chim. Acta*, **9**, 1053 (1926), Rupe and Gugenbühl, *ibid.*, **8**, 358 (1925); **10**, 926 (1927). Rupe and Stocklin, *ibid.*, **7**, 557 (1924). Rupe and Kersten, *ibid.*, **9**, 578 (1926).

[118] Alessandri, *Gazz. chim. ital.*, **38**, 121 (1917).

[119] Heller and Nötzel, *J. prakt. Chem.*, [2] **123**, 257 (1929). Kalb, *Ber.*, **44**, 1455 (1911). Kalb and Berrer, *ibid.*, **57**, 2105 (1924). Martinet, *Compt. rend.*, **166**, 998 (1918).

[120] Laurent, *J. prakt. Chem.*, [1] **26**, 123 (1842); **28**, 337 (1843). Schiff, *Ann.*, **144**, 51 (1867); **210**, 121 (1881); **218**, 192 (1883).

[121] Baumann, *Ber.*, **18**, 890 (1885). Schonberg, Schutz, Arend, and Peter, *ibid.*, **60**, 2344 (1927).

[122] Baeyer and Lazarus, *Ber.*, **12**, 1310 (1879); **18**, 2637 (1885). Liebermann and Dainala, *ibid.*, **40**, 3588 (1907). Candea, *Bull. sect. sci. acad. roumaine*, **8**, 31 (1922-1923); *Chem. Abstr.*, **17**, 1638 (1923). Steopoe, *Ber.*, **60**, 1116 (1927). Gabel and Zubarovskii, *J. Gen. Chem. (U. S. S. R.)*, **7**, 305 (1937); *Chem. Abstr.*, **31**, 4666 (1937).

[123] Wegmann and Dahn, *Helv. Chim. Acta*, **29**, 415 (1946). Sumpter, Unpublished work.

condenses to yield compounds of the type of phenolisatin (XXXIII),[124] which like phenolphthalein finds use as a mild purgative.[125]

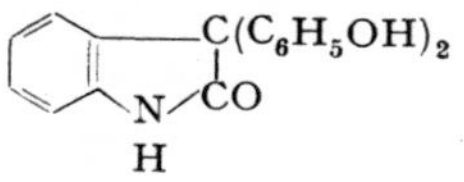

(XXXIII)

The ring expansion reaction of cyclic ketones with diazomethane has its counterpart in isatin chemistry. Thus isatin with this reagent yields both the epoxide (XXXIV), as a simple ketone addition product, or the rearranged 2,3-dihydroxyquinoline (XXXV), usually obtained as the ether

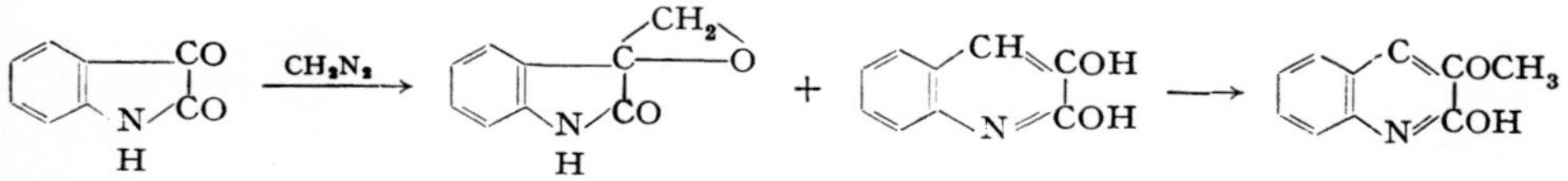

(XXXIV) (XXXV) (XXXVI)

(XXXVI).[126] The reaction has been extended to a number of isatin derivatives.[127]

Application of the Schmidt reaction to the amide linkage of isatin has been made.[128] Loss of a carbon atom occurs to produce anthranilamide. *N*-

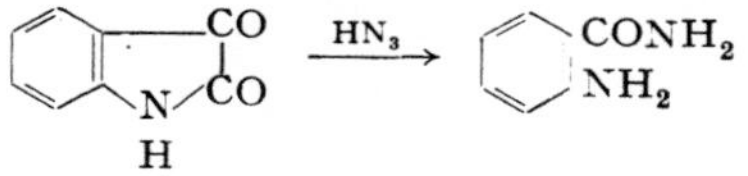

Acetylisatin furnishes the same product, while *N*-ethylisatin yields *o*-ethylaminobenzamide.

Active Methylene Groups. The highly reactive β-carbonyl group of isatin enters into condensation reactions with a number of active methylene

[124] Inagaki, *J. Pharm. Soc. Japan*, **53**, 686, 698 (1933); **58**, 946, 961, 976 (1938); **59**, 1 (1939).

[125] Bergell, *Z. med. Chem.*, **4**, 65 (1926); *Chem. Abstr.*, **21**, 3708 (1927). British patent 523,496. D. R. P. 641,625, 695,691. U. S. patent 2,232,034. Christiansen, *Arch. Pharm. Chem.*, **88**, 47, 69 (1931); *Chem. Abstr.*, **25**, 4264 (1931). D. R. P. 558,238. Silberschmidt, *Chem. Abstr.*, **31**, 2677 (1937). Weiss, *Deut. med. Wochschr.*, **52**, 1343 (1926); *Chem. Abstr.*, **21**, 2738 (1927).

[126] Heller, *Ber.*, **52**, 741 (1919). Heller, Fuchs, Jacobson, Raschig, and Schutze, *ibid.*, **59**, 706 (1926). Arndt, Amende, and Ender, *Monatsh.*, **59**, 202 (1932). Arndt, Eistert, and Ender, *ibid.*, **62**, 44 (1929).

[127] Ault, Hirst, and Morton, *J. Chem. Soc.*, **1935**, 1653.

[128] Caronna, *Gazz. chim. ital.*, **71**, 585 (1941). Edwards and Petrov, *J. Chem. Soc.*, **1948**, 1713.

groups. The number of such reactions reported is too large to consider in detail, but several examples will serve to illustrate the generality of the process and its applications. For more complete lists of these condensations reference should be made to reviews by Heller[129] and by Sumpter.[130]

Isatin condenses with acetone and acetophenone in the presence of mildly basic catalysts such as diethylaniline or piperidine in a typical aldol type reaction.[131] Dehydration may be effected with hydrochloric acid, or the

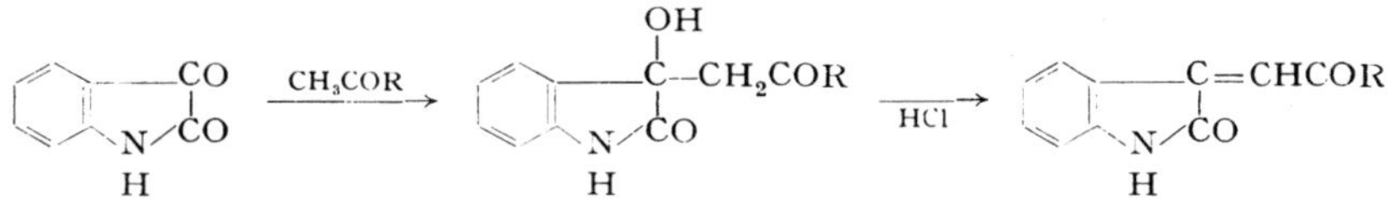

$$\text{isatin} \xrightarrow{CH_3COR} \text{C(OH)}-CH_2COR \text{ (3-hydroxy-oxindole)} \xrightarrow{HCl} \text{C}=CHCOR \text{ (oxindole)}$$

same effect may be accomplished by carrying out the reaction with an acid catalyst. Other compounds possessing active methylene groups react equally well. Ethyl phenylacetate and isatin yield XXXVII,[132] and nitro paraffins such as nitromethane yield aldol type products (XXXVIII).[133]

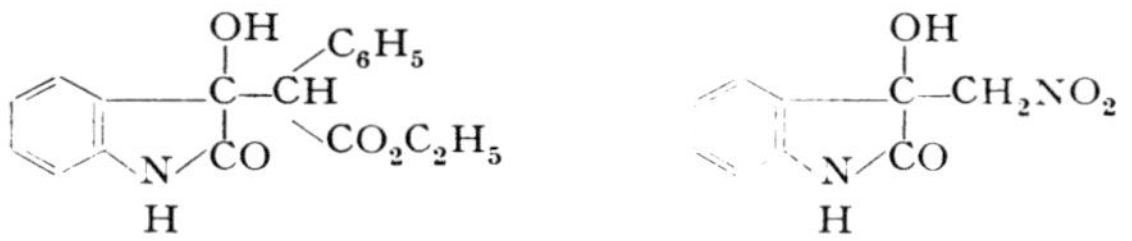

OH, C—CH(C_6H_5)($CO_2C_2H_5$), N—CO, H

(XXXVII)

OH, C—CH_2NO_2, N—CO, H

(XXXVIII)

A number of condensation reactions are not so straightforward, in the aldol sense. Malonic ester and isatin, rather than entering into a clear-cut Knoevenagel reaction, yield XXXIX.[134] Similarly, ethyl cyanoacetate condenses with isatin two to one to give XL,[135] which, however, decomposes at the melting point to produce XLI.[136]

[129] Heller, *Über Isatin, Isatyd, Dioxindol, und Indophenin*, 173 pages, Ahrens Sammlung, Vol. 5 (1931).

[130] Sumpter, *Chem. Revs.*, **34**, 393 (1944).

[131] Lindwall and Braude, *J. Am. Chem. Soc.* **55**, 325 (1933). DuPuis and Lindwall, *ibid.*, **56**, 471, 2716 (1934). Lindwall and Maclennan, *ibid.*, **54**, 4739 (1932).

[132] Zrike and Lindwall, *ibid.*, **57**, 207 (1935); **58**, 49 (1936).

[133] Conn and Lindwall, *ibid.*, **58**, 1236 (1936). Crawford and Lindwall, *ibid.*, **62**, 171 (1940). Compare, however, Colonna, *Bull. sci. facolta chim. ind. Bologna*, **1941,** 89; *Chem. Zentr.*, **1942,** I, 2131.

[134] Lindwall and Hill, *J. Am. Chem. Soc.*, **57**, 735 (1935).

[135] Hill and Samachson, Unpublished dissertation of J. Samachson, Yale University, 1940. Sumpter, Unpublished work.

[136] Yokayama, *J. Chem. Soc. Japan*, **57**, 251 (1936); *Chem. Abstr.*, **30**, 5204 (1936).

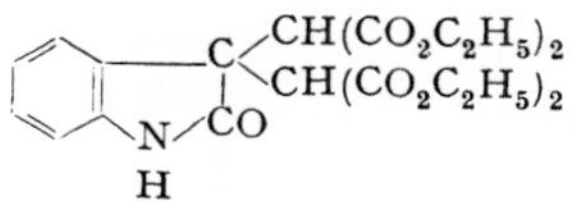

$CH(CO_2C_2H_5)_2$
$CH(CO_2C_2H_5)_2$
N CO
H

(XXXIX)

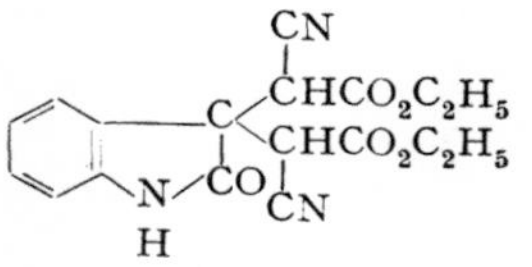

CN
$CHCO_2C_2H_5$
$CHCO_2C_2H_5$
N CO CN
H

(XL)

CN
$C=CCO_2C_2H_5$
N CO
H

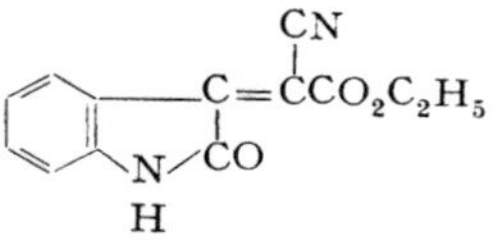

(XLI)

In the presence of strongly basic catalysts, the condensation of isatin with active methylene groups takes another course. As would be expected, strong alkali causes the lactam ring to open, and subsequent closure on acidification produces a six- rather than a five-membered ring, yielding quinoline derivatives. When the active methylene group is furnished in the form of a ketone, the reaction is known as the Pfitzinger reaction[137] and the products are cinchoninic acids. For example, ethyl methyl ketone and isatin under these conditions form 2-ethyl- (XLII) and 2,3-dimethylcinchoninic

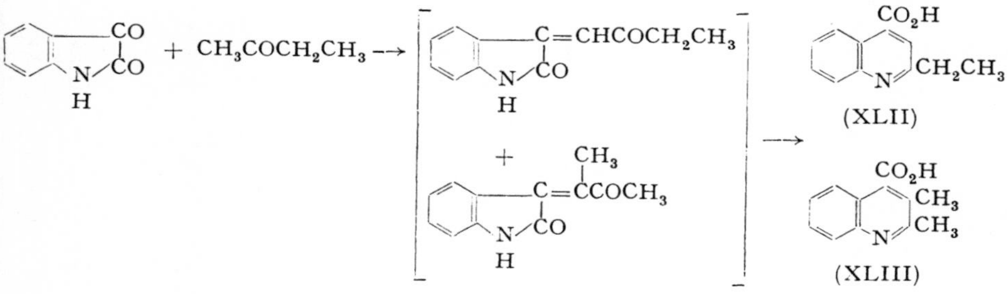

CO CO N H + $CH_3COCH_2CH_3$ → [$C=CHCOCH_2CH_3$ N CO H + $C=C(CH_3)COCH_3$ N CO H] → CO_2H CH_2CH_3 N (XLII); CO_2H CH_3 CH_3 N (XLIII)

acid (XLIII).[138] Isatinic acid behaves in the same fashion[139]; the reaction then is not specific for the isatin molecule. The Pfitzinger reaction has been

[137] Pfitzinger, *J. prakt. Chem.*, [2] **33**, 100 (1886); *ibid.*, **38**, 583 (1888); **56**, 283 (1897); **66**, 263 (1902).
[138] Braun, Gmelin, and Schultheiss, *Ber.*, **56**, 1344 (1923).
[139] Pfitzinger, *J. prakt. Chem.*, [2] **66**, 263 (1902).

studied by a number of workers,[140] and extensive application has been made in the preparation of quinoline derivatives for antimalarial study.[141]

Behavior similar to that in the Pfitzinger reaction has been observed in other cases. The condensation product (XLIV) of isatin and ethyl phenylacetate, for example, on hydrolysis yields the quinolinol (XLV).[142] The

OH, C_6H_5, C—CH, CO, $CO_2C_2H_5$, N, H (XLIV) ⟶ CO_2H, C_6H_5, OH, N (XLV)

same product is obtained directly by using phenylacetic acid.[143] In the same manner, malonic acid and isatin condense to give 2-hydroxycinchoninic acid (XLVI),[144] which is also obtained from *N*-acetylisatin (XLVII) by heating with alkali.[145]

CO, CO, N, H + $CH_2(CO_2H)_2$ ⟶ CO_2H, OH, N (XLVI) ⟵ CO, CO, N, $COCH_3$ (XLVII)

[140] Calaway and Henze, *J. Am. Chem. Soc.*, **61**, 1355 (1939). Gross and Henze, *ibid.*, **61**, 2730 (1939). Lesesne and Henze, *ibid.*, **64**, 1897 (1942). Isbell and Henze, *ibid.*, **66**, 2096 (1944). Crippa and Scevola, *Gazz. chim. ital.*, **67**, 119 (1937). v. Braun, Anton, Haensel, Irmisch, Michaelis, and Teuffert, *Ann.*, **507**, 14 (1933). v. Braun, *Ber.*, **56**, 2343 (1923). Steinkopf and Petersdorf, *Ann.*, **543**, 119 (1940). Steinkopf and Engelmann, *ibid.*, **546**, 205 (1941). Buu-Hoi and Cagniant, *Ber.*, **77**, 118 (1944); *Bull. soc. chim.*, **11**, 343 (1944); *Rec. trav. chim.*, **64**, 214 (1945). Buu-Hoi, *J. Chem. Soc.*, **1946**, 795. Buu-Hoi and Royer, *Compt. rend.*, **223**, 806 (1946). Buu-Hoi and Cagniant, *Bull. soc. chim. France*, **1946**, 123. Buu-Hoi and Royer, *ibid.*, **1946**, 374; *Rec. trav. chim.*, **66**, 300 (1947). Cagniant and Deluzarche, *Compt. rend.*, **223**, 808, 1148 (1946). Nguyen-Hoan and Buu-Hoi, *ibid.*, **224**, 1363 (1947). de Clercq and Buu-Hoi, *ibid.*, **227**, 1251 (1948). Mueller and Stobaugh, *J. Am. Chem. Soc.*, **72**, 1598 (1950).

[141] Bachman and Picha, *J. Am. Chem. Soc.*, **68**, 1599 (1946). Senear, Sargent, Mead, and Koepfli, *ibid.*, **68**, 2695 (1946). Buchman, Sargent, Meyers, and Senear, *ibid.*, **68**, 2692 (1946). Rapport, Senear, Mead, and Koepfli, *ibid.*, **68**, 2697 (1946). Newell and Calaway, *ibid.*, **69**, 116 (1947). Snyder, Freier, Kovacic, and Van Hayningen, *ibid.*, **69**, 371 (1947). Dowell, McCullough, and Calaway, *ibid.*, **70**, 226 (1948). Sublett and Calaway, *ibid.*, **70**, 674 (1948). Huntress and Bornstein, *ibid.*, **71**, 745 (1949).

[142] Zrike and Lindwall, *ibid.*, **57**, 207 (1935); **58**, 49 (1936).

[143] Borsche and Jacobs, *Ber.*, **47**, 354 (1914). Gysae, *ibid.*, **26**, 2484 (1893).

[144] Aeschlimann, *J. Chem. Soc.*, **1926**, 2902.

[145] Camps, *Arch. Pharm.*, **237**, 687 (1899). Panaotovic, *J. prakt. Chem.*, [2] **33**, 58 (1886).

The condensation of isatin with dioxindole in the presence of piperidine yields isatide (XLVIII),[146] the bimolecular reduction product of isatin. It is interesting to note that at one time a quinhydrone structure (XLIX) for isatide was proposed,[147] but has since been discredited.[148]

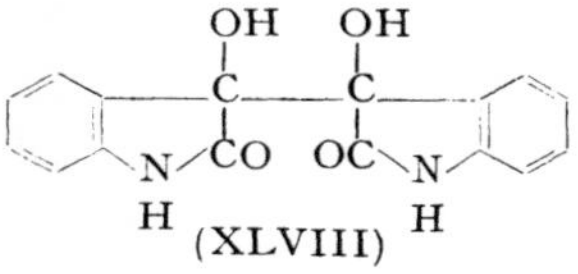

(XLVIII) 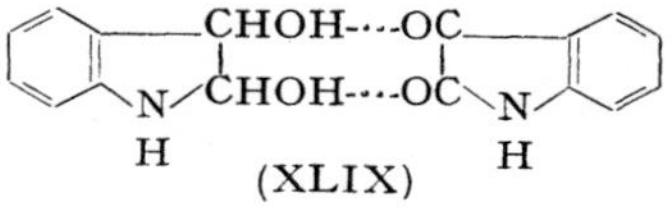 (XLIX)

Isatin condenses with oxindole in acid solution to give isoindigo (L).[149] In the presence of pyridine the dehydration step is prevented and the product is a desoxyisatide, isatane (LI).[150] With indoxyl, isatin forms indirubin (LII).[151]

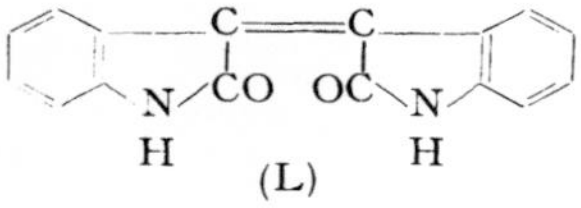

(L) 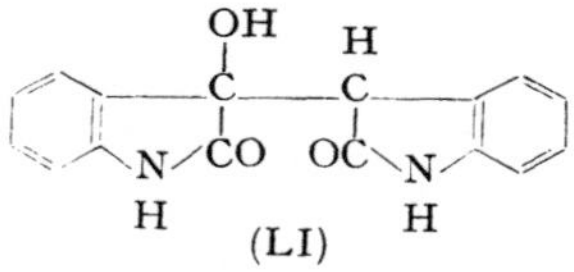 (LI) 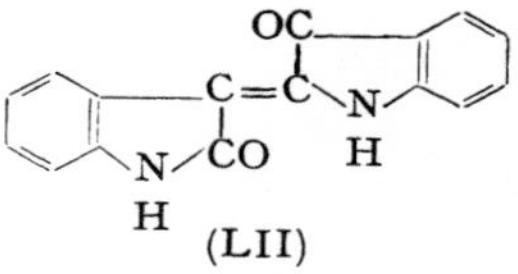 (LII)

Accounts of the use of isatin in the indophenine test,[152] and the subsequent discovery[153] of thiophene are classic. The indophenine reaction

[146] Hansen, *Ann. chim.*, [10] **1,** 94, 126 (1924). Kohn, *Ber.*, **49,** 2514 (1916). Kohn and Klein, *Monatsh.*, **33,** 929 (1912). Kohn and Ostersetzer, *ibid.*, **34,** 789 (1913); **37,** 25 (1916). LeFevre, *Bull. soc. chim.*, **19,** 113 (1916). Stolle and Merkle, *J. prakt. Chem.*, **139,** 329 (1934). Sumpter, *J. Am. Chem. Soc.*, **54,** 2917 (1932). Wahl and Hansen, *Compt. rend.*, **176,** 1070 (1923); **178,** 214, 393 (1924).

[147] Heller, *Ber.*, **49,** 2757 (1916); *J. prakt. Chem.*, **135,** 222 (1932). Heller and Lauth, *Ber.*, **62,** 343 (1929).

[148] Stolle and Merkle, *J. prakt. Chem.*, **139,** 329 (1934).

[149] Fericean, *Bull. Soc. Chim. Roumânia,* **13,** 27 (1931); *Chem. Abstr.*, **26,** 1280 (1932). Hansen, *Ann. chim.*, [10] **1,** 94 (1924). Laurent, *Ann. chim. phys.*, [3] **3,** 372 (1840). Livovschi, *Compt. rend.*, **201,** 217 (1935). Neber and Keppler, *Ber.*, **57,** 788 (1924). Wahl and Bagard, *Compt. rend.*, **148,** 716 (1909); *Bull. soc. chim.*, [4] **5,** 1043 (1909). Wahl and Faivret, *Ann. chim.*, [10] **5,** 314 (1926); *Compt. rend.*, **180,** 589 (1935); **181,** 790 (1925). Wahl and Fericean, *Ann. chim.*, [10] **9,** 277 (1928); *Compt. rend.*, **186,** 378 (1928).

[150] Laurent, *Rev. sci. ind.* (September, 1842).

[151] v. Braun and Hahn, *Ber.*, **56,** 2343 (1923). Ettinger and Friedländer, *ibid.*, **45,** 2074 (1912). Friedländer and Kunz, *ibid.*, **55,** 1597 (1922). Neber and Keppler, *ibid.*, **57,** 788 (1924).

[152] Baeyer, *ibid.*, **12,** 1309 (1879).

[153] Meyer, *ibid.*, **15,** 2893 (1882); **16,** 1465, 2791 (1883).

has been studied by a number of workers,[154] and the structure of the blue product, indophenine (LIII), is regarded as established. That the α-position

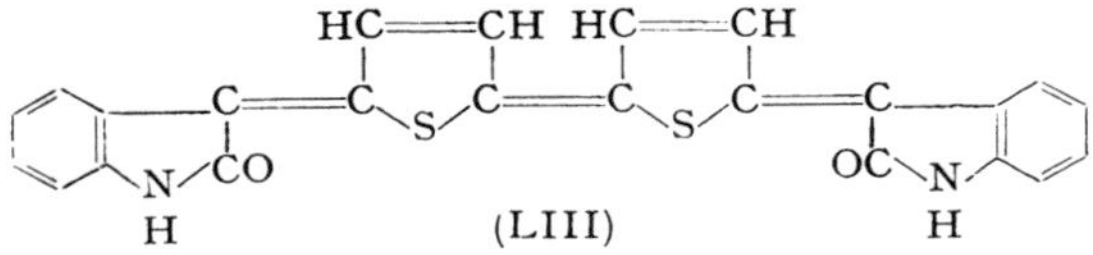

(LIII)

of the thiophene molecule is utilized in the reaction is shown by the fact that the α-methyl derivatives does not give the reaction, while β-methylthiophenes do.[155] β-Derivatives of isatin, on the other hand, do not form the indophenine, while α-derivatives do.[156] Formulations were advanced by both Steinkopf[157] and Heller,[158] but the structure proposed by the latter was accepted as correct.[159]

A number of other dyes have been prepared from isatin. Pyrrole yields pyrrole blue,[160] regarded as analogous to indophenine in structure,[161] though arguments to the contrary have been advanced.[162] With piperidine two products are obtained: isatin monopiperidide and isatin dipiperidide (LIV).[163] On heating, the latter compound is changed to isatin blue, most logically represented as LV.[158]

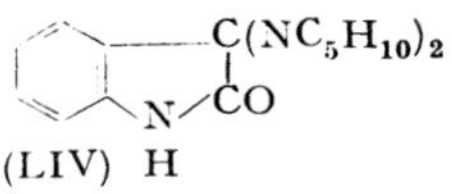

(LIV) 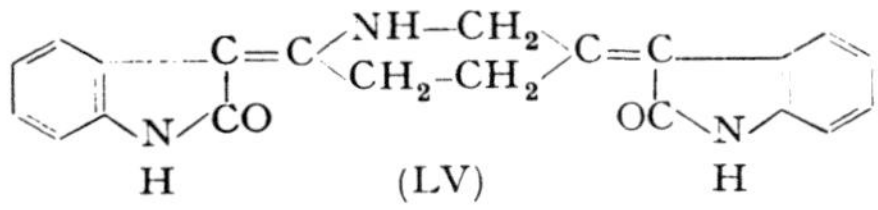 (LV)

Biological Activity of Isatin

Isatin possesses an apparent enzymelike activity in the dehydrogenation of amino acids. Thus, heating α-aminophenylacetic acid with isatin in aqueous solution converts the amino acid into benzaldehyde in good

[154] Bauer, *ibid.*, **37**, 1244, **3182** (1904). Liebermann and Kraus, *ibid.*, **40**, 2492 (1907). Liebermann and Hase, *ibid.*, **38**, 2847 (1905). Liebermann and Pleus, *ibid.*, **37**, 2461 (1904).

[155] Schlenk and Blum, *Ann.*, **443**, 95 (1923). Scheibler and Schmidt, *Ber.*, **54**, 139 (1921).

[156] Heller, *Ber.*, **49**, 1406 (1916); *Z. angew. Chem.*, **29**, I, 415 (1916). Reissert and Hessert, *Ber.*, **57**, 964 (1924).

[157] Steinkopf and Hanske, *Ann.*, **541**, 238 (1939). Steinkopf and Hempel, *Ann.*, **495**, 144 (1932). Steinkopf and Roch, *Ann.*, **482**, 251 (1930).

[158] Heller, *Über Isatin, Isatyd, Dioxindol, und Indophenin*, 173 pages, Ahrens Sammlung, Vol. 5 (1931). Heller, *Z. angew. Chem.*, **37**, 1017 (1924). Heller, *Chem.-Ztg.*, **54**, 585 (1930); **57**, 74 (1933).

yield.[164] Other reports[165] confirm this activity and the following mechanism has been suggested,[166] in which either atmospheric oxygen or methylene blue serves as the hydrogen acceptor, isatin being reduced reversibly in an intermediate step to isatide.

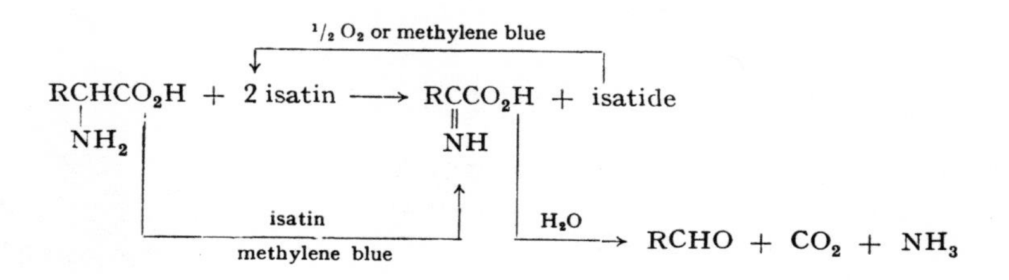

Analytical Uses of Isatin

Isatin can be determined quantitatively by titration with titanous chloride,[167] and has been suggested for use as a microchemical reagent.[168]

[159] Steinkopf and Hanske, *Ann.*, **541,** 238 (1939).

[160] Liebermann and Kraus, *Ber.*, **40,** 2492 (1907). Liebermann and Hase, *ibid.*, **38,** 2847 (1905). Mayer and Stadler, *ibid.*, **17,** 1034 (1884). Ciamician and Silber, *ibid.*, **17,** 142 (1884).

[161] Steinkopf and Wilhelm, *ibid.*, **70,** 2233 (1937); *Ann.*, **546,** 211 (1941).

[162] Pratesi, *Atti accad. Lincei*, **17,** 945 (1933); **18,** 53 (1933); *Ann.*, **504,** 258 (1933); *Ann. chim. applicata*, **25,** 195, 203 (1935); *Chem. Abstr.*, **27,** 4797 (1933). Pratesi and Zanetta, *Atti accad. Lincei*, **22,** 159 (1935).

[163] Schotten, *Ber.*, **24,** 1367 (1897).

[164] Traube, *ibid.*, **44,** 3145 (1911).

[165] Langenbeck, *ibid.*, **60,** 930 (1927); **61,** 942 (1928). Langenbeck, Hellrung, and Juttemann, *Ann.*, **499,** 201 (1932); **512,** 275 (1934). Langenbeck, Hulschenreuter, and Juttemann, *ibid.*, **485,** 53 (1931). Langenbeck and Godde, *Ber.*, **70,** 669 (1937). Langenbeck and Weissenborn, *ibid.*, **72,** 724 (1939). Langenbeck, Weschy, and Godde, *ibid.*, **70,** 672 (1937); Langenbeck and Weschy, **70,** 1039 (1937). Abderhalden, *Z. physiol. Chem.*, **252,** 81 (1938). Dethloff and Schreiber, *Chem. Ber.*, **83,** 157 (1950). Schönberg, Moubasher, and Mostafa, *J. Chem. Soc.*, **1948,** 176; **1950,** 1422. Compare also Giovannini and Portmann, *Helv. Chim. Acta*, **31,** 1361 (1948).

[166] Langenbeck, *Ber.*, **60,** 930 (1927); **61,** 942 (1928). Compare also Schachat, Becker, and McLaren, *J. Org. Chem.*, **16,** 1349 (1951).

[167] Knecht and Hibbert, *J. Chem. Soc.*, **125,** 1537 (1924).

[168] Menke, *Rec. trav. chim.*, **42,** 199 (1923).

CHAPTER IV

Oxindole

In the course of studies on the structure of isatin, it was found[1] that reduction led, in addition to isatide, which had been obtained previously, to dioxindole, $C_8H_7NO_2$, and further, to oxindole, C_8H_7NO. The proposal[2] of a structure for isatin paved the way for correct formulations[3] for the dioxindole (I) and oxindole (II) molecules, and the structure of the latter was confirmed[4] as the lactam of 2-aminophenylacetic acid through its

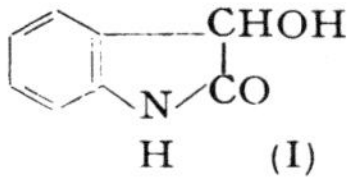

CHOH — CO — N H (I) CH$_2$ — CO — N H (II)

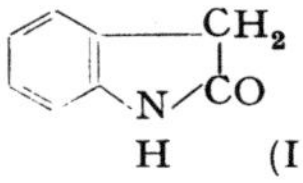

synthesis by the reduction of 2-nitrophenylacetic acid with tin and hydrochloric acid.

Preparation of Oxindoles

Reduction of 2-nitrophenylacetic acid and various derivatives has proved to be an extremely useful method of preparation of oxindoles,[5] and several modifications of the procedure have been made. Using zinc and

[1] Baeyer, *Ber.*, **1,** 17 (1868). Baeyer and Knop, *Ann.*, **140,** 1 (1866).

[2] Kekulé, *Ber.*, **2,** 748 (1869).

[3] Baeyer, *ibid.*, **11,** 582, 1228 (1878).

[4] Baeyer, *loc. cit.*

[5] Wispec, *Ber.*, **16,** 1580 (1883). Smith and MacMullen, *J. Am. Chem. Soc.*, **58,** 629 (1936). Livovschi, *Compt. rend.*, **201,** 217 (1935); **203,** 1265 (1936). Wahl and Livovschi, *ibid.*, **206,** 738 (1937); *Bull. soc. chim.*, [5] **5,** 653 (1938). Ruggli and Grand, *Helv. Chim. Acta*, **20,** 373 (1937). Parks and Aldis, *J. Chem. Soc.*, **1938,** 1841. Wahl and Bagard, *Bull. soc. chim.*, [4] **5,** 1033 (1909). Hahn and Schulz, *Ber.*, **72,** 1308 (1939). Hahn and Tulus, *ibid.*, **74,** 500 (1941). Trinius, *Ann.*, **227,** 274 (1885). Suida, *Ber.*, **11,** 584 (1878). Heller, *ibid.*, **43,** 1097 (1911). Gabriel and Mayer, *ibid.*, **14,** 823 (1881). May and Mosettig, *J. Org. Chem.*, **11,** 429 (1946). Clair, Clarke, Edmuston, and Wiesner, *Can. J. Research*, **28,** 745 (1950).

[6] Reissert, *Ber.*, **30,** 1043 (1897); **41,** 3921 (1908). Reissert and Scherk, *ibid.*, **31,** 393 (1898).

hydrochloric acid both oxindole (II) and 1-hydroxyoxindole (III) result.[6] A method of catalytic hydrogenation, using platinum[7] or Raney nickel,[8] has been developed to offer a convenient method for the preparation of oxindole.

CH_2CO_2H, NO_2 ⟶ CH_2, N, CO, H (II) + CH_2, N, CO, OH (III)

Under certain conditions the use of platinum results in the formation of 1-hydroxyoxindole.[7] The reduction of *o*-nitrophenylacetanilide also yields some oxindole along with *o*-aminophenylacetanilide.[9]

One useful procedure for preparing oxindole is by the reduction of isatin. The original preparation[10] was effected in two steps: reduction to dioxindole through the agency of sodium amalgam in alkaline media, and subsequent reduction of the dioxindole to oxindole with either tin and mineral acids or sodium amalgam in acid solution. A convenient modification[11] involving reduction to dioxindole with sodium hydrosulfite, and then to oxindole with sodium amalgam in a solution saturated with carbon dioxide has been applied in a number of cases.[12]

The reduction of isatin directly to oxindole may be effected by the Wolff-Kishner procedure,[13] although the yield is not large.

CO, N, CO, H ⟶ $C{=}NNH_2$, N, CO, H $\xrightarrow{NaOH}$ CH_2, N, CO, H

A very general method for preparing oxindole and its 1- and 3-alkyl derivatives, as well as those containing nuclear substituents, is that of Stolle.[14] An α-halogenated acid halide is condensed with an aromatic amine. Ring closure with aluminum chloride yields the corresponding oxindole.

[7] DiCarlo, *J. Am. Chem. Soc.*, **66,** 1420 (1944).

[8] Hahn and Tulus, *Ber.*, **74,** 500 (1941). Koelsch, *J. Am. Chem. Soc.*, **66,** 2019 (1944).

[9] Koenig and Reissert, *Ber.*, **32,** 793 (1899).

[10] Baeyer and Knop, *Ann.*, **140,** 1 (1866).

[11] Marschalk, *Ber.*, **45,** 582 (1912); *J. prakt. Chem.*, [2] **88,** 227 (1913).

[12] Wahl and Faivret, *Compt. rend.*, **181,** 790 (1925); *Ann. chim.*, [10] **5,** 314 (1926). Wahl and Ferecean, *Compt. rend.*, **186,** 378 (1928); *Ann. chim.*, [10] **9,** 277 (1928).

[13] Curtius and Thun, *J. prakt. Chem.*, [2] **44,** 187 (1890). Borsche and Mayer, *Ber.*, **54,** 2841 (1921). Shapiro, *ibid.*, **62,** 2133 (1929). Siebert, *Chem. Ber.*, **80,** 494 (1947).

[14] D. R. P. 341,112, 335,673. Stolle, *Ber.*, **47,** 2120 (1914). Stolle, Bergdoll, Luther, Auerhahn, and Wacker, *J. prakt. Chem.*, **128,** 1 (1930). Stolle, Hecht, and Becker, *ibid.*, **135,** 345 (1932).

This method has found extensive application,[15] particularly in the case of

$$C_6H_5NH_2 \xrightarrow{ClCH_2COCl} C_6H_5NH\text{—}CO\text{—}CH_2Cl \xrightarrow{AlCl_3} \text{oxindole (}CH_2\text{, }CO\text{, }NH\text{)}$$

$$C_6H_5NHR + R_2C(Br)\text{—}COBr \longrightarrow C_6H_5NR\text{—}CO\text{—}CR_2Br \xrightarrow{AlCl_3} \text{1-R-3,3-}R_2\text{-oxindole}$$

oxindole derivatives, where ring closure occurs at lower temperatures than in the case of the preparation of oxindole itself. Partial failure is observed in the case of *N*-benzylchloroacetanilide, which loses the benzyl group on heating with aluminum chloride.[16] A related procedure, involving dichloroacetic acid, yields 3-arylaminoöxindoles (IV).[17]

$$2\,C_6H_5NH_2 \xrightarrow{Cl_2CHCO_2H} \text{3-}(CHNHC_6H_5)\text{-oxindole (N—H, CO)} \quad \text{(IV)}$$

Heating *β*-acetylphenylhydrazine (V) with lime at 220° yields oxindole.[18] The method is of general application,[19] particularly for the preparation of 3-alkyl derivatives.

$$C_6H_5NH\text{—}NH\text{—}CO\text{—}CH_3 \text{ (V)} \longrightarrow \text{oxindole (}CH_2\text{, }CO\text{, }NH\text{)}$$

[15] Julian and Pikl, *J. Am. Chem. Soc.*, **57**, 563 (1935). Julian, Pikl, and Boggess, *ibid.*, **56**, 1797 (1934). Livovschi, *Compt. rend.*, **201**, 217 (1935); **203**, 1265 (1936). Porter, Robinson, and Weyler, *J. Chem. Soc.*, **1941**, 620. Wahl and Livovschi, *Compt. rend.*, **205**, 738 (1937); *Bull. soc. chim.* [5] **5**, 653 (1938). Petyunin and Panferova, *J. Gen. Chem.* (*U. S. S. R.*), **17**, 502 (1947). Mathew and Menon, *Proc. Indian Acad. Sci.*, **29**, 361 (1949).

[16] Stolle, Bergdoll, Luther, Auerhahn, and Wacker, *J. prakt. Chem.*, **128**, 1 (1930).

[17] Duisberg, *Ber.*, **18**, 190 (1885). Heller, *Ann.*, **358**, 349 (1907). Heller and Aschkenasi, *ibid.*, **375**, 261 (1910). Mayer, *Ber.*, **16**, 2262 (1883). Paukscli, *ibid.*, **17**, 2800 (1884). Crippa and Pietra, *Gazz. chim. ital.*, **78**, 456 (1948).

[18] Brunner, *Monatsh.*, **18**, 95, 531 (1897); **27**, 1183 (1906); **58**, 369 (1931); **62**, 373 (1933). Brunner and Moser, *ibid.*, **61**, 15 (1932).

[19] D. R. P. 218,477, 218,727. Schwarz, *Monatsh.*, **24**, 572 (1903). Wahl, *ibid.*, **38**, 525 (1918). Stanek and Rybar, *Chem. Listy*, **40**, 173 (1946); *Chem. Abstr.*, **45**, 5147 (1951). Tomicek, *Chem. Listy*, **16**, 1, 35 (1922); *Chem. Abstr.*, **17**, 1467 (1923).

An adaptation of the Fischer indole synthesis is also applicable for the preparation of oxindoles of the 1,3,3-trialkyl type (VI).[20] The intermediate 2-hydroxyindoline is incapable of dehydration to the indole, and oxidation with ammoniacal alcoholic silver nitrate yields the oxindole.

CHCH(CH$_3$)$_2$ N N CH$_3$ $\xrightarrow[ZnCl_2]{HCl}$ C(CH$_3$)$_2$ CHOH N CH$_3$ ⟶ C(CH$_3$)$_2$ CO N CH$_3$ (VI)

A similar preparation[21] utilizes the indole (VII), prepared by the Fischer synthesis. The action of excess methyl iodide forms the indolenium iodide, which loses hydrogen iodide with 20% sodium hydroxide. Potassium permanganate oxidation converts the methylene indole (VIII) into the oxindole.

CH$_3$O C—CH$_2$CH$_2$OC$_6$H$_5$ C—CH$_3$ N H (VII) $\xrightarrow{CH_3I}$ CH$_3$O CH$_3$ C—CH$_2$CH$_2$OC$_6$H$_5$ C—CH$_3$ N⊕ CH$_3$ I⊖ $\xrightarrow{NaOH}$

(VIII) CH$_3$O CH$_3$ C—CH$_2$CH$_2$OC$_6$H$_5$ C=CH$_2$ N CH$_3$ ⟶ CH$_3$O CH$_3$ C—CH$_2$CH$_2$OC$_6$H$_5$ CO N CH$_3$

N-Alkylindoles can be converted into the corresponding oxindoles by treatment with sodium hypobromite to give first the *N*-substituted 3,3-dibromoöxindole, reduction of which yields the *N*-alkyloxindole.[22] Treatment

CH CH N R ⟶ CBr$_2$ CO N R ⟶ CH$_2$ CO N R

of indole with sulfuryl chloride, followed by steam distillation, also produces oxindole.[23]

N-Alkyloxindoles can also be prepared by hydrolysis of the condensation product of a secondary amine and the sodium bisulfite addition product of glyoxal.[24]

[20] Brunner, *Monatsh.*, **17,** 276, 488 (1896); **21,** 173 (1900).

[21] Boyd-Barrett, *J. Chem. Soc.*, **1932,** 321. Boyd-Barrett, and Robinson, *ibid.*, **1932,** 317. King and Robinson, *ibid.*, **1932,** 326, 1433.

[22] Colman, *Ann.*, **248,** 116 (1888). Michaelis, *Ber.*, **30,** 2811 (1897).

[23] Mazzaro and Borgo, *Gazz. chim. ital.*, **35,** II, 320, 563 (1905).

[24] Hinsberg, *Ber.*, **41,** 1367 (1908). Hinsberg and Rosenweig, *ibid.*, **27,** 3253 (1894).

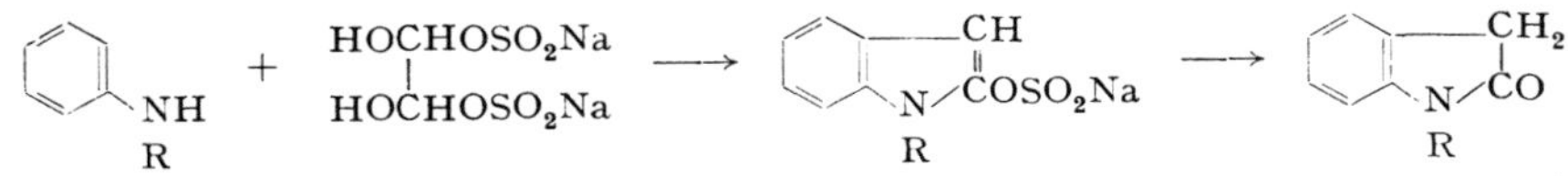

$$\text{C}_6\text{H}_5\text{NHR} + \begin{matrix}\text{HOCHOSO}_2\text{Na}\\ |\\ \text{HOCHOSO}_2\text{Na}\end{matrix} \longrightarrow \text{(indole, N-R) CH}{=}\text{COSO}_2\text{Na} \longrightarrow \text{(oxindole, N-R) CH}_2\text{–CO}$$

An unusual preparation of 4,5,6,7-tetrahydroöxindole-3-propionic acid (IX) has been effected from 2-ketocyclohexane-2-glutaric acid (X) by treatment with ammonia in alcohol.[25]

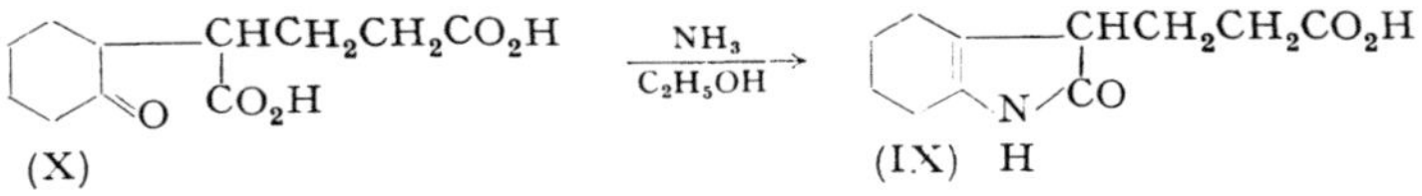

$$\underset{\text{(X)}}{\text{C}_6\text{H}_9\text{O}\text{–CH(CO}_2\text{H)CH}_2\text{CH}_2\text{CO}_2\text{H}} \xrightarrow[\text{C}_2\text{H}_5\text{OH}]{\text{NH}_3} \underset{\text{(IX)}}{\text{(tetrahydroöxindole, NH–CO)–CHCH}_2\text{CH}_2\text{CO}_2\text{H}}$$

Properties of Oxindole

Oxindole crystallizes from water in colorless needles of m.p. 126–127°C., and boils at 195° at 17 mm.[26] and 227° at 73 mm.[27] It is soluble in hot water and the usual organic solvents, the presence of alkali increasing it solubility.

Tautomerism

Although the structure of the oxindole molecule is usually regarded as the lactam (I) of *o*-aminophenylacetic acid, the two enol tautomers (II and III) also represent possible formulations. The absorption spectra of 1-methyl- (IV) and 1,3,3-trimethyloxindole (V) are quite similar to that of oxindole,[28] indicating that the lactam structure is at least predominating. Evidence for the enol structure (III) is observed in the consumption of two moles of Grignard reagent and the evolution of two moles of methane.[29]

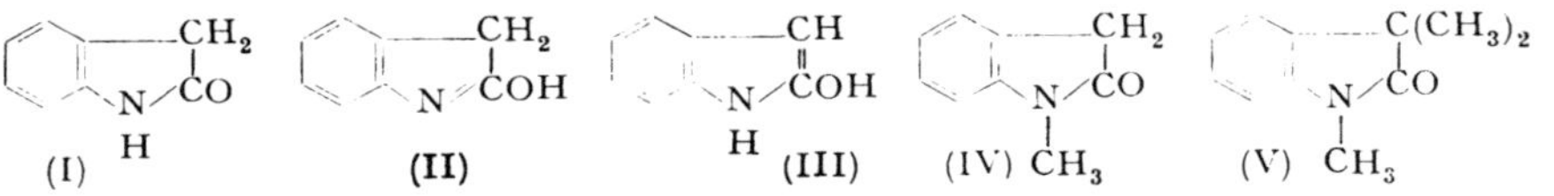

(I) CH_2–CO, N–H; (II) CH_2–COH, N=; (III) CH=COH, N–H; (IV) CH_2–CO, N–CH_3; (V) $C(CH_3)_2$–CO, N–CH_3

Salts

Like isatin, oxindole forms sodium and silver salts. The former is

[25] Kendall, Osterberg, and Mackenzie, *J. Am. Chem. Soc.*, **48**, 1384 (1926). Kendall and Osterberg, *ibid.*, **49**, 2047 (1927).
[26] Wahl and Bagard, *Compt. rend.*, **149**, 132 (1910).
[27] Curtius and Thun, *J. prakt. Chem.*, [2] **44**, 187 (1890).
[28] Ramart-Lucas and Biquard, *Bull. soc. chim.*, [5] **2**, 1383 (1935).
[29] Julian, Pikl, and Wantz, *J. Am. Chem. Soc.*, **57**, 2026 (1935).

obtained with sodium amalgam in warm benzene[30] or with sodium ethoxide.[31] The silver salt is produced with cold ammoniacal silver nitrate.[32] On warming, the silver nitrate is reduced by the oxindole. Oxindole also forms an easily soluble hydrochloride.[32]

The lactam ring of oxindole is opened by heating with barium hydroxide at 150°, yielding the barium salt of 2-aminophenylacetic acid. Acidification of such salts reforms oxindole.[33]

Alkylation of the sodium salts of oxindole produces the *N*-alkyl derivatives.[34] Although *O*-alkyl derivatives of oxindole itself are unknown, they have been obtained from certain of its derivatives. Thus, while the action of methyl iodide and sodium methoxide on 3,3-dimethyloxindole (VII) yields 1,3,3-trimethyl oxindole (VI), treatment of the silver salt of VII with methyl iodide furnishes the lactim ether (VIII).[35] Similar behavior is observed in other cases.[36]

$C(CH_3)_2$, N, CO, CH_3 (VI) $\xleftarrow[NaOCH_3]{CH_3I}$ $C(CH_3)_2$, N, CO, H (VII) $\xrightarrow[AgNO_3]{CH_3I}$ $C(CH_3)_2$, N, $COCH_3$ (VIII)

Reactions of Oxindole

Reduction

Passing the vapor of oxindole over hot zinc effects reduction to indole.[37] A more practical method for the reduction of *N*-methyloxindoles to *N*-methylindoles in good yield has been developed by the use of lithium aluminum hydride.[38] Reduction of the amide carbonyl group of highly

[30] Wheeler, *Am. Chem. J.*, **23,** 465 (1900).

[31] Heller, *Ber.*, **43,** 1907 (1911).

[32] Baeyer and Knop, *Ann.*, **140,** 1 (1866).

[33] Baeyer and Comstock, *Ber.*, **16,** 1705 (1883). Marschalk, *ibid.*, **45,** 582 (1912); *J. prakt. Chem.*, [2] **88,** 227 (1913).

[34] Brunner, *Monatsh.*, **18,** 531 (1897). Julian and Pikl, *J. Am. Chem. Soc.*, **57,** 563 (1935). Julian, Pikl, and Boggess, *ibid.*, **56,** 1797 (1934). Julian, Pikl, and Wantz, *ibid.*, **57,** 2026 (1935).

[35] Brunner, *Monatsh.*, **18,** 95 (1897).

[36] Schwarz, *ibid.*, **24,** 572 (1903).

[37] Baeyer, *Ann.*, **140,** 296 (1866).

[38] Julian and Printy, *J. Am. Chem. Soc.*, **71,** 3206 (1949). Julian and Magnani, *ibid.*, **71,** 3207 (1949).

CH_2, CO, N, CH_3 $\xrightarrow{LiAlH_4}$ CH, CH, N, CH_3

alkylated oxindoles, such as the 1,3,3-trimethyl derivative, takes place with sodium and alcohol, yielding the indolinols (I).[39] This step may be reversed by the oxidative action of ammoniacal silver nitrate.[40]

$C(CH_3)_2$, CO, N, CH_3 $\xrightarrow[C_2H_5OH]{Na}$ / $\xleftarrow[NH_3\ C_2H_5OH]{AgNO_3}$ $C(CH_3)_2$, $CHOH$, N, CH_2 (I)

Halogenation and Nitration

The presence of the acylamino group on the aromatic ring in oxindole activates positions 5 and 7 toward nuclear substitution reactions.[41] In addition, the 3-position, by virture of the influence of the adjacent carbonyl group and aromatic ring, may be acted upon under certain conditions. For example, bromination of certain *N*-substituted oxindoles in aqueous solution yields 5-bromoöxindoles with one molecular proportion of bromine and 5,7-dibromoöxindoles with two equivalents.[42] On the other hand, the bromination of the same compounds in anhydrous carbon tetrachloride solution gives the 3,3-dibromo derivatives.

Early reports[43] of the bromination of oxindole itself led to some confusion. A recent investigation[44] has shown that the product on monobromination in aqueous solution is the expected 5-bromoöxindole (II), and that two equivalents produce the 5,7-dibromo derivative (III). Three molecular proportions of bromine give 3,5,7-tribromoöxindole (IV). As with the *N*-alkyl derivatives, bromination in carbon tetrachloride yields 3,3-dibromo derivatives (V, VI, VII).

[39] Brunner, *Monatsh.*, **17**, 276,488 (1896). Brunner and Moser, *ibid.*, **61**, 15 (1932).

[40] Brunner, *ibid.*, **21**, 173 (1900). Ciamician and Piccinini, *Ber.*, **29**, 2467 (1896).

[41] Brunner, *Monatsh.*, **58**, 369 (1931). Kohn, *ibid.*, **32**, 905 (1911). Martinet, *Ann. chim.*, **11**, 85 (1919). Stolle, Bergdoll, Luther, Auerhahn, and Wacker, *J. prakt. Chem.*, **128**, 1 (1930). Sumpter, *J. Am. Chem. Soc.*, **67**, 1140 (1945). Sumpter, Miller, and Hendrick, *ibid.*, **67**, 1656 (1945). Sumpter, Miller, and Magan, *ibid.*, **67**, 499 (1945).

[42] Stolle, Bergdoll, Luther, Auerhahn, and Wacker, *J. prakt. Chem.*, **128**, 1 (1930).

[43] Baeyer and Knop, *Ann.*, **140**, 1 (1866). Henze and Blair, *J. Am. Chem. Soc.*, **55**, 4621 (1933).

[44] Sumpter, Miller, and Hendrick, *J. Am. Chem. Soc.*, **67**, 1656 (1945).

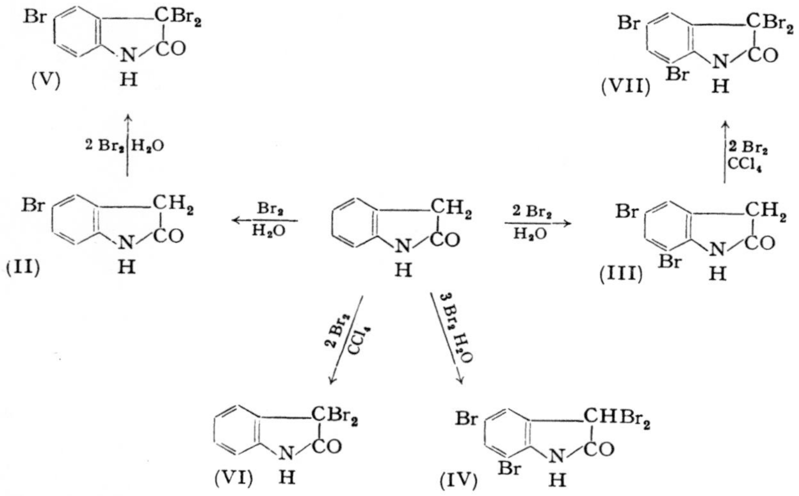

(V) 5-bromo-3,3-dibromooxindole ← $2Br_2$, H_2O ← (II) 5-bromooxindole ← Br_2, H_2O ← oxindole → $2Br_2$, H_2O → (III) 5,7-dibromooxindole → $2Br_2$, CCl_4 → (VII) 5,7-dibromo-3,3-dibromooxindole

oxindole → $2Br_2$, CCl_4 → (VI) 3,3-dibromooxindole

oxindole → $3Br_2$, H_2O → (IV) 5,7-dibromo-3-(dibromomethyl)oxindole ($CHBr_2$)

Prolonged treatment of 5,6-dimethoxyoxindole with bromine in boiling chloroform results in substitution in the 3 and 7 positions, along with cleavage of the 6-methoxyl group.[45]

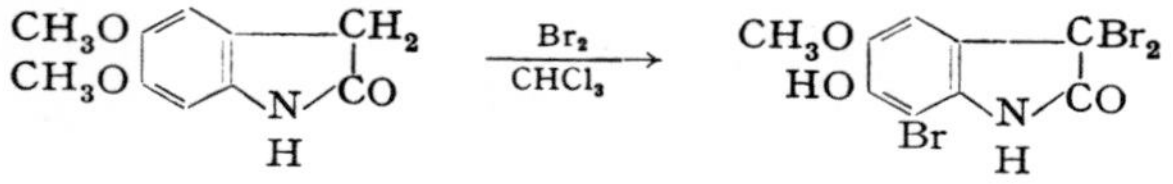

CH_3O, CH_3O (5,6-dimethoxyoxindole, CH_2, CO, NH) → Br_2, $CHCl_3$ → CH_3O, HO, Br (CBr_2, CO, NH)

The products of the chlorination of oxindole derivatives have been assumed[46] to be the 5,7-dichloro derivatives by analogy with the structures of the bromination products. Iodination may be accomplished directly by the use of iodine, potassium iodide, and potassium iodate in acetic acid.[46] This product has been shown to be 5-iodoöxindole by synthesis from the 5-amino derivative.

N-Alkyloxindoles yield 3,3-dichloro derivatives through the action of calcium hypochlorite[47] or sodium hypochlorite.[48] Such compounds are also readily prepared through the Stolle synthesis from the appropriate trichloroacetyl-*N*-alkylanilide (VIII), or from isatin or *N*-alkylisatins by the agency of phosphorus pentachloride.[49] One reported preparation of 3,3-di-

[45] Hahn and Tulus, *Ber.*, **74,** 500 (1941).

[46] Brunner, *Monatsh.*, **58,** 369 (1931).

[47] Stolle, Bergdoll, Luther, Auerhahn, and Wacker, *J. prakt. Chem.*, **128,** 1 (1930).

[48] Colman, *Ann.*, **248,** 116 (1888). Fischer and Hess, *Ber.*, **17,** 564 (1884). Michaelis, *ibid.*, **30,** 2811 (1897).

[49] Baeyer, *Ber.*, **12,** 456 (1879); Hantzsch, *ibid.*, **54,** 1221, 1257 (1921).

chloroöxindoles from isatin by the combined action of chlorosulfonic acid

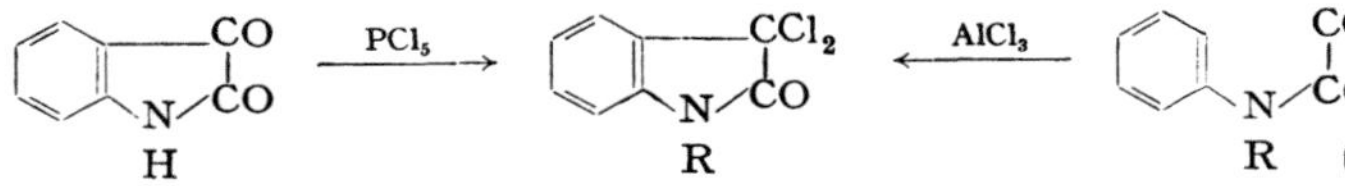

CO, N/CO, H $\xrightarrow{PCl_5}$ CCl_2, N/CO, R $\xleftarrow{AlCl_3}$ CCl_3, N/CO, R (VIII)

and hydrochloric acid[50] has been shown[51] to yield products containing sulfur, probably as the result of sulfonation.

Treatment of oxindole with phosphorus pentachloride yields a product designated as chloroöxindole chloride (2,3-dichloroindole) (IX),[52] also obtainable from dioxindole with the same reagent.

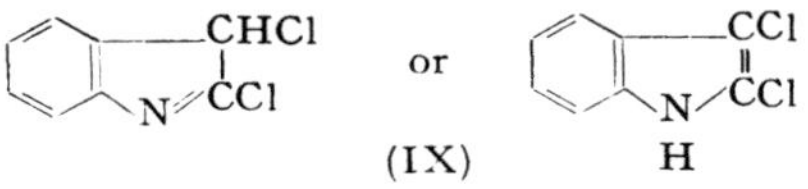

CHCl, N=CCl or CCl, N/CCl, H

(IX)

The mononitration of oxindole[53] has been shown[54] to furnish the 5-nitro derivative. Similarly, the nitration of 3,3-dimethyloxindole gives two mononitro derivatives, 5- and 7-nitro-3,3-dimethyloxindole.[55]

The only study of the direct sulfonation of an oxindole derivative has been made with 3,3-dimethyloxindole.[56] Concentrated sulfuric acid introduces a sulfonic acid group into position 5, while the 5,7-disulfonic acid is produced with fuming sulfuric acid.

Condensation Reactions

The activity of the 3-position of oxindole is observed in the large number of condensations, of both the aldol and Claisen types, into which oxindole enters. For example, oxindole reacts readily with benzaldehyde and substituted benzaldehydes in the presence of pyridine to give benzaloxindoles (X).[57] A number of similar condensations have been effected.[58]

[50] D. R. P. 694,044.

[51] Sumpter, Unpublished work. U. S. patent 2,335,273.

[52] Baeyer, *Ber.*, **12,** 456 (1879).

[53] Baeyer, *ibid.*, **12,** 1312 (1879). Borsche, Weussman, and Fritzsche, *ibid.*, **57,** 1149 (1924).

[54] Sumpter, Miller, and Magan, *J. Am. Chem. Soc.*, **67,** 499 (1945).

[55] Brunner, *Monatsh.*, **58,** 369 (1931).

[56] Brunner, *ibid.*, **62,** 373 (1933).

[57] Wahl and Bagard, *Bull. soc. chim.*, [4] **5,** 1033 (1909); *Compt. rend.*, **148,** 716 (1909); **149,** 132 (1910).

[58] Armit and Robinson, *J. Chem. Soc.*, **127,** 1604 (1925). Borsche, Wagner-Roemmich, and Barthenheier, *Ann.*, **550,** 160 (1942). Horner, *ibid.*, **548,** 117 (1941).

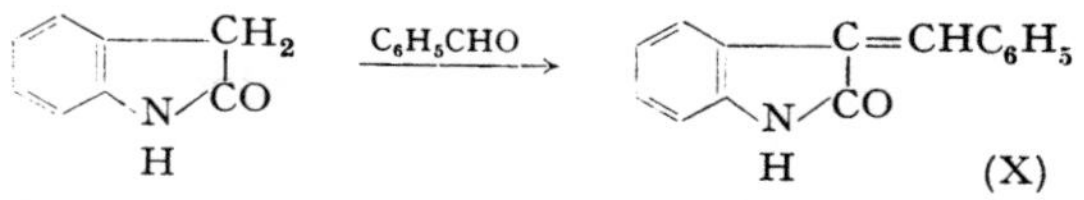

CH_2, NH, CO $\xrightarrow{C_6H_5CHO}$ $C{=}CHC_6H_5$, NH, CO (X)

With nitrous acid oxindole nitrosates in the 3-position to give isatin-β-oxime (XI).[59] Nitrosobenzene[59,60] and *p*-nitrosodimethylaniline[61] condense with oxindole and nuclear substituted oxindoles to give derivatives of isatin-β-anil (XII).

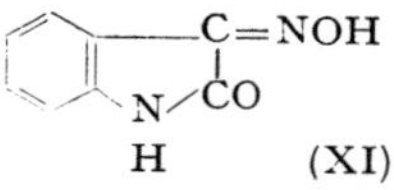

$C{=}NOH$, NH, CO (XI) $C{=}NC_6H_5$, NH, CO (XII)

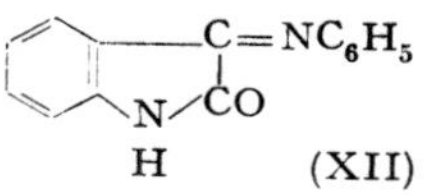

The condensation of oxindole with isatin forms isoindigo (XIII)[62] in acid media and isatane (XIV) in the presence of pyridine.[63] With isatin-α-chloride, indirubin (XV) is produced.[64]

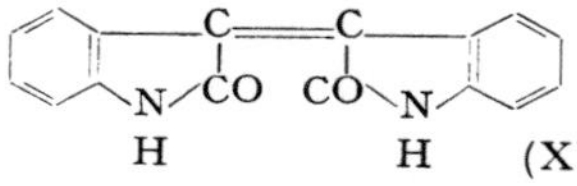

C═C, NH, CO, CO, NH (XIII)

Kirchner, *Nachr. kgl. Ges. Wiss. Göttingen, Math.-physik. Kl.*, **1921**, 154; *Chem. Abstr.*, **17**, 1012 (1923). Kliegl and Schmalenbach, *Ber.*, **56**, 1517 (1923). Neber, *ibid.*, **55**, 826 (1922). Neber and Rocker, *ibid.*, **56**, 1710 (1923). Stolle, Bergdoll, Luther, Auerhahn, and Wacker, *J. prakt. Chem.*, **128**, 1 (1930). Wahl and Faivret, *Compt. rend.*, **181**, 790 (1925); *Ann. chim.*, [10] **5**, 314 (1926). Hahn and Ferecean, *Compt. rend.*, **186**, 378 (1928); *Ann. chim.*, [10] **9**, 277 (1928). Windaus and Eickel, *Ber.*, **57**, 1871 (1924).

[59] Baeyer and Knop, *Ann.*, **140**, 1 (1866). Borsche, Weussmann, and Fritzsche, *Ber.*, **57**, 1149 (1924). Gabriel, *ibid.*, **16**, 518 (1883). Hahn and Schulz, *ibid.*, **72**, 1309 (1939). Hahn and Tulus, *ibid.*, **74**, 500 (1941). Kohn and Ostersetzer, *Monatsh.*, **34**, 1714 (1913).

[60] Neber and Keppler, *Ber.*, **57**, 778 (1924).

[61] Stolle, Bergdoll, Luther, Auerhahn, and Wacker, *J. prakt. Chem.*, **128**, 1 (1930).

[62] Fericean, *Bull. soc. chim. România*, **13**, 27 (1931); *Chem. Abstr.*, **26**, 1280 (1932). Hansen, *Ann. chim.*, [10] **1**, 94, 126 (1924). Langenbeck, Hellrung, and Juttemann, *Ann.*, **499**, 201 (1932). Livovschi, *Compt. rend.*, **201**, 217 (1935). Neber and Keppler, *Ber.*, **57**, 778 (1924). Wahl and Bagard, *Bull. soc. chim.*, [4] **5**, 1033 (1909); *Compt. rend.*, **148**, 716 (1909). Wahl and Faivret, *ibid.*, **180**, 589 (1925); **181**, 790 (1925). Wahl and Ferecean, *Compt. rend.*, **186**, 378 (1928). Wahl and Hansen, *ibid.*, **176**, 1070 (1923); **178**, 214, 393 (1924).

[63] Hansen, *Ann. chim.*, [10] **1**, 94, 126 (1924). Laurent, *Rev. sci. ind.* (September, 1842). Lefevre, *Bull. soc. chim.*, **19**, 113 (1916). Livovschi, *Compt. rend.*, **201**, 217 (1935). Neber and Keppler, *Ber.*, **57**, 778 (1924). Wahl and Faivret, *Compt. rend.*, **181**, 790 (1925); *Ann. chim.*, [10] **5**, 314 (1926). Wahl and Hansen, *Compt. rend.*, **176**, 1070 (1923); **178**, 214, 393 (1924).

[64] Wahl and Bagard, *Compt. rend.*, **148**, 716 (1909); *Bull. soc. chim.*, [4] **5**, 1043 (1909).

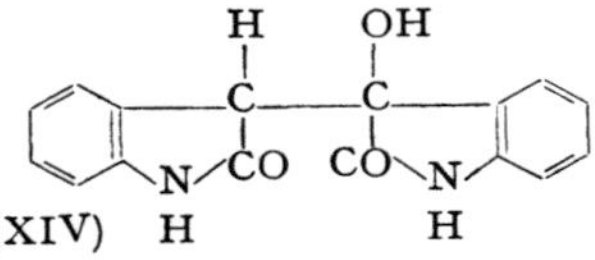

(XIV) 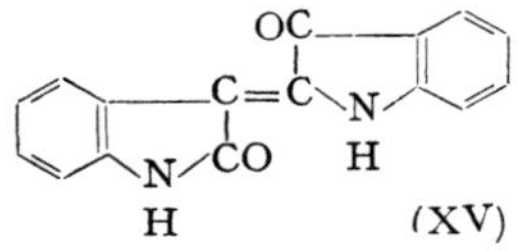 (XV)

Acyl derivatives of the 3-position of oxindole are readily prepared by the Claisen condensation. The first preparation of this type was of 3-formyl-1-phenyloxindole (XVI), through the condensation of ethyl formate with 1-phenyloxindole.[65] This type of compound had been prepared previously

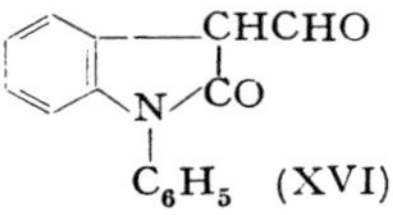

(XVI)

by the hydrolysis, with alcoholic sodium hydroxide, of the dye, Thioindigo Scarlet R[66] (see page 194).

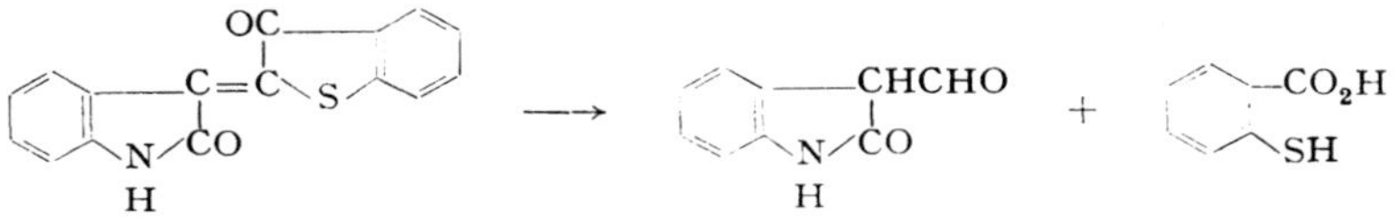

The reaction was generalized for the formation of 3-acyl-1-alkyl-oxindoles as intermediates in the preparation of 3-alkyl derivatives (XVII) by catalytic reduction,[67] although it has been reported that 3-acyloxindoles unsubstituted on nitrogen are not reduced in this manner.

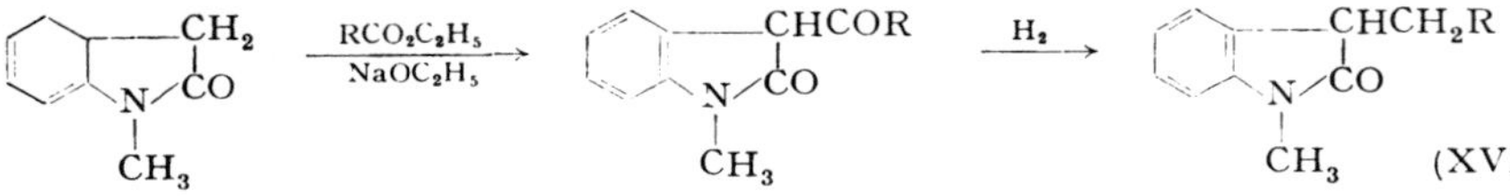

$\xrightarrow[NaOC_2H_5]{RCO_2C_2H_5}$ $\xrightarrow{H_2}$ (XVII)

The above acylation procedure has been extended to oxindole itself by condensation with a number of esters[68]: ethyl acetate, ethyl oxalate, ethyl malonate, and certain others. It is noteworthy that the vinylogous 3-ethyl-ideneoxindole (XVIII) possesses a sufficiently reactive position to undergo

[65] Stolle, Hecht, and Becker, *J. prakt. Chem.*, **135**, 345 (1932).

[66] Friedländer and Kielbasinski, *Ber.*, **44**, 3098 (1911). D. R. P. 246,338. Friedländer and Schwenck, *Ber.*, **43**, 1974 (1910). Kalb and Berrer, *ibid.*, **57**, 2105 (1924).

[67] Julian, Pikl, and Boggess, *J. Am. Chem. Soc.*, **56**, 1797 (1934). Julian, Pikl, and Wantz, *ibid.*, **57**, 2026 (1935). Porter, Robinson, and Weyler, *J. Chem. Soc.*, **1941,** 620.

[68] Horner, *Ann.*, **548,** 117 (1941).

[69a] Julian *et al.*, in Elderfield, *Heterocyclic Compounds*, Vol. 3, Wiley, New York, 1952, p. 167; *J. Am. Chem. Soc.*, **75**, 5301, 5305 (1953).

$$\text{3-acetyloxindole (CHCOCH}_3\text{)} \xleftarrow[\text{NaOC}_2\text{H}_5]{\text{CH}_3\text{CO}_2\text{C}_2\text{H}_5} \text{oxindole (CH}_2\text{)} \xrightarrow[\text{NaOC}_2\text{H}_5]{(\text{CO}_2\text{C}_2\text{H}_5)_2} \text{CHCOCO}_2\text{C}_2\text{H}_5$$

$$\text{oxindole} \xrightarrow[\text{CH}_2(\text{CO}_2\text{C}_2\text{H}_5)_2]{\text{NaOC}_2\text{H}_5} \text{CHCOCH}_2\text{COCH (bis-oxindole)}$$

this same condensation.

$$\text{C=CHCH}_3 \text{ (XVIII)} \xrightarrow[\text{NaOC}_2\text{H}_5]{(\text{CO}_2\text{C}_2\text{H}_5)_2} \text{C=CHCH}_2\text{COCO}_2\text{C}_2\text{H}_5$$

3-Formyloxindole will undergo a Knoevenagel reaction with malonic acid to yield oxindole-3-acrylic acid (XIX), reducible to oxindole-3-propionic acid (XX).[69]

$$\text{CHCHO} \xrightarrow{\text{CH}_2(\text{COOH})_2} \text{CHCH=CHCO}_2\text{H (XIX)} \longrightarrow \text{CHCH}_2\text{CH}_2\text{CO}_2\text{H (XX)}$$

Oxindole also enters into a Michael condensation with two moles of ethyl acrylate to give a product originally designated as the 1,3-derivatives,[68] but recently shown to be the 3,3-isomer (XXI).[68a]

$$\text{CH}_2 + 2\,\text{CH}_2\text{=CHCO}_2\text{C}_2\text{H}_5 \longrightarrow \text{C(CH}_2\text{CH}_2\text{CO}_2\text{C}_2\text{H}_5)_2 \text{ (XXI)}$$

The reduction of ethyl oxindole-3-glyoxalate (XXII) under Clemmensen conditions, with zinc amalgam and hydrochloric acid[68] does not yield the expected oxindole-3-acetic acid (XXIII), but rather undergoes a rearrangement to produce the quinoline derivative (XXIV).[70] Reduction of the ester with zinc and acetic acid does yield ethyl oxindole-3-acetate (XXIII)[68,70] but hydrolysis of the ester causes rearrangement to XXIV.[70]

[69] D. R. P. 451,957. U. S. patent 1,656,239.

[70] Sumpter, Miller, and Hendrick, *J. Am. Chem. Soc.*, **67**, 1037 (1945). The preparation of oxindole-3-acetic acid through catalytic reduction of the condensation product of oxindole with dibenzyl oxlate under conditions precluding ring opening and enlargement has been announced as this volume goes to press. See Julian *et al.*, *J. Am. Chem. Soc.*, **75**, 5305 (1953).

(XXII) $\xrightarrow[HCl]{Zn}$ (XXIV)

(XXII) $\xrightarrow{Zn, CH_3CO_2H}$ (XXIII) $\xrightarrow{H_2O}$ (XXIV)

Alkylation and Acylation

Direct alkylation of the oxindole nucleus can be effected in positions 1 and 3 by treating oxindole with alkyl halides in the presence of sodium alkoxides. Thus, 3-methyloxindole (XXV) yields 1,3,3-trimethyloxindole (XXVI) with methyl iodide and sodium methoxide.[71] Apparently it is

(XXV) $\xrightarrow[NaOC_2H_5]{2\ CH_3I}$ (XXVI)

necessary that there be either an alkyl or acyl group in the 3-position for alkylation to take place, since it was found impossible to introduce an alkyl group into the the 3-position of oxindole itself or of 1-methyloxindole.[72]

Alkylation of 1-methyl-3-formyloxindole (XXVII) proceeds stepwise, with elimination of the acyl group.[72,73] On the other hand, treatment of the

(XXVII) $\xrightarrow{CH_3I}$ 1,3-dimethyloxindole $\xrightarrow{CH_3I}$ 1,3,3-trimethyloxindole

(XXVII) $\xrightarrow{2\ CH_3I}$ 1,3,3-trimethyloxindole

[71] Brunner, *Monatsh.*, **18**, 531 (1897).

[72] Julian, Pikl, and Wantz, *J. Am. Chem. Soc.*, **57**, 2026 (1935). Compare also Rutenberg and Horning, Abstracts, Atlantic City Meeting, Am. Chem. Soc., p. 2L, 1949.

sodium salt of the formyl derivative in acetone with methyl iodide yields the *O*-methyl ether (XXVIII).[72a]

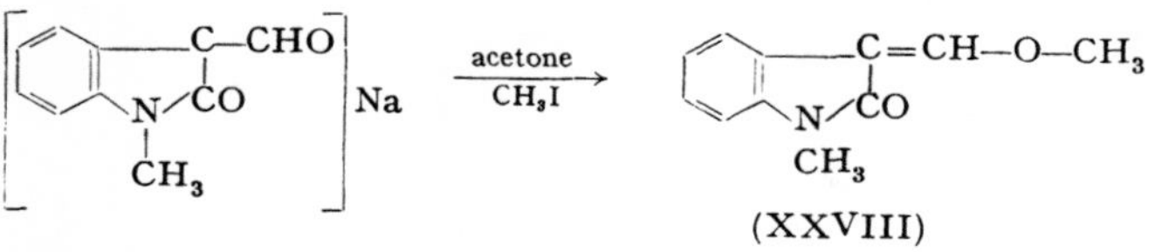

(XXVIII)

This same general alkylation technique has been used for the introduction of the acetonitrile and acetaldehyde groups into position 3 of the oxindole nucleus.[72,74]

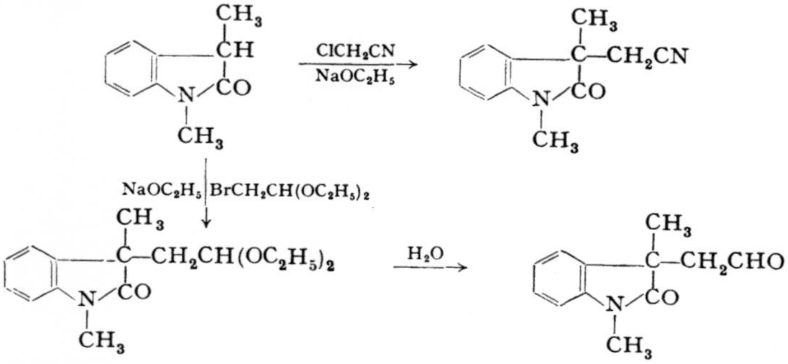

Cyanoethylation of *N*-methyloxindole has also been accomplished,[74]

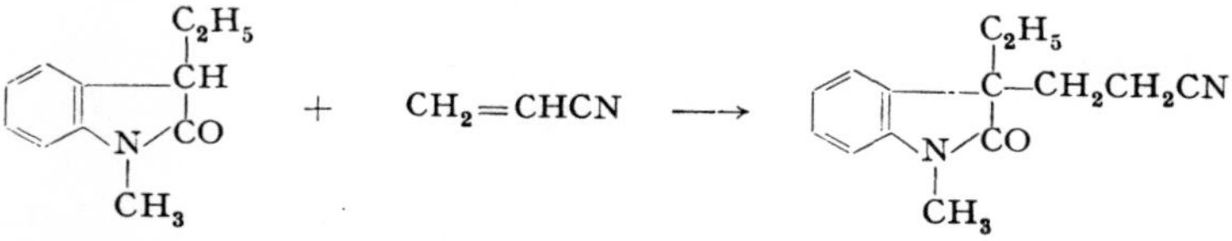

and introduction of the dialkylaminomethyl group at the 3-position has been effected through the Mannich reaction.[74]

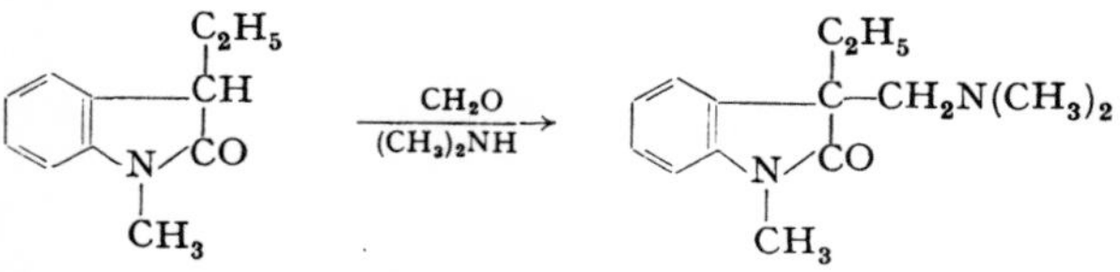

Alkylation at the 2-position of 1,3,3-trialkyloxindoles with the Grignard reagent produces indolinol derivatives (XXIX).[75]

[72a] Wenkert, Bose, and Reid, *J. Am. Chem. Soc.*, **75**, 5514 (1953).

[73] Julian and Pikl, *J. Am. Chem. Soc.*, **57**, 563 (1935). Julian, Pikl, and Boggess, *ibid.*, **56**, 1797 (1934).

[74] Horning and Rutenberg, *ibid.*, **72**, 3534 (1950).

[75] Brunner, *Monatsh.*, **26**, 1359 (1905). Jenisch, *ibid.*, **27**, 1223 (1906).

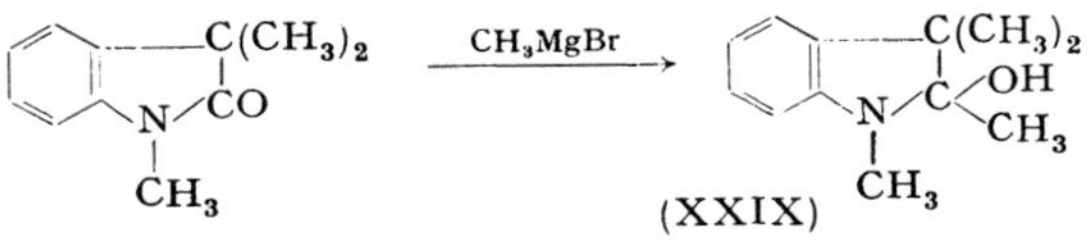

$C(CH_3)_2$, CO, N, CH_3 $\xrightarrow{CH_3MgBr}$ $C(CH_3)_2$, C, OH, CH_3, N, CH_3 (XXIX)

The acidic amide hydrogen atom of oxindole may be replaced by acyl groups directly by the action of acid anhydrides on oxindole[76] or its derivatives[77] or by the treatment of the sodium salt of oxindole with acid chlorides.[78]

Derivatives

Hydroxyoxindoles

Dioxindole. 3-Hydroxyoxindole (dioxindole) (I) was first obtained by the reduction of isatin with sodium amalgam.[79] It is usually prepared in this manner, either by the agency of zinc and acetic acid,[80] or better, sodium hydrosulfite.[81] The latter method appears to be quite general, having been

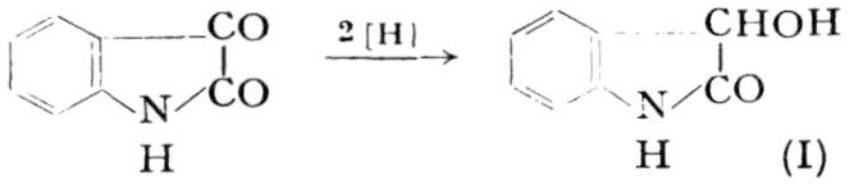

CO, CO, N, H $\xrightarrow{2[H]}$ CHOH, CO, N, H (I)

employed for the preparation of a number of dioxindole derivatives.[82] The sodium amalgam procedure has been used for the reduction of isatin-4-carboxylic acid,[83] and *N*-alkyldioxindoles have been obtained from the corresponding *N*-alkylisatin with zinc and hydrochloric acid.[84] 5-Amino-

[76] Suida, *Ber.*, **12,** 1326 (1879).

[77] Brunner, *Monatsh.*, **18,** 95, 531 (1897). Mazzaro and Borgo, *Gazz. chim. ital.*, **35,** II, 320, 563 (1905). Reissert and Scherk, *Ber.*, **31,** 393 (1898). Schwarz, *Monatsh.*, **24,** 572 (1903). Wahl, *ibid.*, **38,** 525 (1918).

[78] Heller, *Ber.*, **43,** 1907 (1911).

[79] Baeyer and Knop, *Ann.*, **140,** 1 (1866).

[80] Heller, *Ber.*, **37,** 943 (1904).

[81] Martinet, *Compt. rend.*, **166,** 851 (1918). Marschalk, *J. prakt. Chem.*, [2] **88,** 227 (1913). Kalb, *Ber.*, **44,** 1455 (1911).

[82] Hill and Sumpter, Unpublished dissertation of W. C. Sumpter, Yale University, 1930. Sumpter, *J. Am. Chem. Soc.*, **67,** 1140 (1945). Wahl and Faivret, *Compt. rend.*, **181,** 790 (1925); *Ann. chim.*, [10] **5,** 314 (1926).

[83] v. Braun and Hahn, *Ber.*, **56,** 2343 (1923).

[84] Colman, *Ann.*, **248,** 116 (1888). Michaelis, *Ber.*, **30,** 2811 (1897). Sumpter, Unpublished work.

dioxindole has been prepared by the hydrogenation of 5-nitroisatin[85] in the presence of a nickel catalyst.[86]

Another method for the preparation of dioxindole and its derivatives is the Martinet procedure.[87] Aniline or a substituted aniline is condensed with oxomalonic ester to give a dioxindole carboxylic ester (II). Although hydrolysis of the ester in the presence of oxygen yields an isatin, exclusion of oxygen during the final step produces the dioxindole. The reaction has been employed by a number of workers.[88]

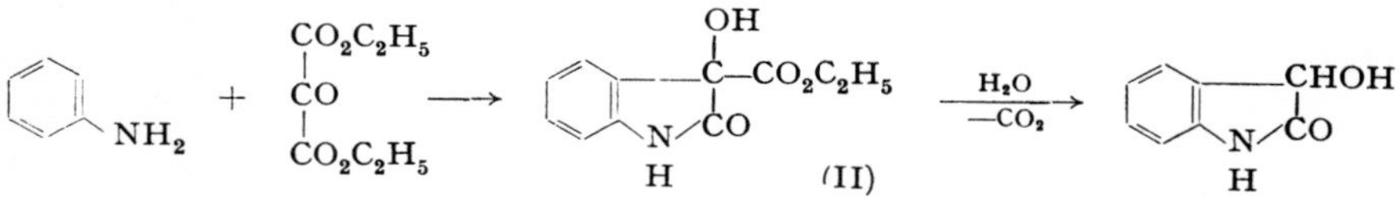

$$C_6H_5NH_2 + CO(CO_2C_2H_5)_2 \longrightarrow \text{(II)} \xrightarrow[-CO_2]{H_2O} \text{dioxindole (CHOH, CO, NH)}$$

(II)

3-Alkyldioxindoles are readily prepared by the action of Grignard reagents on isatin.[89] The Reformatsky reaction may also be employed in the

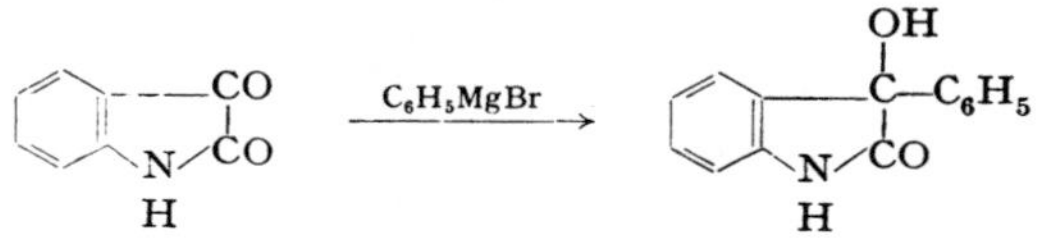

$$\text{isatin} \xrightarrow{C_6H_5MgBr} \text{3-hydroxy-3-}C_6H_5\text{-oxindole}$$

case of *N*-alkylisatins to give the dioxindole acetic esters (III).[90] Hydrolysis of the ester does not yield the acid, but rather the quinolone carboxylic acid (IV).

[85] Sumpter and Jones, *J. Am. Chem. Soc.*, **65**, 1802 (1943).

[86] Rupe and Apotheker, *Helv. Chim. Acta*, **9**, 1049 (1926). Hartmann and Panizzon, *ibid.*, **19**, 1327 (1936).

[87] Bonnefoy and Martinet, *Compt. rend.*, **172**, 220 (1921). Guyot and Martinet, *ibid.*, **156**, 1625 (1913). Martinet, *ibid.*, **166**, 851, 998 (1918); *Ann. chim.*, **11**, 85 (1919). Martinet and Vacher, *Bull. soc. chim.*, **31**, 435 (1922). Compare also Heller and Notzel, *J. prakt. Chem.*, [2] **77**, 145 (1907).

[88] Halberkann, *Ber.*, **54**, 3079 (1921). Hinsberg, *ibid.*, **21**, 117 (1888). Kalb, *ibid.*, **44**, 1455 (1911). Kalb and Berrer, *ibid.*, **57**, 2105 (1924). Langenbeck, Hellrung, and Juttemann, *Ann.*, **499**, 201 (1932); **512**, 276 (1934).

[89] Hill and Sumpter, Unpublished dissertation of W. C. Sumpter, Yale University, 1930. Kohn, *Monatsh.*, **31**, 747 (1910); **32**, 905 (1911). Kohn and Ostersetzer, *ibid.*, **34**, 789, 1714 (1913); **37**, 25 (1916). Myers and Lindwall, *J. Am. Chem. Soc.*, **60**, 2135 (1938). Steinkopf and Hanske, *Ann.*, **541**, 238 (1939). Steinkopf and Hempel, *ibid.*, **495**, 144 (1932). Steinkopf and Wilhelm, *Ber.*, **70**, 2233 (1937). Stolle, Hecht, and Becker, *J. prakt. Chem.*, **135**, 345 (1932). Sumpter, *J. Am. Chem. Soc.*, **64**, 1736 (1942). Sumpter, *Trans. Kentucky Acad. Sci.*, **9**, 61 (1941).

[90] Myers and Lindwall, *J. Am. Chem. Soc.*, **60**, 644 (1938).

$$\text{1-methylisatin} + BrCH_2CO_2C_2H_5 \xrightarrow{Zn} \text{(III)} \longrightarrow \text{(IV)}$$

(III): OH, $C\text{—}CH_2CO_2C_2H_5$, N, CO, CH_3

(IV): CO_2H, $C{=}CH$, N, CO, CH_3

A large number of dioxindole derivatives have been prepared by aldol condensations of isatins with compounds containing active methylene groups, such as phenylacetonitrile to yield V,[91] and nitromethane to form VI.[92]

(V): OH, $C\text{—}CH(CN)C_6H_5$, N, CO, H

(VI): OH, $C\text{—}CH_2NO_2$, N, CO, H

Dioxindole is obtained as crystals, m.p. 167–168°C., from alcohol. It is soluble in ether, but difficulty so in benzene. By virture of the asymmetric β-carbon atom, dioxindole is resolvable into its enantiomorphs.[93]

Acylation of dioxindole is readily effected with benzoyl chloride and acetic anhydride, yielding the 3-acyl derivatives under ordinary conditions. Thus, the benzoyl derivative, prepared, by the direct action of benzoyl chloride on dioxindole has been shown[93,94] to be 3-benzoyldioxindole (VII), rather than the 1-isomer (VIII). An isomeric substance prepared by the action of benzoyl chloride on the sodium salt of dioxindole is most likely the *N*-benzoyl derivative (VIII).[95] 1,3-Dibenzoyldioxindole has been prepared by the Schotten-Baumann technique.[93,94]

(VII): $CHOCOC_6H_5$, N, CO, H

(VIII): CHOH, N, CO, COC_6H_5

[91] Hill and Samachson, Unpublished dissertation of J. Samachson, Yale University, 1930.

[92] Conn and Lindwall, *J. Am. Chem. Soc.*, **58**, 1236 (1936). Crawford and Lindwall, *ibid.*, **62**, 171 (1940). For other preparations see: Braude and Lindwall, *ibid.*, **55**, 325 (1933). DuPuis and Lindwall, *ibid.*, **56**, 471, 2716 (1934). Lindwall and Maclennan, *ibid.*, **54**, 4739 (1932). Zrike and Lindwall, *ibid.*, **57**, 207 (1935). Baumann, *Ber.*, **18**, 890 (1885). Schonberg, Schütz, Arend, and Peter, *ibid.*, **60**, 2344 (1927). Sumpter, *Chem. Revs.*, **34**, 393 (1944).

[93] McKenzie and Stewart, *J. Chem. Soc.*, **1935**, 104.

[94] Heller, *Ber.*, **37**, 943 (1904). Heller and Lauth, *ibid.*, **62**, 343 (1929).

[95] Hill and Sumpter, Unpublished dissertation of W. C. Sumpter, Yale University, 1930. Hellman and Renz, *Chem. Ber.*, **84**, 901 (1951).

The action of acetic anhydride on dioxindole yields a product[96] described[97] as the 3-acetyl derivative. The isomeric 1-acetyldioxindole, which has the same melting point (127°), has been prepared by the reduction of 1-acetylisatin.[98]

Reaction of dioxindole with phenylhydrazine yields the β-phenylhydrazone of isatin (IX),[99] affording a convenient route for comparison of the two series.

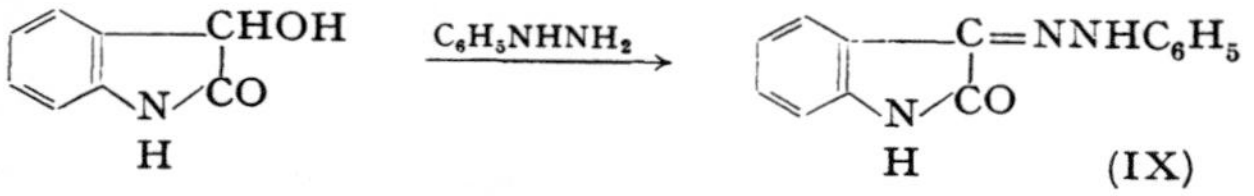

CHOH — N(H)—CO $\xrightarrow{C_6H_5NHNH_2}$ C=$NNHC_6H_5$ — N(H)—CO (IX)

A number of direct substitution reactions of dioxindole have been effected,[100] but most of these early results have not been confirmed. In the case of the bromination of dioxindole, it has been found[101] that the constants reported[100] are in error. As would be expected, the first bromine enters the 5-position, and the dibromination product is 5,7-dibromodioxindole.[101]

1-Hydroxyoxindole. This compound was first obtained, along with oxindole, by the reduction of *o*-nitrophenylacetic acid with zinc and hydrochloric acid.[102] Under certain conditions, catalytic hydrogenation of *o*-nitrophenylacetic acid yields 1-hydroxyoxindole (X), along with oxindole.[103]

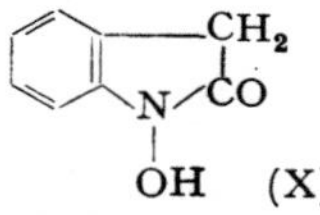

CH_2 — N(OH)—CO (X)

Conversion to 1-acetoxyoxindole and 1-methoxyoxindole may be effected with acetic anhydride and methyl sulfate, respectively.[104] The acetoxy compound may be reduced to oxindole with zinc dust and acetic acid.[105]

Aminoöxindoles

Although reduction of *o*-nitrophenylacetic acid ordinarily gives oxindole,

[96] Suida, *Ber.*, **11**, 584 (1878); **12**, 1326 (1879).

[97] Heller, *Über Isatin, Isatyd, Dioxindol, und Indophenin*, 173 pages, Ahrens Sammlung, Vol. 5, 1931, p. 18. Stolle and Merkle, *J. prakt. Chem.*, **139**, 329 (1934).

[98] Sumpter, Unpublished work.

[99] Heller, *Ber.*, **37**, 943 (1904). Martinet, *Compt. rend.*, **168**, 689 (1919).

[100] Baeyer and Knop, *Ann.*, **140**, 1 (1866).

[101] Sumpter, *J. Am. Chem. Soc.*, **67**, 1140 (1945).

[102] Reissert, *Ber.*, **30**, 1043 (1897); **41**, 3921 (1908).

[103] DiCarlo, *J. Am. Chem. Soc.*, **66**, 1420 (1944).

[104] Heller, *Ber.*, **39**, 2345 (1906). Reissert, *ibid.*, **41**, 3921 (1908).

[105] Heller, *ibid.*, **43**, 1907 (1911).

under certain conditions, *o*-aminophenylacetic acid (XI) can be obtained. When the latter compound is diazotized, followed by reduction to the hydrazine with stannous chloride, and the product quickly distilled, 1-amino-oxindole (XII) is produced.[106]

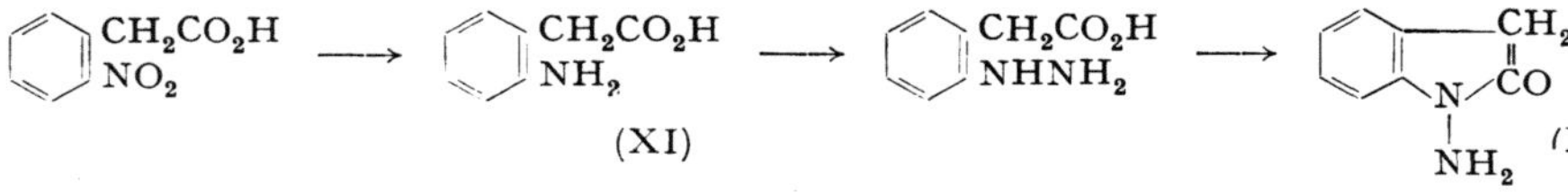

$$C_6H_4(CH_2CO_2H)(NO_2) \longrightarrow C_6H_4(CH_2CO_2H)(NH_2)\ \text{(XI)} \longrightarrow C_6H_4(CH_2CO_2H)(NHNH_2) \longrightarrow \text{1-aminooxindole (N–}NH_2\text{)}\ \text{(XII)}$$

A number of 3-aminoöxindoles have been obtained[107] by the original method of reduction of isatin-β-oxime,[108] produced either by the action of nitrous acid on oxindole or hydroxylamine on isatin.

$$\text{isatin-}\beta\text{-oxime (C=NOH, CO, NH)} \longrightarrow \text{3-aminooxindole (CHNH}_2\text{, CO, NH)}$$

6-Aminoöxindole has been prepared by several workers[109] from 2,4-dinitrophenylacetic acid, or its ester, by reduction.

$$O_2N\text{–}C_6H_3(CH_2CO_2H)(NO_2) \longrightarrow H_2N\text{–oxindole (}CH_2\text{, CO, NH)}$$

3,3-Diaryloxindoles

A variety of 3,3-diaryl derivatives of oxindole (XIII and XIV) have been obtained by the condensation of isatin with activated aromatic nuclei such as those of phenols and anilines.[110] Benzene condenses under the agency

[106] Neber, *Ber.*, **55,** 826 (1922). Neber and Keppler, *ibid.*, **57,** 778 (1924).

[107] Langenbeck, Hutschenreuter, and Juttemann, *Ann.*, **485,** 53 (1931). Langenbeck, Hellrung, and Juttemann, *ibid.*, **499,** 201 (1932). Langenbeck and Weissenborn, *Ber.*, **72,** 724 (1939). DiCarlo and Lindwall, *J. Am. Chem. Soc.*, **67,** 199 (1945).

[108] Baeyer, *Ber.*, **11,** 582, 1228 (1878). Baeyer and Knop, *Ann.*, **140,** 1 (1866).

[109] Gabriel and Mayer, *Ber.*, **14,** 823 (1881). Parks and Aldis, *J. Chem. Soc.*, **1938,** 1841. Ruggli and Grand, *Helv. Chim. Acta,* **20,** 373 (1937). Kishi, *J. Pharm. Soc. Japan,* **546,** 667 (1927); *Chem. Abstr.*, **22,** 421 (1928).

[110] Baeyer and Lazarus, *Ber.*, **12,** 1310 (1879); **18,** 2637 (1885). Liebermann and Danaila, *ibid.*, **40,** 3588 (1907). Inagaki, *J. Pharm. Soc. Japan,* **53,** 686, 698 (1933); **58,** 946, 961, 976 (1938); **59,** 1 (1939). Myers and Lindwall, *J. Am. Chem. Soc.*, **60,** 2153 (1938). Sumpter, *ibid.*, **54,** 3736 (1932). Sumpter, Miller, and Hendrick, *ibid.*, **67,** 1656 (1945). Petyunin *et al.*, *J. Gen. Chem. (U.S.S.R.),* **21,** 1853, 1859, 2016, 2019 (1951); *Chem. Abstr.*, **46,** 6638, 6639 (1952). *J. Gen. Chem. (U.S.S.R.),* **21,** 1699, 1703, 2193 (1951); *Chem. Abstr.*, **46,** 8070, 8071 (1952). *J. Gen. Chem. (U.S.S.R.),* **22,** 190, 296 (1952); *Chem. Abstr.*, **46,** 11161, 11162 (1952).

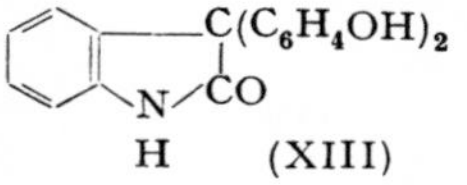

$C(C_6H_4OH)_2$ — N—CO, H (XIII)

$C(C_6H_4NH_2)_2$ — N—CO, H (XIV)

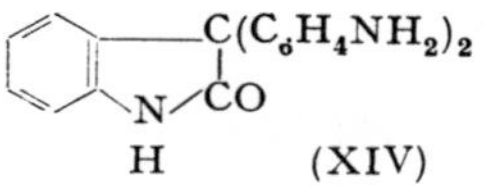

of aluminum chloride with isatin[111] or 3,3-dibromoöxindole.[112] 3,3-Diphenyl-1-methyloxindole (XV) is also produced, with other products, by the action of excess phenylmagnesium bromide on *N*-methylisatin,[113] and from *N*-methylbenzoylformanilide (XVI) with the same Grignard reagent.[114]

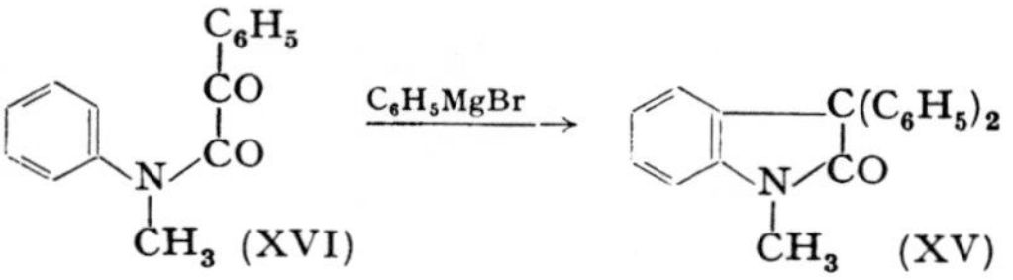

C_6H_5—CO—CO—N—CH_3 (XVI) $\xrightarrow{C_6H_5MgBr}$ $C(C_6H_5)_2$—CO—N—CH_3 (XV)

[111] Sumpter, Unpublished work. Wegmann and Dahn, *Helv. Chim. Acta*, **29**, 415 (1946).

[112] Sumpter, Miller, and Hendrick, *J. Am. Chem. Soc.*, **67**, 1656 (1945).

[113] Kohn and Ostersetzer, *Monatsh.*, **34**, 1714 (1913). Myers and Lindwall, *J. Am. Chem. Soc.*, **60**, 2153 (1938). Witkop and Ek, *ibid.*, **73**, 5664 (1951).

[114] Reeves and Lindwall, *J. Am. Chem. Soc.*, **64**, 1086 (1942).

CHAPTER V

Isatogens

The first members of this group of compounds were prepared by Baeyer[1] during the course of his classic researches on indigo. Ethyl isatogenate (I) was obtained by the rearrangement of ethyl *o*-nitrophenylpropiolate under the influence of cold concentrated sulfuric acid. Diisatogen was prepared

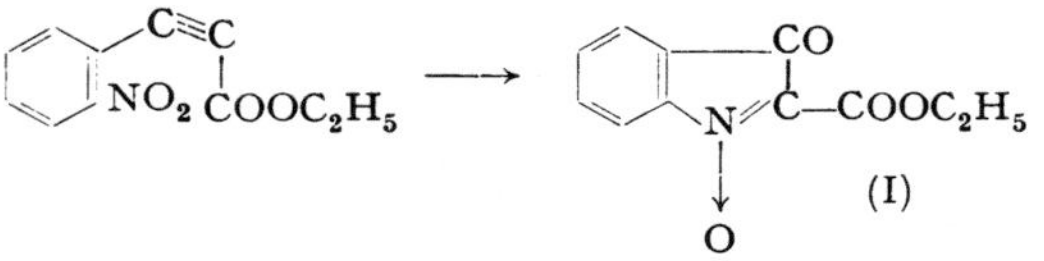

(I)

similarly from o,o′-dinitrodiphenylbutadiyne through the agency of cold sulfuric acid. The isatogens were originally regarded as having the cyclic

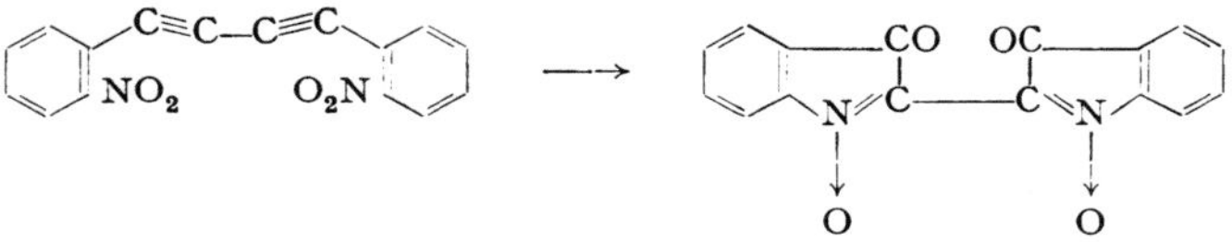

structure (II) but are now generally considered to have the nitrone struc-

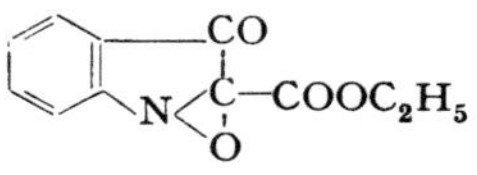

(II) 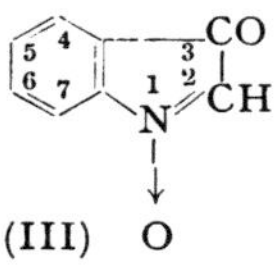 (III)

ture (I). The parent compound of this series (III) is unknown. The isatogens are isomeric with, and closely related to, the isatins. The system of numbering currently used is shown in III and is that of Ruggli.[2] A different system of numbering was employed by Pfeiffer[3,4] in earlier work.

[1] Baeyer, *Ber.*, **13**, 2258 (1880); **14**, 1741 (1881); **15**, 50 (1882).

[2] Ruggli, *ibid.*, **52**, 1 (1919); *Helv. Chim. Acta*, **4**, 626 (1921); **6**, 594 (1923).

[3] Pfeiffer, *Ber.*, **44**, 1209 (1911); **45**, 1825 (1912); **46**, 3661 (1913).

[4] Pfeiffer, Braude, Fritsch, Halberstadt, Kirchhoff, Kleber, and Wittkopp, *Ann.*, **411**, 72 (1916).

While formula I is the one usually written for the isatogens it should be recognized that these compounds are probably in reality resonance hybrids and that several other structures also contribute.

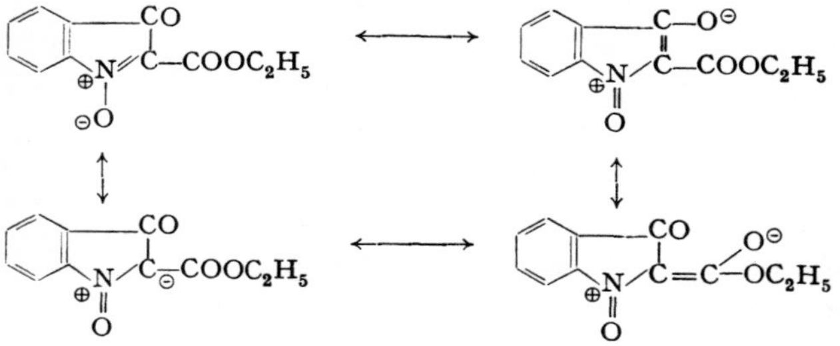

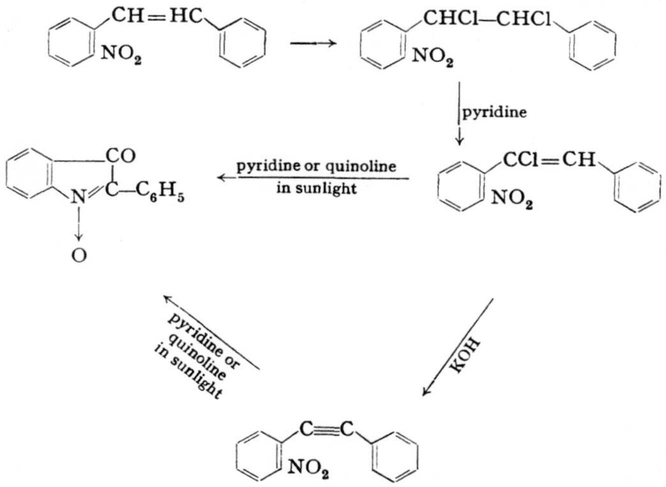

Pfeiffer[4] found that the isatogens could be prepared quite readily from *o*-nitrostilbene derivatives and from *o*-nitrotolanes by means of a rearrangement which takes place in pyridine solution in the presence of sunlight. The synthesis of 2-phenylisatogen by this method is illustrated in the accompanying equations. The addition of chlorine to *o*-nitrostilbene gives the dichloride which on warming with pyridine loses one molecule of hydrogen chloride to give the chlorostilbene. The solution of the chlorostilbene in pyridine on exposure to sunlight gradually becomes orange in color, then red, and finally deposits orange-colored crystals of 2-phenylisatogen in good yield. Alternately, treatment of the chloronitrostilbene with alcoholic potassium hydroxide gives *o*-nitrotolane which with pyridine (or quinoline) in sunlight likewise gives the phenylisatogen. A number of isatogens have been prepared by Pfeiffer and by Ruggli in this manner. For example, 2-phenyl-6-nitroisatogen was prepared by the above procedure from 2,4-dinitrotolane. The role of pyridine (or quinoline) as a catalyst seems to be a specific one since other solvents are without effect.

Another procedure for converting *o*-nitrotolane into 2-phenylisatogen has been developed by Ruggli.[5] In this procedure a solution of *o*-nitrotolane in chloroform is treated with nitrosobenzene. After standing nineteen days a 57% yield of 2-phenylisatogen had separated. The action of the nitrosobenzene seems to be catalytic.

That the isatogens are compounds of the indole series is shown by the fact that on reduction with zinc and acetic acid they yield indoxyl derivatives:

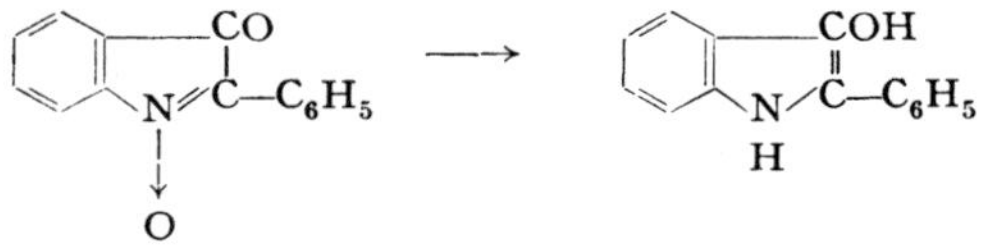

The isatogen molecule contains the two reactive groups: $>C=O$, $>N \rightarrow O$

This is shown by the action of hydroxylamine hydrochloride in boiling alcohol. Two isomeric oximes are obtained; one is yellow, m.p. 172°, while the other is orange and melts at 236°. The orange oxime is identical with the compound obtained by Angelo and Angelico[6] by the action of amyl nitrite on *N*-hydroxy-2-phenylindole and must be the *C*-oxime. Reduction of this oxime yields 3-amino-2-phenylindole. If the yellow isomer was simply

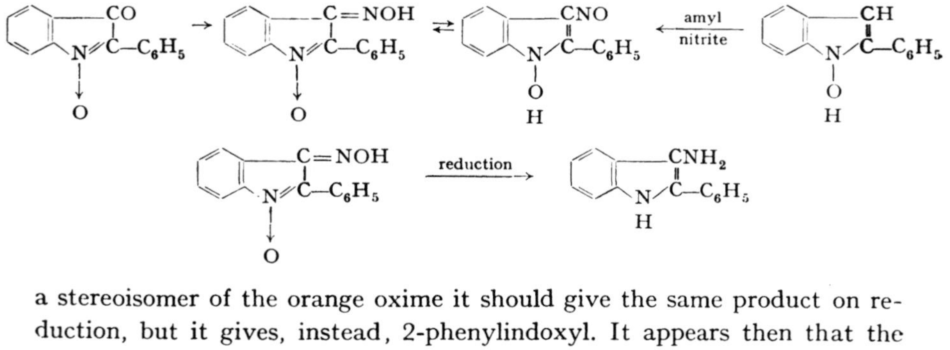

amyl nitrite

reduction

a stereoisomer of the orange oxime it should give the same product on reduction, but it gives, instead, 2-phenylindoxyl. It appears then that the

[5] Ruggli, Caspar, and Hegedus, *Helv. Chim. Acta*, **20**, 250 (1937). See also: Króhnke and Meyer-Delius, *Chem. Ber.*, **84**, 932 (1951); Campbell, Shavel, and Campbell, *J. Am. Chem. Soc.*, **75**, 2400 (1953).

[6] Angeli and Angelico, *Atti accad. Lincei*, [5] **15**, II, 761 (1907); *Chem. Zentr.* **1907**, I, 732.

yellow oxime must be either the *N*-oxime or the nitroso compound. The

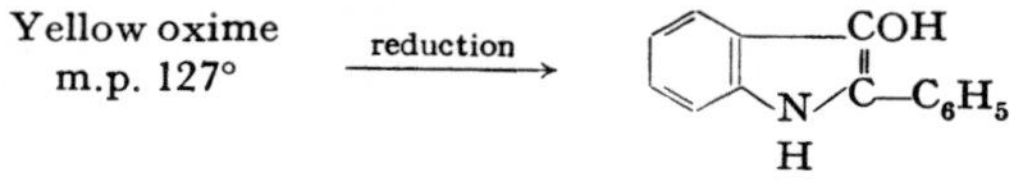

Yellow oxime m.p. 127° $\xrightarrow{\text{reduction}}$ COH, C—C_6H_5, N, H

oxime formula is indicated because the acetyl derivative on reduction gives

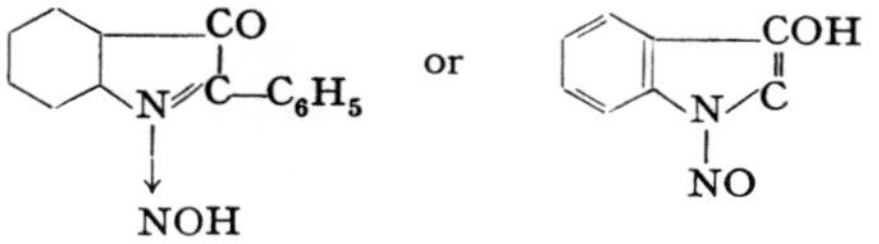

CO, N—C—C_6H_5, ↓ NOH or COH, N—C, NO

2-phenylindoxyl rather than 3-acetyl-2-phenylindoxyl. It thus seems that

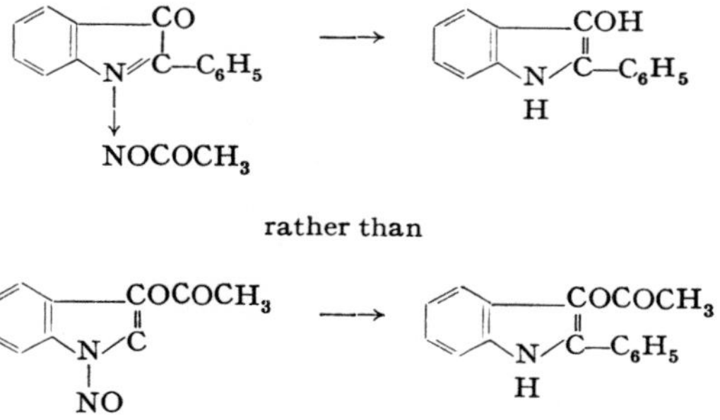

CO, N—C—C_6H_5, ↓ $NOCOCH_3$ $\longrightarrow$ COH, N—C—C_6H_5, H

rather than

$COCOCH_3$, N—C, NO $\longrightarrow$ $COCOCH_3$, N—C—C_6H_5, H

the $\geqslant N \rightarrow O$ group can be converted into an oxime and hence is somewhat like the carbonyl group. The structure written above for the *N*-oxime seems at first sight to be subject to serious question because of the negative charge on the oxime nitrogen atom, formula A being identical with that in which the arrow is used for the coordinate bond. This objection can be overcome by assuming that the predominating resonance forms are probably A and B with minor contributions being made to the hybrid by structures having the negative charge on carbon atom 2 and on the para and two ortho carbon atoms in the phenyl group attached to position 2. It seems probable that the

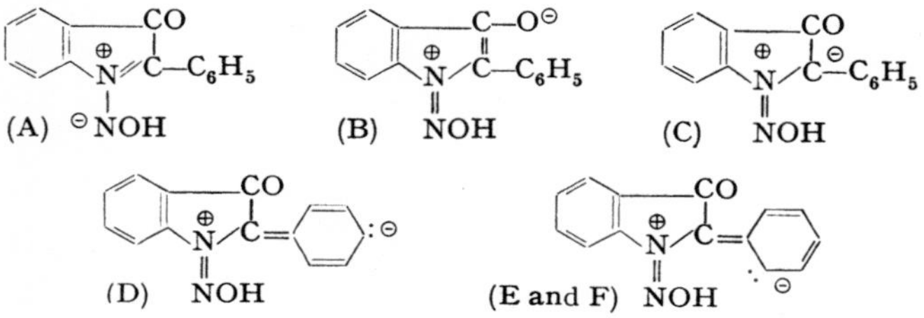

(A) $^{\ominus}$NOH (B) NOH (C) NOH

(D) NOH (E and F) NOH

N-oxime is best represented as a resonance hybrid of A and B with recognition of the fact that structures C, D, E, and F may also contribute.[7]

The formula previously written for the *N*-oxime is completely un-

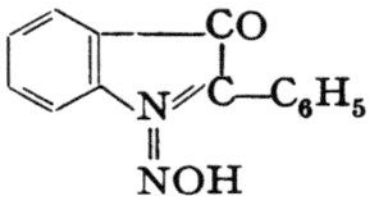

CO
$N \!=\! C—C_6H_5$
NOH

tenable, containing, as it does, pentacovalent nitrogen. It seems strange that while previous writers[8] have realized that the similar formula for the isatogen

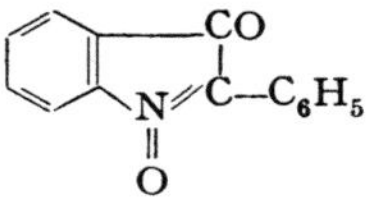

CO
$N \!=\! C—C_6H_5$
O

is unsatisfactory they have continued to write the equally unsatisfactory formula for the *N*-oxime.

The isatogens exhibit the characteristics of quinones. They are deeply colored, they liberate iodine from hydriodic acid, and they form quinhydrones with indoxyls (the indoxyls corresponding to the hydroquinones). Thus 6-cyano-2-phenylisatogen (orange) and 6-cyano-2-phenylindoxyl (colorless) give a black crystalline addition compound with is separated easily into its components.

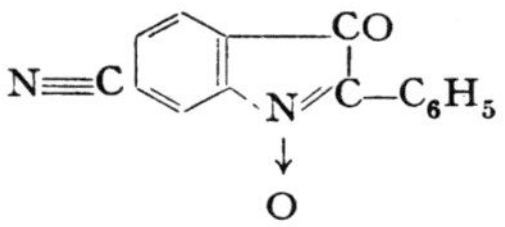

N≡C — CO — $N \!=\! C—C_6H_5$ — ↓ O

Orange

N≡C — COH — $N \!=\! C—C_6H_5$ — H

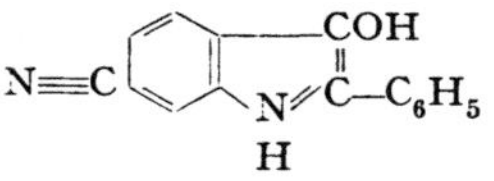

Colorless

When 2-phenylisatogen is treated with alcoholic hydrochloric (or sulfuric) acid a yellow isomer,[9] 2-phenylisoisatogen, m.p. 93°, is obtained. The structure of this isomer has not been definitely established but the compound is regarded by Ruggli as having the oxide structure which was originally assigned to the isatogens by Baeyer. 2-Phenylisoisatogen reacts,

[7] The authors are indebted to Professor Lawrence H. Amundsen of the University of Connecticut for suggesting that the several resonating structures pictured above contribute to the stability of the *N*-oxime which appears to be an improbable structure if written as formula A alone.

[8] Smith, *Chem. Revs.*, **23,** 248 (1938). Morton, *The Chemistry of Heterocyclic Compounds*, McGraw-Hill, New York, 1946, p. 140.

[9] Ruggli, *Ber.*, **52,** 1 (1919). Ruggli and Bolliger, *Helv. Chim. Acta*, **4,** 626 (1921). Ruggli, Caspar and Hegedus, *ibid.*, **22,** 140 (1939).

with hydroxylamine[9] to give two products, 2-phenylketoindolenine oxime m.p. 260°, and 2-phenylisatogen-*N*-oxime formed through the opening of the -N-O-C- ring. The isoisatogen does not liberate iodine from potassium

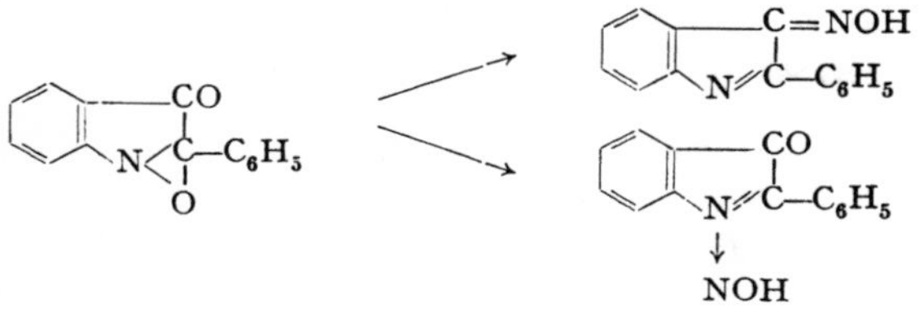

iodide and exhibits no quinoid properties. Reduction of both the isatogen and the isoisatogen gives the same indoxyl. When heated above its melting point, or in glacial acetic acid solution, or with phenylisocyanate, the iso-isatogen reverts to the normal isatogen. The normal isatogen is reduced by phenylhydrazine[10] to the corresponding indoxyl while the isoisatogen yields a compound corresponding in composition to the formula:

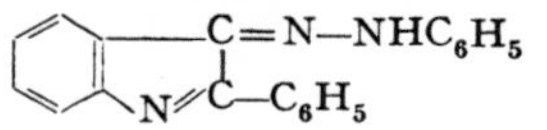

The isatogens add a variety of reagents, the addition seeming to be over a 1,3-system. Thus, 6-nitro-2-phenylisatogen on refluxing with acetyl chloride gives a yellow addition compound,[11] which decomposes into its components when heated strongly but is fairly stable at room temperature

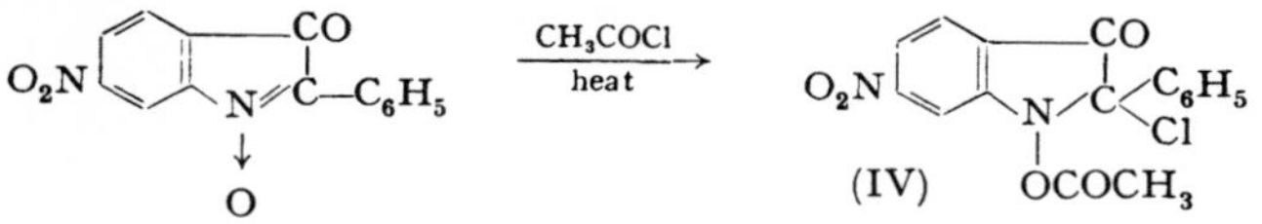

if kept dry. The compound reacts with aniline and with phenylhydrazine to give the original isatogen. With less active bases such as *p*-nitroaniline the chlorine is replaced yielding the secondary amine (V) while with methanol the ether (VI) is obtained. If the solution of the chloro compound in acetyl

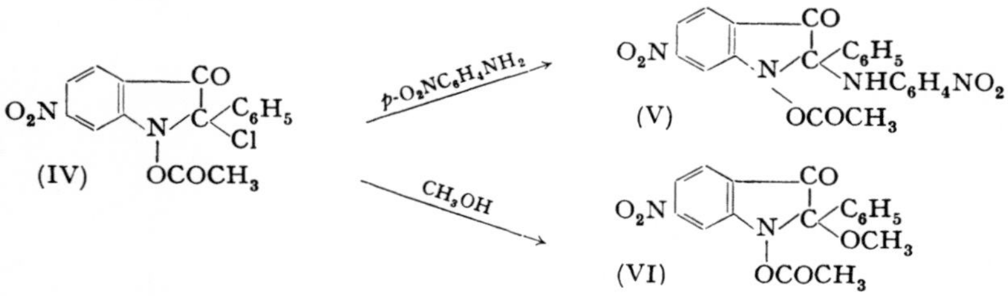

[10] Ruggli and Disler, *Helv. Chim. Acta*, **10**, 938 (1927).

[11] Ruggli, Bolliger, and Leonhardt, *ibid.*, **6**, 594 (1923).

chloride is poured into water the compound is hydrolysed and transformed into the isatogen by the heat evolved. However, if the solution is poured onto ice the yellow precipitate of VII can be isolated. The compound (VII) dissociates into acetic acid and the isatogen at 125°. The action of cold

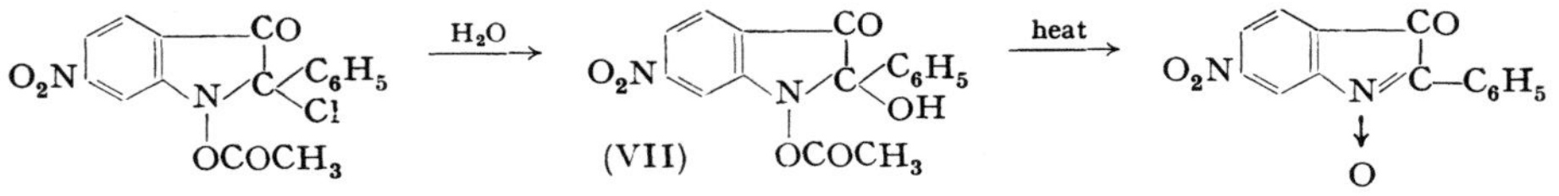

O_2N CO N C C_6H_5 Cl $OCOCH_3$ —H_2O→ O_2N CO N C C_6H_5 OH (VII) $OCOCH_3$ —heat→ O_2N CO N=C—C_6H_5 ↓ O

alcoholic hydrochloric acid on the isatogen gives an addition product (VIII) while hot alcoholic hydrochloric acid gives the isoisatogen.

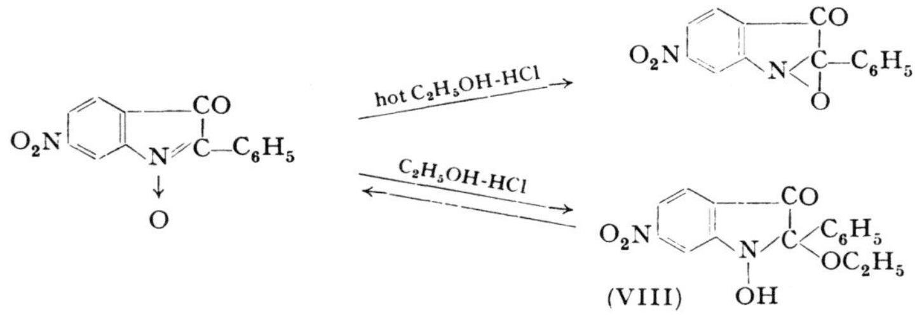

O_2N CO N=C—C_6H_5 ↓ O —hot C_2H_5OH-HCl→ O_2N CO N C—C_6H_5 O

—C_2H_5OH-HCl⇄ O_2N CO N C C_6H_5 OC_2H_5 (VIII) OH

With acetic anhydride in the presence of sulfuric acid the isatogen yields a triacetate[12] (IX) which is partially hydrolyzed by hydrogen chloride to the compound (X) which on treatment with sulfuric acid or zinc chloride gives

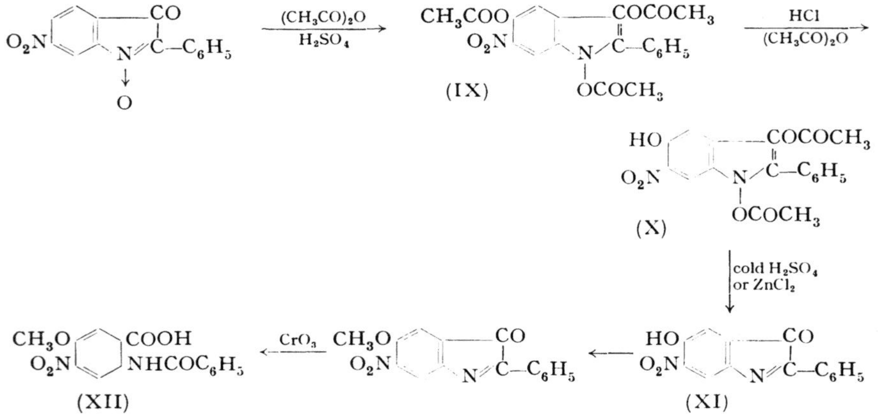

O_2N CO N=C—C_6H_5 ↓ O —$(CH_3CO)_2O$, H_2SO_4→ CH_3COO O_2N $COCOCH_3$ N C—C_6H_5 $OCOCH_3$ (IX) —HCl, $(CH_3CO)_2O$→

HO O_2N $COCOCH_3$ N C—C_6H_5 $OCOCH_3$ (X)

cold H_2SO_4 or $ZnCl_2$ ↓

CH_3O O_2N COOH $NHCOC_6H_5$ (XII) ←CrO_3— CH_3O O_2N CO N=C—C_6H_5 ← HO O_2N CO N=C—C_6H_5 (XI)

the ketoindolenine (XI) which can be converted into the acid XII. Both XI and XII have been synthesized by conventional methods. In the preparation

[12] Ruggli and Leonhardt, *ibid.*, **7,** 689 (1924). Ruggli, Hegedus, and Caspar, *ibid.*, **22,** 411 (1939).

of the triacetate (IX) *N*-phenylisatin is also obtained as a by-product through a Beckmann type rearrangement.[13]

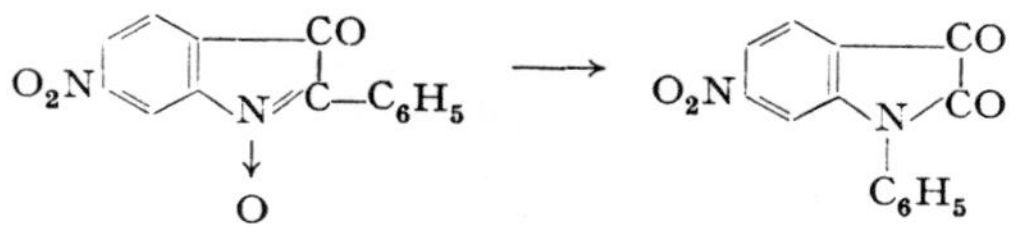

O_2N–(isatogen, N→O, C–C_6H_5) ⟶ O_2N–(N-C_6H_5 isatin)

The first product in the reduction of the isatogens is an indoxyl. Using the methyl ester of isatogenic acid Ruggli, Schmid, and Zimmermann[14] were able to show that this reduction takes place stepwise with a ketoindolenine and a "quinhydrone" being intermediates.

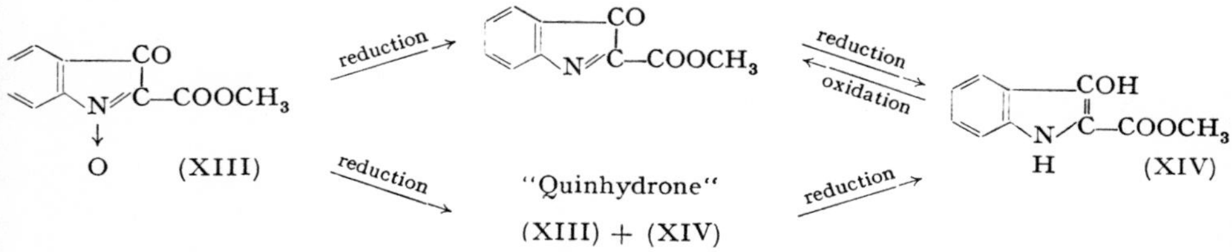

(XIII) —reduction→ ketoindolenine (CO, N=C–$COOCH_3$) ⇌ (reduction / oxidation) (XIV) (COH, N(H)–C–$COOCH_3$)

(XIII) —reduction→ "Quinhydrone" (XIII) + (XIV) —reduction→ (XIV)

Isatogens derived from anthraquinone have been prepared[15] while pyridylisatogens are also known.[16] The isatogen-indoxyl derivative (XV)[17] has been prepared while attempts to prepare the diisatogen (XVI) were unsuccessful,[18] ring closure being effected on only one side of the benzene

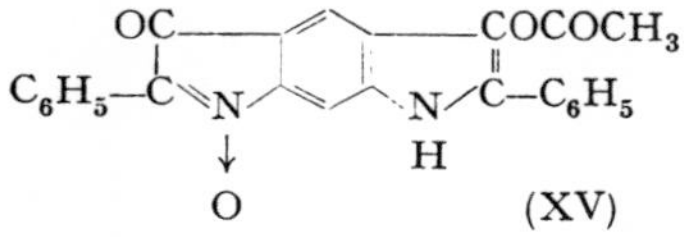

(XV) 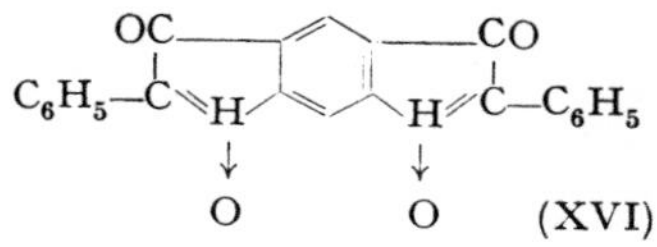 (XVI)

ring. The diisatogen (XVII) having two isatogen rings with a *p*-phenylene group common to both positions 2 has been prepared.[19]

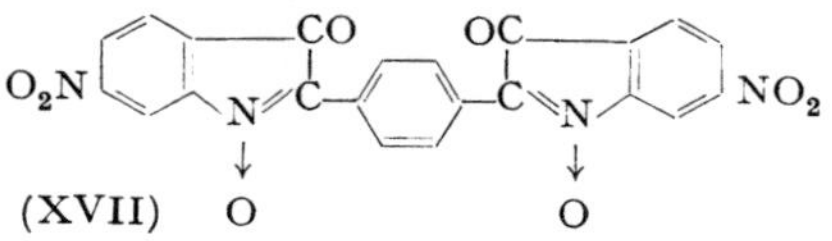

(XVII)

[13] Ruggli and Leonhardt, *ibid.*, **7**, 898 (1924).

[14] Ruggli, Schmid, and Zimmermann, *ibid.*, **17**, 1328 (1934). Ruggli, Zaeslin, and Grand, *ibid.*, **21**, 33 (1938).

[15] Ruggli and Disler, *ibid.*, **10**, 938 (1927).

[16] Ruggli and Cuenin, *ibid.*, **27**, 649 (1944).

[17] Ruggli and Zimmermann, *ibid.*, **16**, 69 (1933).

[18] Ruggli, Zimmermann, and Thouvay, *ibid.*, **14**, 1256 (1931).

[19] Ruggli and Wolff, *ibid.*, **19**, 5 (1936).

2-Phenylisatogen reacts with phenylmagnesium bromide, the product being formulated by Ruggli[20] as shown below.

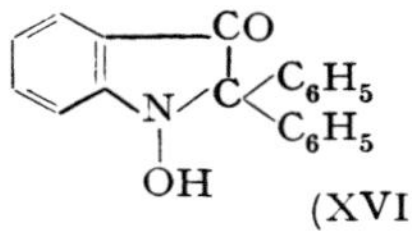

$$\text{(benzo ring)}-CO-C(C_6H_5)(C_6H_5)-N(OH)-\text{(benzo ring)}$$

(XVIII)

[20] Ruggli, Hegedus, and Caspar, *ibid.*, 22, 411 (1939).

CHAPTER VI

Indoxyl

Introduction

Indoxyl, 3-hydroxyindole (I) or 3-keto-2,3-dihydroindole (II), is an important intermediate in the synthesis of indigo. The compound is also of biochemical importance because of its occurrence in the form of its glucoside in certain plants and its excretion by animals as the sulfuric acid ester. The compound was first obtained[1] through hydrolysis of the naturally occurring

COH / CH / N H (I) CO / CH_2 / N H (II)

indoxyl sulfuric acid. The compound is also obtained as an intermediate in the preparation of indigo from the indoxyl glucoside obtained from the indigo-bearing plants as well as in the synthesis of indigo.

Synthesis of Indoxyl

Indoxyl has been synthesized through the fusion of *N*-phenylglycine[2] with alkali. The use of *N*-phenylglycine-*o*-carboxylic acid in place of *N*-

NH_2 $\xrightarrow{ClCH_2COOH}$ COOH / CH_2 / N H $\xrightarrow{-H_2O}$ CO / CH_2 / N H

phenylglycine yields in similar fashion indoxylic acid or at higher temperatures a mixture of indoxylic acid and indoxyl.[3] Many modifications of these

[1] Baumann and Tiemann, *Ber.*, **12,** 1099, 1192 (1879); **13,** 415 (1880).

[2] Heumann, *ibid.*, **23,** 3043 (1890). Feuchter, *Chem.-Ztg.*, **38,** 273 (1914); *Chem. Abstr.*, **8,** 3780 (1914). German Patent 54,626; *Friedl.*, **2,** 100 (1892).

[3] German patent 56,273; *Friedl.*, **3,** 281 (1896). German patent 85,071; *Friedl.*, **4,** 1032 (1899). German patent 152,548; *Friedl.*, **7,** 267 (1905). German patent 232,986, *Friedl.*, **10,** 345 (1913).

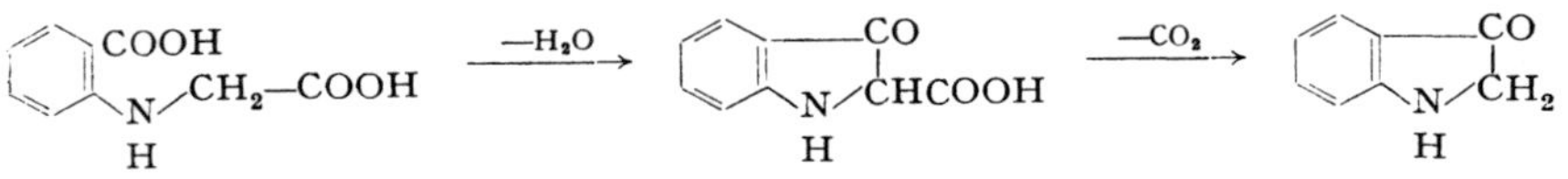

COOH N CH$_2$—COOH H $\xrightarrow{-H_2O}$ CO N CHCOOH H $\xrightarrow{-CO_2}$ CO N CH$_2$ H

two procedures have been proposed,[4] the most important of which is the use of sodium amide[5] in place of the alkali hydroxides in the fusion of *N*-phenylglycine or *N*-phenylglycine-*o*-carboxylic acid.[6] With this condensing agent ring closure takes place at lower temperatures and with good yields. Fusion of phenylglycylphenylglycine with sodium amide and alkali hydroxides yields indoxyl[7] which is also obtained by heating *N*-phenyl-*N*-acetylglycine with $AlCl_3$ in absence of air.[8] The compound is further formed from 3-methylanthranil[9] through rearrangement on heating. Indoxyl can be prepared through decarboxylation of indoxylic acid even on heating

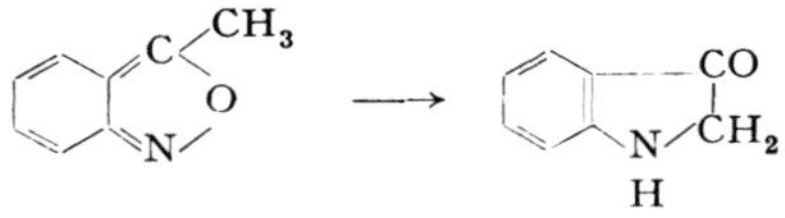

C CH$_3$ O N $\longrightarrow$ CO N CH$_2$ H

with water,[10] as well as by boiling *o*-nitrophenylacetylene with ammonium acid sulfite and treating the resulting compound with zinc dust and ammonia.[11] The compound is also obtained from β-anilinoethanol, β-ethylanilinoethanol, bis-(β-hydroxyethyl)-aniline or from 2-(β-hydroxyethylamino)-benzoic acid on fusion with alkali.[12] Similarly heating 2-cyanoanilinoacetic acid with aqueous alkali yields indoxyl.[13] Indoxyl has also been

[4] Vorländer, *Ber.*, **35**, 1683 (1902). Vorländer and Pfeiffer, *ibid.*, **52**, 325 (1919). Also patents too numerous to list here.

[5] German patent 137,955; *Chem. Zentr.*, **1903**, I, 110. German patent 141,749; *Chem. Zentr.*, **1903**, I, 1323. Feuchter, *Chem. Ztg.*, **38**, 273 (1914); *Chem. Abstr.*, **8**, 3780 (1914).

[6] German patent 79,409; *Friedl.*, **4**, 1031 (1899). German patent 137,208; *Chem. Zentr.*, **1903**, I, 110. German patent 139,393; *Chem. Zentr.*, **1903**, I, 745. German patent 145,601; *Chem. Zentr.*, **1903**, II, 1225. Fränkel and Spiro, *Ber.*, **28**, 1685 (1895).

[7] German patent 141,749; *Friedl.*, **6**, 1316 (1909).

[8] German patent 188,436; *Friedl.*, **9**, 514 (1911).

[9] Bamberger and Elger, *Ber.*, **36**, 1624 (1903). Camps, *ibid.*, **32**, 3232 (1899).

[10] Baeyer, *ibid.*, **14**, 1744 (1881). German patent 17,656; *Friedl.*, **1**, 135 (1888). Vorländer and Drescher, *Ber.*, **34**, 1856 (1901); *ibid.*, **35**, 1702 (1902). German patent 131,400; *Friedl.*, **6**, 551 (1904).

[11] Baeyer, *Ber.*, **15**, 56 (1882).

[12] German patent 171,172; *Friedl.*, **8**, 412 (1908).

[13] German patent 206,903; *Friedl.*, **9**, 516 (1911).

prepared from 1-phenylhydantoin through alkali fusion.[14] Indoxyl and derivatives of indoxyl have been prepared by heating ethylenedianiline, $C_6H_5NHCH_2CH_2NHC_6H_5$, and its derivatives with alkali.[15] The compound can be detected as an intermediate in the oxidation of indole to indigo by hydrogen peroxide.[16] Indoxyl results (reversibly) in poor yields from indigo through heating with molten potassium hydroxide or with strong solutions of potassium hydroxide.[17]

Physical Properties

Indoxyl is a yellow crystalline substance, melting at 85°, possessing a strong fecal odor and soluble in hot water with a yellow-green fluorescence which disappears on addition of either potassium hydroxide or hydrochloric acid.[18]

Reactions

The phenolic character of indoxyl is shown by its solubility in aqueous alkali as well as by the development of a dark red color when a solution of ferric chloride is added to an alcoholic solution of indoxyl. In alkaline solution indoxyl is converted through strong oxidation into indigo.[19] In this oxidation some indirubin is also formed through oxidation of part of the

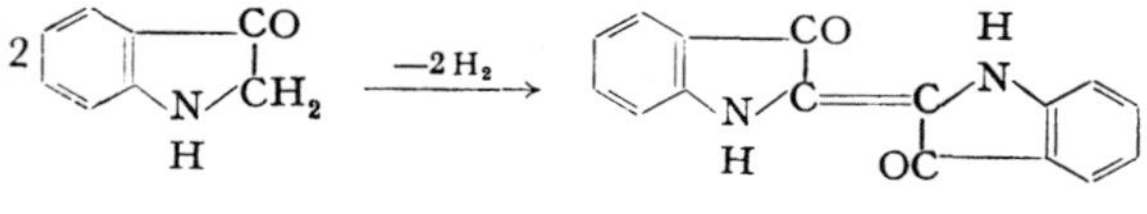

$\xrightarrow{-2\,H_2}$

Indigo

indoxyl to isatin which then condenses with unchanged indoxyl.[20] Isatin can become the principal product in the oxidation of indoxyl if manganese

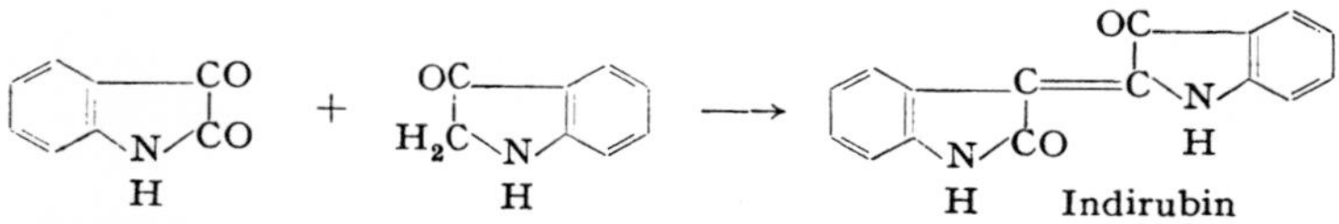

Indirubin

[14] German patent 132,477; *Friedl.*, **6**, 568 (1904).

[15] German patent 220,172; *Chem. Abstr.*, **4**, 2187 (1910).

[16] Porcher, *Bull. soc. chim.*, [4] **5**, 527 (1909).

[17] Heumann and Bachofen, *Ber.*, **26**, 225 (1893). Friedländer and Schwenck, *ibid.*, **43**, 1972 (1910).

[18] Vorländer and Drescher, *ibid.*, **34**, 1856 (1901); **35**, 1701 (1902).

[19] Baeyer, *ibid.*, **14**, 1744 (1881).

[20] Thomas, Bloxam, and Perkin, *J. Chem. Soc.*, **95**, 842 (1909). Perkin, *ibid.*, **95**, 847 (1909).

dioxide and a solution of sodium hydroxide are employed as the oxidizing agent.[21] On the other hand, indoxyl can be reduced to indole by sodium amalgam or by zinc dust and alkali.[22]

The ketonic character of indoxyl (formula II) is shown in those reactions involving the condensation of the active methylene group with the carbonyl groups of aromatic aldehydes or of ketonic acids which condense easily to yield indogenides.[23] Similarly isatin and indoxyl condense easily

$$C_6H_4\langle{}^{CO}_{NH}\rangle CH_2 + OCH{-}C_6H_4NO_2 \longrightarrow C_6H_4\langle{}^{CO}_{NH}\rangle C{=}CHC_6H_4NO_2$$

to give indirubin.[24] Indoxyl condenses with isatin chloride to give indigo[25] and with isatin-α-anil in absence of air to give indirubin anil.[26] Indoxyl also condenses with aromatic nitroso compounds to give characteristic derivatives.[27] These condensation reactions, especially the indirubin reaction, serve

$$C_6H_4\langle{}^{CO}_{NH}\rangle CH_2 + ON{-}C_6H_4{-}N(CH_3)_2 \longrightarrow C_6H_4\langle{}^{CO}_{NH}\rangle C{=}N{-}C_6H_4{-}N(CH_3)_2$$

for the identification and determination of indoxyl.[28] Indoxyl reacts with benzenediazonium chloride to give 2-phenylazoindoxyl,[29] a compound for

$$C_6H_4\langle{}^{CO}_{NH}\rangle CH_2 + C_6H_5N_2Cl \longrightarrow C_6H_4\langle{}^{CO}_{NH}\rangle CHN{=}NC_6H_5$$

[21] German patent 107,719; *Chem. Zentr.*, **1900,** I, 1112.

[22] Vorländer and Apelt, *Ber.*, **37,** 1134 (1904). Baeyer, *ibid.*, **14,** 1741 (1881).

[23] Baeyer, *ibid.*, **16,** 2196 (1883). Perkin and Thomas, *J. Chem. Soc.*, **95,** 795 (1909).

[24] Baeyer, *Ber.*, **14,** 1765 (1881); **16,** 2200 (1883). German patent 17,656; *Friedl.*, **1,** 136 (1888).

[25] Wahl and Bagard, *Bull. soc. chim.*, [4] **7,** 1100 (1910).

[26] Pummerer and Göttler, *Ber.*, **43,** 1385 (1910).

[27] Pummerer and Göttler, *ibid.*, **43,** 4269 (1909).

[28] Orchardson, Wood, and Bloxam, *Chem. Zentr.*, **1907,** I, 847. Porcher, *Bull. soc. chim.*, [4] **5,** 528 (1909). Thomas, Bloxam, and Porcher, *J. Chem. Soc.*, **95,** 842 (1909).

[29] Baeyer, *Ber.*, **16,** 2190 (1883). Bamberger and Elger, *ibid.*, **36,** 1625 (1903). Heller, *ibid.*, **40,** 1298 (1907).

[30] Martinet and Dornier, *Compt. rend.*, **170,** 592 (1920). *Rev. gén. mat. color.*, **28,** 65 (1923); *Chem. Abstr.*, **17,** 3031 (1923).

which several other tautomeric structures can also be written.[30] With nitrous

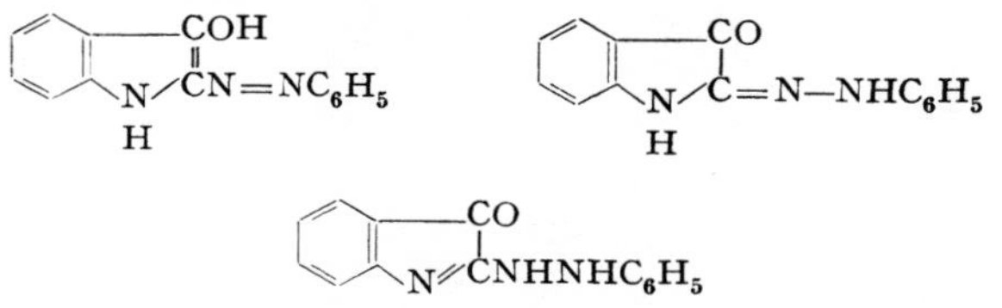

COH, $CN{=}NC_6H_5$, N, H CO, $C{=}N{-}NHC_6H_5$, N, H

CO, $CNHNHC_6H_5$, N

acid indoxyl gives the *N*-nitroso derivative while on heating under pressure

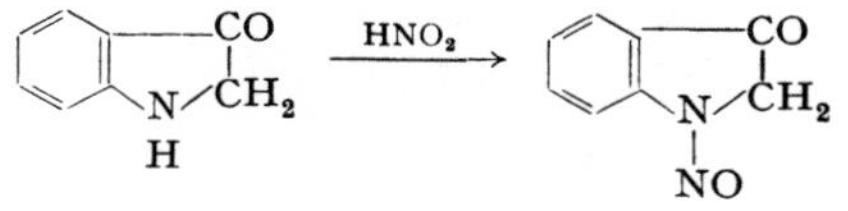

CO, CH_2, N, H $\xrightarrow{HNO_2}$ CO, CH_2, N, NO

with 10% sodium hydroxide and subsequent oxidation by air indoxyl yields indoxyl red.[31]

Methylation of indoxyl with methyl sulfate gives 3-methoxyindole while methylation with methyl iodide yields 2,2-dimethylindoxyl.[31a]

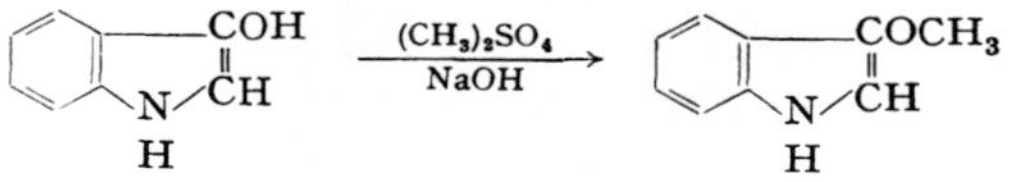

COH, CH, N, H $\xrightarrow[NaOH]{(CH_3)_2SO_4}$ $COCH_3$, CH, N, H

3-Methoxyindole, m.p. 69°

CO, CH_2, N, H $\xrightarrow[CH_3I]{NaOH}$ CO, C, CH_3, CH_3, N, H

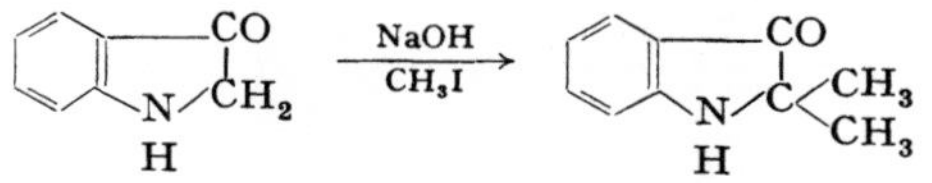

2,2-Dimethylindoxyl, m.p. 88°

Acyl Derivatives of Indoxyl

Indoxyl and acetic anhydride[31a, 32] react in the cold to give the alkali-soluble *N*-acetylindoxyl, m.p. 136°. On heating the diacetyl derivative, m.p. 82°, is formed. Shaking an aqueous solution of indoxyl with acetic anhydride

[31] Schmitz-Dumont, Hamann, and Geller, *Ann.*, **504**, 4 (1933). German patent 255,691; *Chem. Zentr.*, **1913**, I, 481.
[31a] Etienne, *Compt. rend.*, **225**, 124 (1947); *Bull. soc. chim. France*, **1948**, 651.
[32] Vorländer and Drescher, *Ber.*, **34**, 1857 (1901); **52**, 325 (1919). German patent 113,240; *Friedl.*, **5**, 940 (1901). German patent 133,144; *Friedl.*, **6**, 554 (1904). German patent 131,400; *Friedl.*, **6**, 553 (1904). Spencer, *J. Soc. Chem. Ind.*, **50**, 63 (1931).

yields *O*-acetylindoxyl, m.p. 126°. The latter compound reacts with sodium

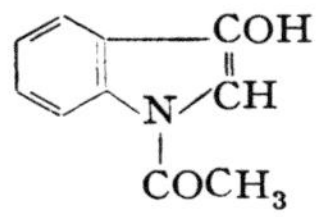

N-Acetylindoxyl 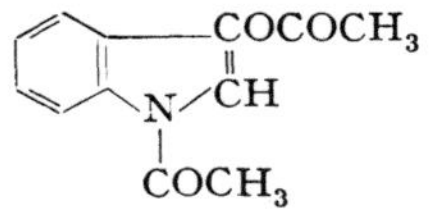 1,3-Diacetylindoxyl 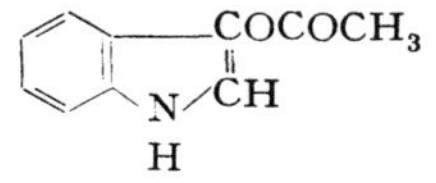 *O*-Acetylindoxyl

nitrite in acetic acid to give a *N*-nitroso derivative, m.p. 83°. *N*-Acetylindoxyl is also obtained by heating indoxylic acid with acetic anhydride. The propionyl and 3-benzoyl derivatives of indoxyl have also been prepared.

Indoxyl Sulfuric Acid

Indoxyl sulfuric acid is found as its potassium salt in the urine of mammals.[33] The action of acids and the subsequent oxidation by air converts the compound into indigo. It was originally supposed that this substance was identical with indican (indoxyl glucoside) from plants but its nature as the potassium salt of indoxyl sulfuric acid was definitely established by Baumann.[34] The salt was prepared[35] through the action of potassium pyrosulfate on an alkaline solution of indoxyl as well as through the action of chlorosulfonic acid on *N*-acetylindoxyl[36] and subsequent treatment with potassium hydroxide. The detection and quantitative determination of indoxyl in urine is based on the hydrolysis of the potassium indoxyl sulfate and subsequent oxidation of indoxyl to indigo or condensation with isatin to indirubin.[37] Indoxyl is found in urine not only as indoxyl sulfuric acid but also as an easily hydrolyzable compound, possibly indoxylglucuronic acid.[38]

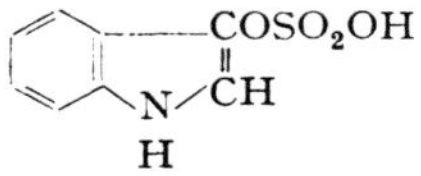

Indican

Indican, indoxyl glucoside, is the substance found in "indigo bearing"

[33] Schunck, *Jahresber.*, **1857**, 564. Hoppe-Seyler, *ibid.*, **1863**, 656.

[34] Baumann, *Z. physiol. Chem.*, **1**, 66 (1877). Baumann and Brieger, *ibid.*, **3**, 254 (1879). Baumann and Tiemann, *Ber.*, **12**, 1098, 1193 (1879); **13**, 408 (1880).

[35] Baeyer, *Ber.*, **14**, 1745 (1881).

[36] Schwenck and Jolles, *Biochem. Z.*, **60**, 467 (1915); *Chem. Abstr.*, **9**, 2883 (1915).

[37] Kozlowski, *Chem. Zentr.*, **1911**, I, 1257. Jolles, *Z. physiol. Chem.*, **87**, 310 (1913). Stanford, *ibid.*, **88**, 47 (1913).

[38] Baumann, *Z. physiol. Chem.*, **1**, 68 (1877). Hoppe-Seyler, *ibid.*, **7**, 425 (1883). Mayer and Neuber, *ibid.*, **29**, 272 (1900).

plants which is finally converted into indigo through hydrolysis to indoxyl and subsequent oxidation to indigo.[39] Indican crystallizes from water with three molecules of water of crystallization as colorless needles of m.p. 57–58° and from mixture of alcohol and benzene as water-free crystals, m.p. 180°; $[\alpha]_D^{15}$ in water —66.17° (*c.* 1.7). Indican is hydrolyzed by dilute acids or by the enzyme "Indimulsin" to indoxyl and glucose. Indican has been synthesized by Robertson.[40] 1-Acetyl-3-tetraacetyl-β-glucosidoxyindole was prepared from 1-acetyl-3-hydroxyindole and tetraacetyl-α-glucosidyl bromide. Anhydrous ammonia in methanol converts this product into 3-β-glucosidoxyindole (indican). 6-Bromoindican was prepared similarly.[41]

Indoxylic Acid and Indoxyl-2-aldehyde

One of the most important derivatives of indoxyl is indoxyl-2-carboxylic acid, indoxylic acid, which owes its importance to the fact that it is an intermediate in the synthesis of indigo.[42] The compound was first prepared from *o*-nitrophenylpropiolic acid.[43] Indoxylic acid has also been

C≡C, NO_2 $COOC_2H_5$ —(cold H_2SO_4)→ CO, N(→O)=C—$COOC_2H_5$ —(reduction)→ COH, NH—C—$COOC_2H_5$

reduction

prepared[44] by fusing phenylglycine-*o*-carboxylic acid with molten alkali.

COOH, NH—CH_2COOH ⟶ CO, NH—CHCOOH

[39] Schunck, *Jahresber.*, **1855,** 659; **1858,** 465. Schunck and Römer, *Ber.*, **12,** 2311 (1879). Hoogewerff and ter Meulen, *Rec. trav. chim.*, **19,** 166 (1900). ter Meulen, *ibid.*, **24,** 444 (1905); **28,** 339 (1909). Perkin and Bloxam, *J. Chem. Soc.*, **91,** 1715 (1907). Perkin and Thomas, *ibid.*, **95,** 793 (1909). Thomas, Bloxam, and Perkin, *ibid.*, **95,** 824 (1909).

[40] Robertson, *J. Chem. Soc.*, **1927,** 1937.

[41] Robertson and Waters, *ibid.*, **1931,** 72.

[42] Baeyer, *Ber.*, **14,** 1742 (1881); **15,** 775 (1882). Forrer, *ibid.*, **17,** 976 (1884). Heumann, *ibid.*, **23,** 3434 (1890). Blank, *ibid.*, **31,** 1814 (1898). Vorländer and Schilling, *Ann.*, **301,** 349 (1898). Vorländer and Weissbrenner, *Ber.*, **33,** 555 (1900). Vorländer and Drescher, *ibid.*, **34,** 1854 (1901). Conrad and Reinbach, *ibid.*, **35,** 524 (1902). Vorländer, *ibid.*, **35,** 1683 (1902). Perkin, *J. Chem. Soc.*, **95,** 847 (1909). v. Auwers, *Ann.*, **393,** 338, 379 (1912). German patent 226,689; *Chem. Zentr.*, **1910,** II, 1257. Orendorff and Nichols, *Am. Chem. J.*, **48,** 486 (1912).

[43] Baeyer, *Ber.*, **13,** 2260 (1880).

[44] Heumann, *ibid.*, **23,** 3434 (1890).

Another synthesis has been accomplished by heating anilino malonic ester.[45]

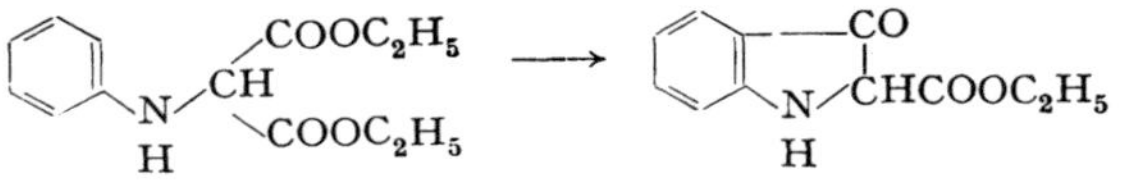

C_6H_5–NH–CH($COOC_2H_5$)$_2$ ⟶ indoxylic ester (N–H, CO, $CHCOOC_2H_5$)

Indoxylic acid is obtained in the form of colorless crystals melting at 122–123° with gas evolution. Even on warming with water the compound loses carbon dioxide, giving indoxyl. In dilute alkaline solution the compound is easily oxidized to indigo. The ethyl ester of indoxylic acid melts at 116°, is alkali soluble and very resistant to saponification by alkali. The compound is oxidized by strong oxidizing agents (chromic acid) to ethyl oxalyl-

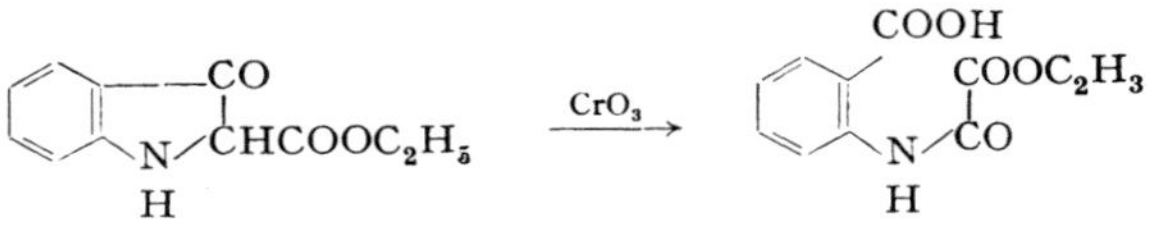

CO, N–H, $CHCOOC_2H_5$ $\xrightarrow{CrO_3}$ COOH, $COOC_2H_3$, CO, N–H

anthranilic acid. As a by-product there also results the ester of indoxanthin-carboxylic acid[46] which on reduction again gives indoxylic ester. Treatment of the potassium salt of indoxylic ester with ethyl iodide gives the

CO, N–H, C(OH)($COOC_2H_5$) ⟶ CO, N–H, $CHCOOC_2H_5$

O-ethyl derivative, m.p. 98°, which is easily saponified to *O*-ethylindoxylic acid which on melting decomposed to give *O*-ethylindoxyl.

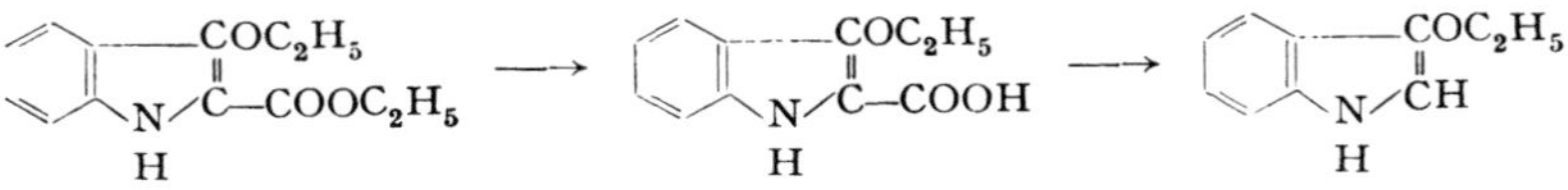

COC_2H_5, N–H, C–$COOC_2H_5$ ⟶ COC_2H_5, N–H, C–COOH ⟶ COC_2H_5, N–H, CH

Indoxyl-2-aldehyde (2-formyl-3-hydroxyindole) is obtained along with anthranilic acid on strong heating of indigo with strong alkali to 150°.[47] Its anil melts at 195° while the phenylhydrazone melts at 116° with decomposition.

[45] Blank, *ibid.*, **31**, 1814 (1898).

[46] Kalb, *ibid.*, **44**, 1458 (1911).

[47] Friedländer and Kielbasinski, *ibid.*, **44**, 3104 (1911). Friedländer and Schwenck, *ibid.*, **43**, 1972 (1910).

CHAPTER VII

Indigo, $\Delta^{2,2'}$-Bipseudoindoxyl

Introduction

Early investigations in the indole series received their impetus largely from interest in indigo which was obtained from certain plants and which had long played an important role in dyeing. The classical researches of Baeyer[1] led to the formula (I) for indigo.

CO OC

N—C=C—N

H H

(I)

The composition of indigo was determined by Dumas[2] while the molecular weight in the vapor state was determined by v. Sommaruga[3] with further determinations by other investigators.[4-6] These studies led to the molecular formula, $C_{16}H_{10}O_2N_2$, for indigo.

In 1841 Erdmann and Laurent discovered isatin as a product of the oxidation of indigo. The studies of Baeyer on the reduction of isatin led to the discovery of oxindole and of dioxindole and in 1870 to the formation of indigo along with indirubin by the reduction of isatin chlorine.[7] The ready preparation of isatin by the oxidation of indigo and of indigo through the oxidation of indoxyl and the reduction of isatin chloride led to the conclusion that indigo contains two indole nuclei. The question of the positions through which the two indole nuclei are united was settled by the synthesis of indigo from *o,o'*-dinitrodiphenylbutadiyne.[8] This compound is converted through

1 Baeyer, *Ber.*, **13**, 2254 (1880).

2 Dumas, *Ann. chim. phys.*, [2] **63**, 265 (1836); [3] **2**, 205 (1841).

3 v. Sommaruga, *Ann.*, **195**, 306 (1879).

4 Beckmann and Gabel, *Ber.*, **39**, 2611 (1906).

5 Vaubel, *ibid.*, **39**, 3587 (1900).

6 Scholl and Berblinger, *ibid.*, **36**, 3430 (1903).

7 Baeyer and Emmerling, *ibid.*, **3**, 514 (1870). Baeyer, *ibid.*, **12**, 456 (1879).

8 Baeyer, *ibid.*, **15**, 50 (1882).

the agency of concentrated sulfuric acid into diisatogen which is quantitatively reduced by ammonium sulfide to indigo.

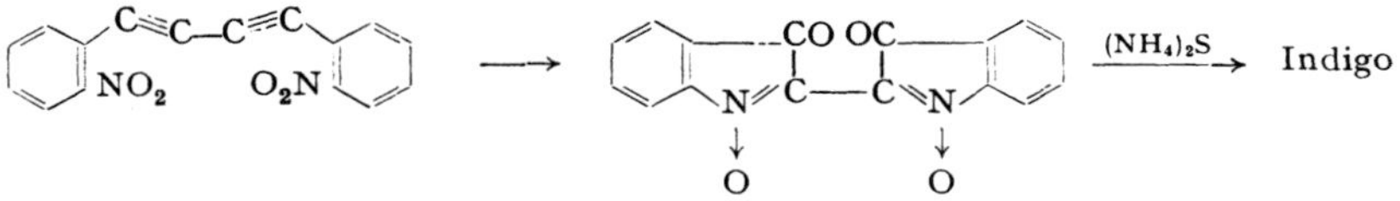

$\xrightarrow{(NH_4)_2S}$ Indigo

Since the Baeyer formula (I) for indigo contains an ethylenic double bond it follows that it should permit of the existence of two geometrically isomeric forms. Falk and Nelson[9] regarded indigo itself as having the *cis* structure and diacetylindigo as having the *trans* structure. They felt that

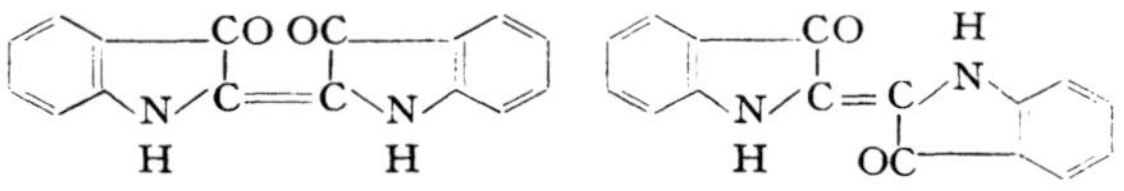

Cis *Trans*

the *trans* form of indigo should be red rather than blue. Robinson,[10] however, regarded indigo as possessing the *trans* structure stabilized by secondary valence forces. On the basic of a comparison of the absorption spectrum of indigo with those of certain condensation products which are definitely *trans* derivatives, Posner[11] likewise concluded that indigo must possess the *trans* structure. X-ray crystallographic studies[12] have established the fact that the indigo molecule possesses a center of symmetry which is possible only if the *trans* configuration is correct. These studies have also established the fact that there are two indigo molecules in the unit cell. Kuhn[13] suggested that indigo be assigned the polar formula II. Gill and Stonehill[14] suggested the hydrogen bonded structure III for indigo, this structure being a resonance hybrid of structure IV (p. 173). To these

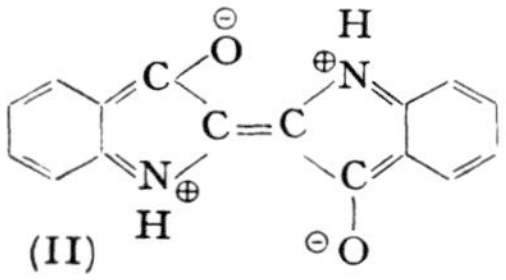

(II)

[9] Falk and Nelson, *J. Am. Chem. Soc.*, **29**, 1739 (1907); **30**, 143 (1908).

[10] Robinson, *J. Soc. Dyers Colourists*, **37**, 77 (1921).

[11] Posner, *Ber.*, **59**, 1799 (1926).

[12] Reis and Schneider, *Z. Krist.*, **68**, 543 (1929); *Chem. Abstr.*, **23**, 2083 (1929); v. Eller, *Acta Cryst.*, **5**, 142 (1952).

[13] Kuhn, *Naturwissenschaften*, **20**, 618 (1932).

[14] Gill and Stonehill, *J. Soc. Dyers Colourists*, **60**, 183 (1944).

recommended resonance forms of Kuhn and of Gill and Stonehill several

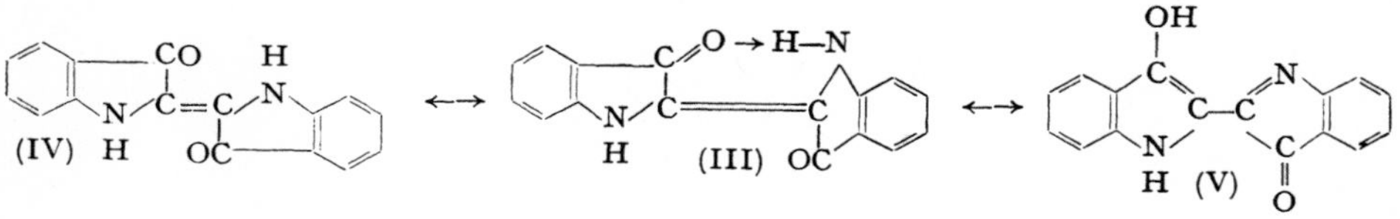

(IV) ⟷ (III) ⟷ (V)

other resonating structures have been added in the proposals of Hodgson[15] and of Van Alphen.[16] The suggestions of these several workers constitute essentially a translation of the views of Robinson into modern electronic terms.

It seems evident that the *trans* modification of the Baeyer formula is the most useful representation to give to this compound which must be the hybrid of several resonating structures. It should be observed, however, that the formula offers no adequate explanation for the fact that in solvents such as aniline and decalin indigo is monomolecular and red, but is blue when the molecules are associated.

Indigo condenses with phenylacetyl chloride to give Ciba lake red B, a *trans* derivative. Indigo likewise condenses with malonic ester to give the

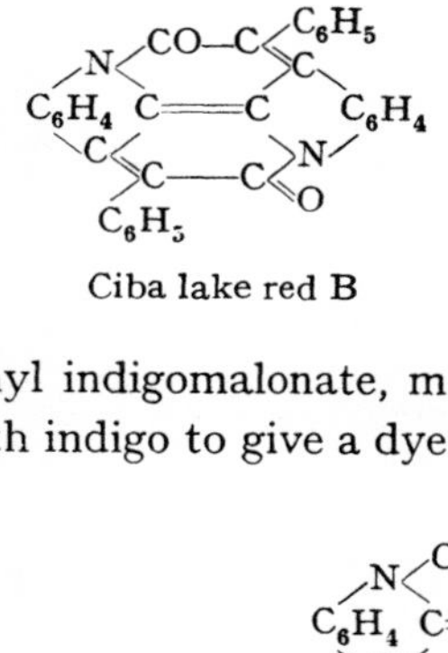

Ciba lake red B

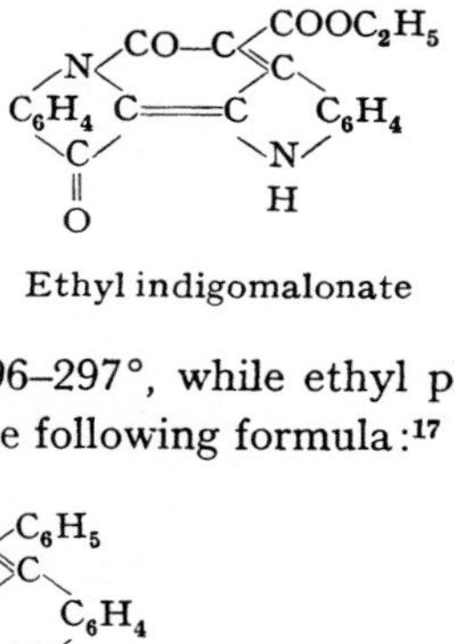

Ethyl indigomalonate

red-violet ethyl indigomalonate, m.p. 296–297°, while ethyl phenylacetate condenses with indigo to give a dye of the following formula:[17]

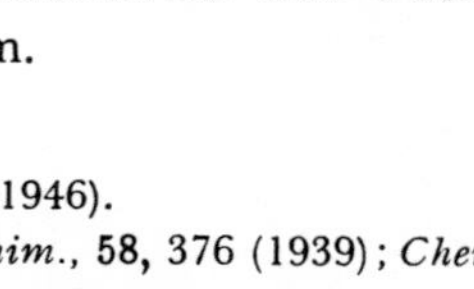

It was on the basis of the comparison of the absorption spectrum of indigo with those of these *trans* derivatives that Posner concluded that indigo itself exists in the *trans* form.

[15] Hodgson, *ibid.*, **62,** 176 (1946).

[16] Van Alphen, *Rec. trav. chim.*, **58,** 376 (1939); *Chem. Weekblad,* **35,** 435 (1938).

[17] Posner and Kemper, *Ber.*, **57,** 1311 (1924).

On the other hand, indigo condenses with oxalyl chloride or with ethoxyoxalyl chloride to give oxalyl indigo[18] which is a *cis* derivative and

```
        O           O
        ‖           ‖
       ⁄C⟍        ⁄C⟍
   C6H4  C═════C    C6H4
       ⟍N⟨       ⟩N⁄
          CO—CO
```

yellow in color. It has been reported[19] that the *cis* form of indigo has been isolated but is very unstable and changes rapidly to the *trans* form. It should be said, however, that the evidence supporting this claim does not appear to be wholly convincing.

Occurrence of Indigo

Indigo does not occur as such in the indigo-bearing plants, but as the glucoside, indican. This glucoside is split easily by hydrolysis into glucose and indoxyl. Indican is found to the extent of 1.5 to 2% in different species of the genus *Indigofera*. For example, the glucoside is found in *Indigofera tinctoria*, *I. arrecta*, and *I. leptostachya*.[20] Indican or other indigo-producing substances are also found in *Isatis tinctoria* (Waid)[21] and in some orchids[22] as well as in *Crotalaria retusa* L.[23] and *Lonchocarpus cyanesceus*.[24]

Indigo is obtained from the plants by first allowing the leaves to soak in vats to ferment. In this operation the indican is extracted from the leaves and hydrolyzed by an enzyme contained in the plant. The solution is then agitated in contact with air to effect the conversion of indoxyl to indigo.[25] The crude indigo obtained in this manner is contaminated with by-products,

[18] Friedländer and Sander, *ibid.*, **57**, 648 (1924). Van Alphen, *ibid.*, **72**, 525 (1939); *Rec. trav. chim.*, **58**, 378 (1939). Pummerer and Reuss, *Chem. Ber.*, **80**, 242 (1947).

[19] Heller, *Ber.*, **69**, 563 (1936); **72**, 1858 (1939); **77**, 163 (1944).

[20] Robiquet, *Jour. de Pharmacie*, **25**, 62 (1839). Hervy, *ibid.*, **26**, 322 (1840). Girardin and Preisser, *ibid.*, **26**, 344 (1840).

[21] Marchlewski, *Ber.*, **35**, 4338 (1902).

[22] Calvert, *J. pharm. chim.*, [3] **6**, 198 (1844).

[23] Greshoff, *Ber.*, **23**, 3540 (1890).

[24] Perkin, *J. Soc. Chem. Ind.*, **28**, 389 (1909).

[25] Schunck, *J. prakt. Chem.*, [1] **66**, 321 (1855); *Jahresber.*, **1855**, 659; **1858**, 465. Alvarez, *Compt. rend.*, **105**, 1887 (1887). Breaudat, *ibid.*, **127**, 769 (1898); **129**, 1478 (1899). Hazewinkel, *Chem.-Ztg.*, **24**, 409 (1900). ter Meulen, *Rec. trav. chim.*, **24**, 471 (1905); **28**, 339 (1909). Bergtheil, *J. Chem. Soc.*, **85**, 807 (1904). Orchardson, Wood, and Bloxam, *J. Soc. Chem. Ind.*, **26**, 8 (1907). Gaunt, Thomas, and Bloxam, *ibid.*, **26**, 1181 (1907). Thomas, Bloxam, and Perkin, *J. Chem. Soc.*, **95**, 826 (1909). Hoogewerf and ter Meulen, *Rec. trav. chim.*, **19**, 169 (1900).

among them indirubin.[26] Purification of the crude indigo can be accomplished by reduction to indigo white and subsequent reoxidation to indigo.

Synthesis of Indigo

The first synthesis of indigo from materials not obtained from the compound itself was accomplished by Baeyer[27] in 1880, by boiling *o*-nitrophenylpropiolic acid in alkaline solution with glucose.

C≡C, NO_2 COOH ⟶ CO CH_2, NO_2 COOH ⟶ CO, N≠C—COOH, ↓ O ⟶ CO, N≠CH ⟶ Indigo

The first commercially practical synthesis of indigo came in 1897.[28] The process started with naphthalene as the raw material and proceeded through well-known reactions by way of phthalic acid through phthalimide and anthranilic acid to phenylglycine-*o*-carboxylic acid. The latter compound was converted through alkali fusion to indoxylic acid which was in turn oxidized to indigo. In its last two phases this process consists of steps discovered by Heumann[29] for the preparation of indoxylic acid and its con-

⟶ COOH COOH ⟶ CO NH CO $\xrightarrow[NaOH]{Cl_2}$ COOH NH_2 $\xrightarrow{ClCH_2COOH}$

COOH NH—CH_2COOH $\xrightarrow[\text{fusion}]{\text{alkali}}$ COH ‖ N—C—COOH H $\xrightarrow[\text{oxidation}]{\text{air}}$ Indigo

version to indigo. In 1901 another synthetic process was developed which was also based on the work of Heumann.[29, 30] In this process aniline is con-

NH_2 $\xrightarrow{ClCH_2COOH}$ COOH CH_2 N H $\xrightarrow[\text{fusion}]{NaNH_2}$ COH ‖ N—CH H $\xrightarrow[\text{oxidation}]{\text{air}}$ Indigo

[26] Perkin and Bloxam, *J. Chem. Soc.*, **91,** 279 (1907). Perkin and Thomas, *ibid.*, **95,** 803 (1909).

[27] Baeyer, *Ber.*, **13,** 2260 (1880). Muller, *Ann.*, **212,** 143 (1882). German patents 11,857, 11,858; *Friedl.*, **1,** 129, 131 (1888).

[28] Brunck, *Ber.*, **33,** Sonderheft LXXI (1900).

[29] Heumann, *ibid.*, **23,** 3431 (1890).

[30] Biedermann and Lepetit, *ibid.*, **23,** 3289 (1890).

verted into phenylglycine which yields indoxyl on fusion with sodium amide. Oxidation of the indoxyl yields indigo. Small amounts of indirubin result also due to oxidation of indoxyl to isatin and condensation of the isatin with indoxyl.

Another synthesis of indigo which is due to Sandmeyer[31] is from isatin-α-anilide which in turn is prepared from thiocarbanilide. A method

C_6H_5—NH—CS—NH—C_6H_5 $\xrightarrow[\text{and KCN in water}]{\text{boil with PbCO}_3}$ C_6H_5—NH—C(CN)=NC_6H_5 $\xrightarrow{(NH_4)_2S}$ C_6H_5—NH—C($CSNH_2$)=NC_6H_5 $\xrightarrow{H_2SO_4}$ (CO, N(H), C=NC_6H_5 ring) $\xrightarrow{H_2S}$ (CO, N(H), CS ring) $\xrightarrow[\text{alkali}]{\text{weak}}$ Indigo

of preparation due to Lepetit[32] starts with aniline, formaldehyde, and sodium bisulfite. Indigo has also been prepared from ω-bromoacetanilide[33] through

C_6H_5—NH—CO—CH_2Br $\xrightarrow[\text{air oxidation}]{\text{alkali fusicn}}$ Indigo

alkali fusion and air oxidation. Indigo is also obtained when isatin chloride

(CO, N=CCl ring) + (CO, N(H), CH_2 ring) ⟶ Indigo

is condensed with indoxyl.[34] The reduction of ω-chloro-2-nitroacetophenone by zinc and acetic acid[35] and of ω-bromo-2-nitroacetophenone and ω,ω-dibromo-2-nitroacetophenone (or the ω,ω-dichloro derivative) by heating with alcoholic ammonium sulfide[36] gives indigo. Indigo has also been

[31] Sandmeyer, *Z. Farben Textilchem.*, **2,** 134 (1903).

[32] Lepetit, *Chimie & Industrie*, **14,** 852 (1925); *Chem. Abstr.*, **20,** 2585 (1926).

[33] Flimm, *Ber.*, **23,** 57 (1890). Kuhara and Chikashigi, *Am. Chem., J.*, **24,** 167 (1900).

[34] Wahl and Bagard, *Bull. soc. chim.*, [4] **7,** 1100 (1910).

[35] Brodinius, *Chem.-Ztg.*, **40,** 326 (1916); *Chem. Abstr.*, **10,** 2150 (1916).

[36] Gevokoht, *Ann.*, **221,** 331 (1883). German patent 23,785; *Friedl.*, **1,** 139 (1888). Ruggli and Reichwein, *Helv. Chim. Acta*, **20,** 905 (1937).

prepared from *o*-nitroacetophenone on heating with zinc dust and soda lime according to the following scheme:[37]

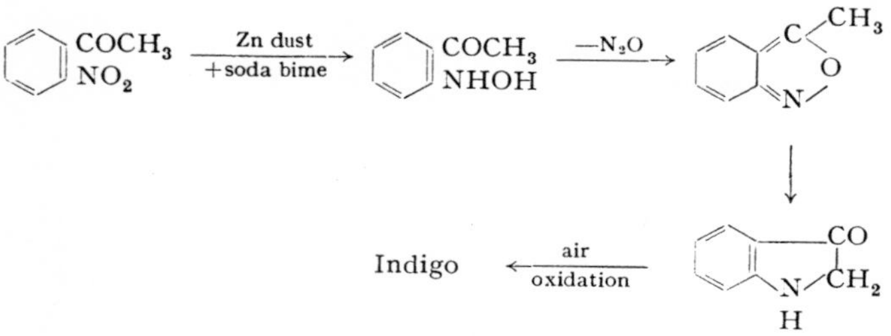

$$C_6H_4(COCH_3)(NO_2) \xrightarrow[+\text{soda bime}]{\text{Zn dust}} C_6H_4(COCH_3)(NHOH) \xrightarrow{-N_2O} \ldots \rightarrow C_6H_4\langle CO, CH_2, NH \rangle \xrightarrow[\text{oxidation}]{\text{air}} \text{Indigo}$$

The Baeyer-Drewson[38] synthesis of indigo consists of treating a solution of *o*-nitrobenzaldehyde in acetone with aqueous alkali. The dye has also

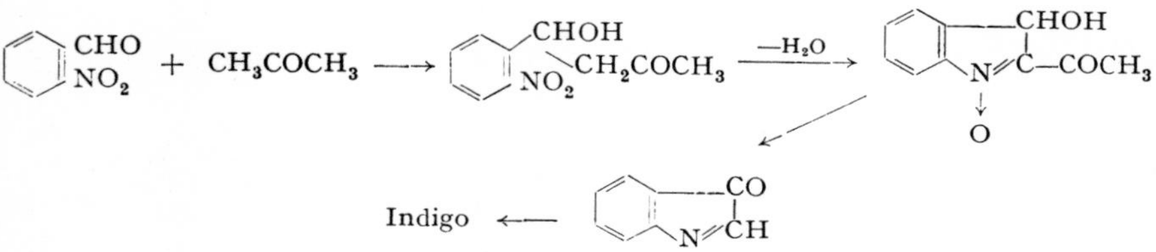

$$C_6H_4(CHO)(NO_2) + CH_3COCH_3 \longrightarrow C_6H_4(NO_2)CHOH{-}CH_2COCH_3 \xrightarrow{-H_2O} \ldots \rightarrow \text{Indigo}$$

been prepared from dianilinomaleic acid through fusion with sodium amide,[39]

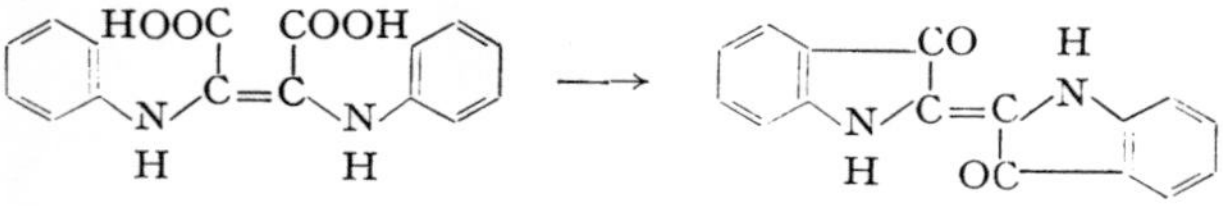

as well as from oxal-*o*-toluidide according to the following scheme of reactions:[40]

[37] Emmerling and Engler, *Ber.*, **3,** 885 (1870); **9,** 1422 (1876). Engler, *ibid.*, **28,** 309 (1895). Camp, *ibid.*, **32,** 3232 (1899). Wichelhaus, *ibid.*, **9,** 1106 (1876). Bamberger and Engler, *ibid.*, **36,** 1611 (1903).

[38] Baeyer and Drewson, *ibid.*, **15,** 2856 (1882). German patent 238,381; *Chem. Zentr.*, **1911,** II, 1187.

[39] Salmony and Simons, *Ber.*, **38,** 2581 (1905).

[40] Madelung, *ibid.*, **45,** 1131 (1912); *Ann.*, **405,** 58 (1914). German patent 262,347; *Chem. Zentr.*, **1913,** II, 553.

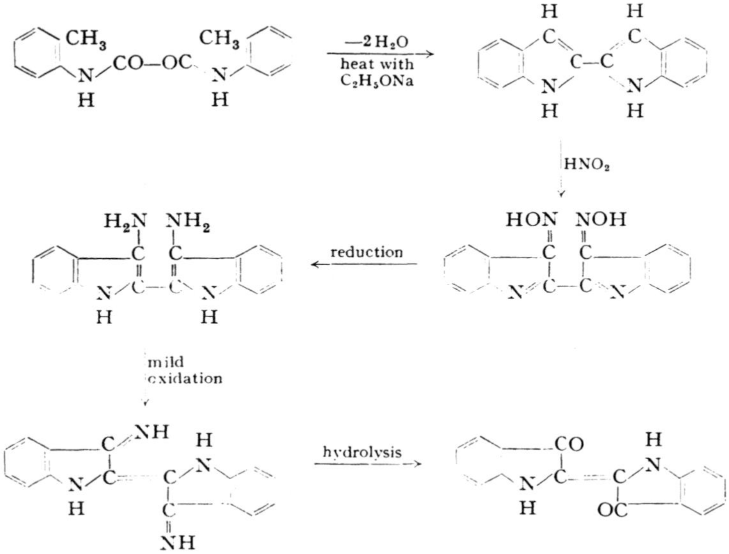

CH_3 CH_3 CO—OC N H N H

$-2\,H_2O$, heat with C_2H_5ONa →

H H C C C—C N N H H

HNO_2 ↓

HON NOH C C N C—C N

← reduction

H_2N NH_2 C C N C—C N H H

↓ mild oxidation

C NH H N N C C C NH H

hydrolysis →

CO H N N C C OC H

Indigo is obtained from indole by the action of ozone,[41] of air in sodium sulfite or sodium bisulfite solution,[42] of hydrogen peroxide,[43] of monopersulfuric acid,[44] or of iodine in sodium carbonate solution.[45] The oxidation of indole-3-carboxylic acid with ozone gives indigo[46] as does also the oxidation of *N*-hydroxyindole-α-carboxylic acid by concentrated sulfuric acid.[47] Indigo is also obtained by the oxidation of indoxyl in alkaline or in ammoniacal solution by air[48] or through treatment with ferric chloride and hydrochloric acid.[49]

Physical Properties of Indigo

Indigo is a dark blue powder possessing a copper-red iridescence,

[41] Nencki, *Ber.*, **8**, 721 (1875).
[42] German patent 130,629; *Chem. Zentr.*, **1902**, I, 1084.
[43] Porcher, *Bull. soc. chim.*, [4] **5**, 526 (1909).
[44] German patent 132,405; *Chem. Zentr.*, **1902**, II, 173.
[45] Pauly and Gunderman, *Ber.*, **41**, 4007 (1908).
[46] Weissgerber, *ibid.*, **46**, 658 (1913). German patent 230,542; *Chem. Abstr.*, **5**, 2734 (1911).
[47] Reissert, *Ber.*, **30**, 1048 (1897).
[48] Baeyer, *ibid.*, **14**, 1744 (1881). German patent 17,656; *Friedl.*, **1**, 135 (1888). Thomas, Bloxam, and Perkin, *J. Chem. Soc.*, **95**, 842 (1909).
[49] Baumann and Tiemann, *Ber.*, **13**, 413 (1880).

of low solubility in most organic solvents. However, it can be crystallized from chloroform, nitrobenzene, aniline, and certain other solvents. In a sealed tube it melts at 390–392° to a purple-red liquid which decomposes rapidly.[50] The vapor of indigo is fiery red with a violet tinge.[51]

Chemical Properties of Indigo

Formation of Salts

Indigo combines with mineral acids to give salts which are easily decomposed by water. It dissolves readily in acetic acid, benzene, or chloroform when dry hydrogen chloride is passed through a suspension of the dye in those solvents.[52] The hydrochloride, $C_{16}H_{10}O_2N_2 \cdot HCl$, is precipitated from the acetic acid solution by the addition of ether. The monosulfate, $C_{16}H_{10}O_2N_2 \cdot H_2SO_4$, is formed when indigo is digested with a solution of concentrated sulfuric acid in acetic acid and the resulting solution is cooled.[53] The disulfate, $C_{16}H_{10}O_2N_2 \cdot 2\,H_2SO_4$, is obtained when indigo is similarly treated with sulfuric acid of density 1.71.[52,54]

When indigo is shaken for half an hour with warm alcoholic sodium hydroxide it is changed to a green powder which has the composition, $C_{16}H_{10}O_2N_2 \cdot NaOH$,[55] and probably has the structure:

```
      ONa HO
   C       | CO
C6H4 C------C    C6H5
   N        NH
   H
```

On heating with strong alkali at 145–150° indigo is split yielding anthranilic acid and indoxylaldehyde.[56] On the other hand, the fusion of indigo with

[50] Michael, *ibid.*, **28**, 1632 (1895).

[51] For absorption spectra of indigo see Vogel, *ibid.*, **11**, 1364 (1878). Kruss and Oekonomides, *ibid.*, **16**, 2054 (1883). Grandmougin, *ibid.*, **42**, 4218 (1909). Schwalbe and Jochheim, *ibid.*, **41**, 2798 (1908). Posner, *ibid.*, **59**, 1799 (1926).

[52] Binz and Kufferath, *Ann.*, **325**, 196 (1902).

[53] Bloxam, *J. Chem. Soc.*, **87**, 977 (1905). DeAguir and Bayer, *Ann.*, **157**, 367 (1871). Binz and Kufferath, *ibid.*, **325**, 197 (1902).

[54] German patent 121,450; *Friedl.*, **6**, 598 (1904).

[55] Binz, *Z. angew. Chem.*, **19**, 1415 (1906). Binz and Schadel, *Ber.*, **45**, 587 (1912). Friedländer, *ibid.*, **41**, 1036 (1908). German patent 158,625; *Chem. Zentr.*, **1905**, I, 787. Madelung, *Ber.*, **57**, 234 (1924). Kunz and Günther, *ibid.*, **56**, 2029 (1925).

[56] Friedländer and Schwenck, *Ber.*, **43**, 1971 (1910). Friedländer and Kielbasinski, *ibid.*, **44**, 3098 (1911). Friedländer and Sander, *ibid.*, **57**, 648 (1924).

$$\text{Indigo} \xrightarrow{\text{NaOH}} C_6H_4\begin{matrix}C(ONa)\\NH\end{matrix}C{-}C(HO)\begin{matrix}CO\\NH\end{matrix}C_6H_5 \longrightarrow C_6H_5\begin{matrix}C(OH)\\NH\end{matrix}C{-}CHO + C_6H_4\begin{matrix}COOH\\NH_2\end{matrix}$$

potassium hydroxide yields indoxyl.[57]

Oxidation of Indigo

Strong oxidation of indigo with nitric acid or with chromic acid accomplishes the separation of the two indole fragments with the formation of isatin.[58] This is the industrial method for the preparation of isatin (see p. 113). Indigo and ozone yield an ozonide which yields isatin on treat-

$$C_6H_4\begin{matrix}CO\\NH\end{matrix}C{=}C\begin{matrix}NH\\CO\end{matrix}C_6H_4 \longrightarrow 2\,C_6H_4\begin{matrix}CO\\NH\end{matrix}CO$$

ment with water. With excess ozone isatinic anhydride is obtained.[59] Through the action of chlorine or bromine water on indigo halogenated derivatives of isatin are obtained (see p. 110) [60]

Dehydroindigo is obtained when a suspension of indigo in carbon tetrachlorine is oxidized by chlorine and the solution treated with calcium hydroxide or when a suspension of indigo in boiling benzene is oxidized with lead oxide.[61] Dehydroindigo crystallizes in dark yellowish-red six-sided plates. It is more soluble in organic solvents than is indigo. Being readily

$$C_6H_4\begin{matrix}CO\\N\end{matrix}\!\!\!=C{-}C\!=\!\!\!\begin{matrix}OC\\N\end{matrix}C_6H_4$$

reduced to indigo, dehydroindigo functions as an oxidizing agent. It liberates iodine from potassium iodide and oxidizes hydroquinone to quinone. It also oxidizes indigo white to indigo while being itself reduced to indigo. Boiling with water likewise gives indigo while simultaneously an equivalent of

$$2\,C_6H_4\begin{matrix}CO\\N\end{matrix}\!\!\!=C{-}C\!=\!\!\!\begin{matrix}OC\\N\end{matrix}C_6H_4 + 2H_2O \longrightarrow \text{Indigo} + 2\,C_6H_4\begin{matrix}CO\\NH\end{matrix}CO$$

[57] Heumann and Bachofen, *ibid.*, **26**, 225 (1893).

[58] Erdmann, *J. prakt. Chem.*, **24**, 1 (1841). Laurent, *ibid.*, **25**, 434 (1842).

[59] Van Alphen, *Rec. trav. chim.*, **57**, 911–914 (1938).

[60] Erdmann, *J. prakt. Chem.*, **19**, 330 (1840).

[61] Kalb, *Ber.*, **42**, 3642 (1909); **44**, 1455 (1911); **45**, 2136 (1912). Kalb and Bayer, *ibid.*, **45**, 2155 (1912). German patent 216,886; *Chem. Zentr.*, **1910**, I, 308.

dehydroindigo is oxidized to isatin. In acetic acid solution dehydroindigo combines with the acid to give the diacetate which is also obtained when a suspension of indigo in glacial acetic acid is shaken with powdered potassium permanganate.[62] The bisulfite addition compound of dehydroindigo, $C_{16}H_8O_2N_2 \cdot 2\,NaHSO_3$, crystallizes in bright canary-yellow crystals, has an intensely sweet taste, and is decomposed by mineral acids to give indigo.

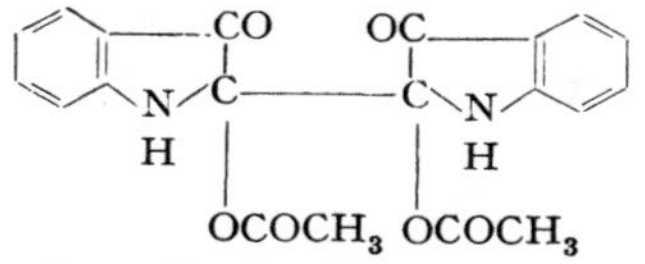

The diimide of dehydroindigo results from indigodiimide through oxidation by lead peroxide and crystallizes in orange-yellow needles melting at 193° (dec.).[63]

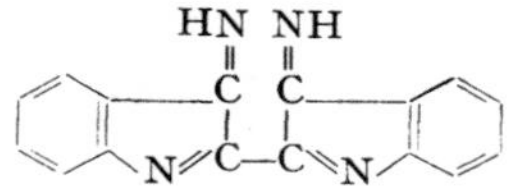

α,α-Diindolyl reacts with nitrous acid in acetic acid to give the hydrochloride of the monoisonitroso derivative (I) in the presence of hydrogen chloride, and in the absence of mineral acids to give the diisonitroso derivative (II), which is the dioxime of dehydroindigo.[64] Diisatogen can be regarded as the *N,N'*-dioxide of dehydroindigo.[64]

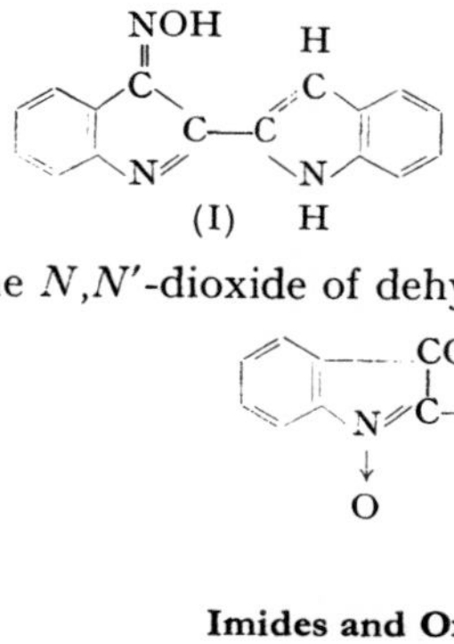

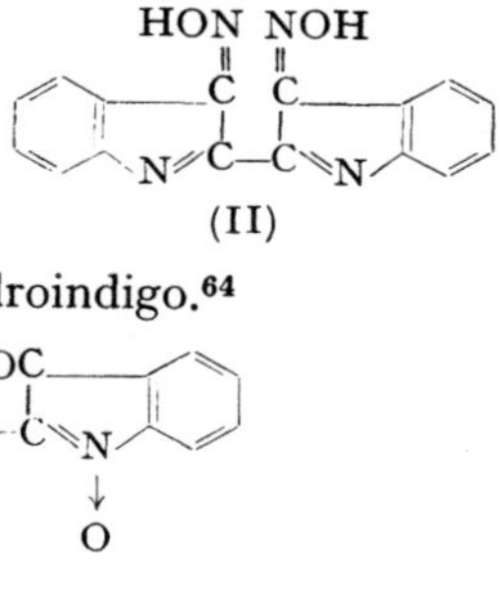

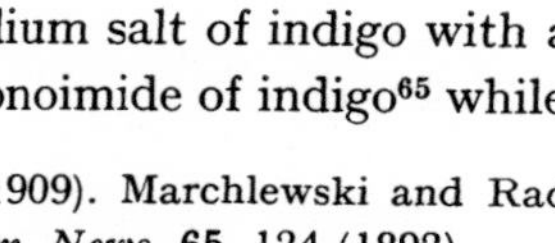

Imides and Oximes of Indigo

Treatment of the sodium salt of indigo with a solution of zinc chloride in ammonia yields the monoimide of indigo[65] while fusion of indigo with the

[62] Kalb, *Ber.*, **42,** 3642 (1909). Marchlewski and Radcliffe, *J. prakt. Chem.*, [2] **58,** 102 (1898). O'Neill, *Chem. News*, **65,** 124 (1892).

[63] Madelung, *Ann.*, **405,** 82 (1914).

[64] Baeyer, *Ber.*, **15,** 52 (1882). Pfeiffer, *ibid.*, **45,** 1821 (1912).

[65] Binz and Lange, *ibid.*, **46,** 1691 (1913).

zinc chloride ammonia complex yields the diimide.[66] The diimide can further

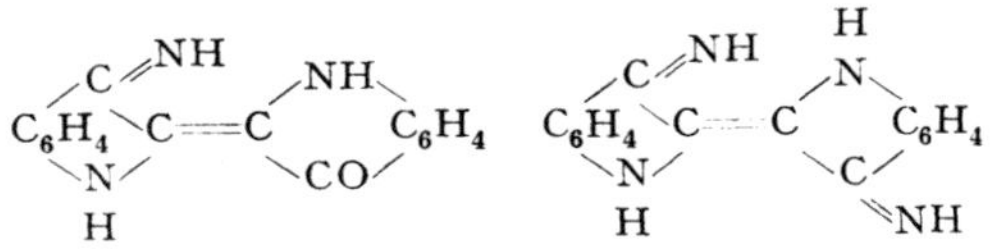

```
      C=NH       NH              C=NH       H
     /          /  \            /           N
C6H4  C====C      C6H4   C6H4  C====C     /  \
     \N/        \CO/            \N/      \C/   C6H4
      H                          H          \\NH
```

be prepared through oxidation of diaminodiindolyl as well as through the oxidation of 3-aminoindole. The di-imide crystallizes from warm benzene in blue needles, m.p. 215°. Boiling indigo with aniline in the presence of anhydrous boric acid yields indigo anil, deep blue needles.[67]

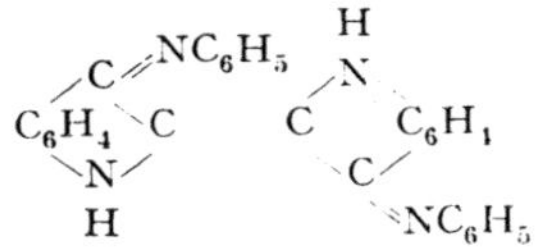

```
        C=NC6H5      H
       /             N
  C6H4  C        C     C6H4
       \N/         \C/
        H            \\NC6H5
```

The monooxime of indigo,[68] m.p. 205° (dec.), crystallizes from alcohol in brownish-violet needles, is formed by warming indigo with an alkaline solution of hydroxylamine. The dioxime[69] is obtained from the diimide through the action of a boiling alcoholic solution of hydroxylamine hydrochloride.

Reduction of Indigo

Indigo white or leucoindigo[70] is prepared from indigo through the agency

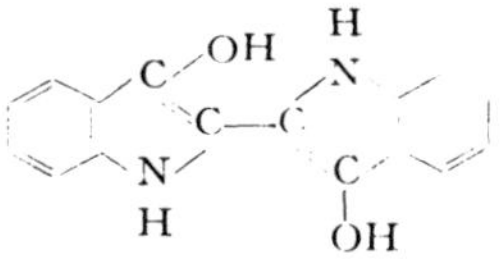

```
           C-OH      H
          /          N
  (C6H4)     C—C         (C6H4)
          \N/        C
           H         |
                     OH
```

of a variety of reducing agents. The conversion is effected by ferrous sulfate[71]

[66] Madelung, *ibid.*, **46,** 2259 (1913); *Z. angew. Chem.*, **34,** 482 (1921); *Ann.*, **405,** 58 (1914). Madelung and Wilhelm, *Ber.*, **57,** 234 (1924).

[67] Grandmougin and Dessoulavy, *Ber.*, **42,** 3636, 4401 (1909). Grandmougin, *ibid.*, **43,** 1317 (1910); *Compt. rend.*, **174,** 1175 (1922).

[68] Thiele and Pickard, *Ber.*, **31,** 1252 (1898).

[69] Madelung, *Ann.*, **405,** 58 (1914).

[70] Dumas, *ibid.*, **22,** 75 (1837). Ullgren, *ibid.*, **136,** 100 (1865). Baeyer, *Ber.*, **12,** 1600 (1879). Baumann and Tiemann, *ibid.*, **13,** 408 (1880). German patent 276,808; *Chem. Zentr.*, **1914,** II, 517.

[71] Berzelius, *Ann. phys.*, **10,** 119 (1827). Erdmann, *J. prakt. Chem.*, [1] **19,** 326 (1840). Dumas, *Ann. chim. phys.*, [3] **2,** 209 (1841); **22,** 75 (1837); **48,** 257 (1843). Ullgren, *Ann.*, **136,** 97 (1865). Binz and Marx, *Z. angew. Chem.*, **21,** 529 (1908). German patent 230,306; *Chem. Abstr.*, **5,** 2733 (1911).

in the presence of alkali as well as by iron powder[72] in the presence of alkali or in neutral or weakly acidic media,[73] by reduction with zinc dust[74] or with silicon.[75] The reduction of indigo to indigo white can also be effected electrolytically,[76] by the action of glucose[77] or by catalytic methods.[78] An especially useful agent for the reduction of indigo to indigo white is sodium hydrosulfite ($Na_2S_2O_4$). This agent is widely used in preparing the "vat" used in dyeing with indigo.[79] Indigo white can also be prepared by the incomplete oxidation of indoxyl in hot alkaline solution by air or by heating indoxyl with indigo in alkaline medium.[80] Indigo white is precipitated from its alkaline solution by the addition of mineral acids.[81]

Indigo white is a crystalline substance, somewhat soluble in boiling water, soluble with a yellow color in alcohol and ether as well as in alkali and alkali carbonates. It is insoluble in dilute acids but is soluble in concentrated sulfuric acid yielding a purple solution. Moist indigo white is readily oxidized to indigo by the oxygen of the air.[82] This oxidation serves to regenerate the dye in the fibers which have been dipped in the "vat."

The reduction of halogenated indigos[83] yields both the halogenated indigo white derivatives and indigo white itself.

Prolonged heating of indigo white with barium hydroxide and zinc dust yields a compound, $C_{15}H_{10}N_2$, quindolin,[84] apparently through splitting of

[72] German patent 165,429; *Chem. Zentr.*, **1906**, I, 106. German patent 171,785; *Chem. Zentr.*, **1906**, II, 374.

[73] German patent 199,375; *Chem. Zentr.*, **1908**, II, 375.

[74] German patent 131,118; *Chem. Zentr.*, **1902**, I, 1287. German patent 131,245; *Chem. Zentr.*, **1902**, I, 1287. Binz and Rung, *Z. angew. Chem.*, **12**, 489 (1899); **13**, 413 (1900); **19**, 1417 (1906). German patent 204,568; *Chem. Abstr.*, **3**, 1467 (1909).

[75] German patent 262,833; *Chem. Abstr.*, **7**, 3368 (1913).

[76] Binz, *Z. Elektrochem.*, **5**, 5 (1898). Binz and Hagenbach, *ibid.*, **6**, 261 (1899).

[77] Fritzsche, *Ann.*, **44**, 290 (1842); *J. prakt. Chem.*, [1] **28**, 193 (1843). Ullgren, *Ann.*, **136**, 97 (1865).

[78] Brochet, *Compt. rend.*, **160**, 306 (1915); *Rev. gén. mat. color.*, **27**, 131 (1922); *Chem. Abstr.*, **17**, 636 (1923).

[79] Grandmougin, *J. prakt. Chem.*, [2] **76**, 142 (1907). German patent 204,568; *Chem. Zentr.*, **1909**, I, 114. German patent 275,121; *Chem. Zentr.*, **1914**, II, 179.

[80] German patent 164,509; *Chem. Zentr.*, **1905**, II, 1753

[81] Binz and Rung, *Z. angew. Chem.*, **13**, 416 (1900). Binz., *J. prakt. Chem.*, [2] **63**, 504 (1901).

[82] Berzelius, *Ann. phys.*, **10**, 119 (1827). Manchot and Herzog, *Ann.*, **316**, 318 (1901).

[83] German patent 176,617; *Chem. Zentr.*, **1906**, II, 1791.

[84] Schutzenberger, *Compt. rend.*, **85**, 147 (1877); *Jahresber.*, **1877**, 511. Girard, *Jahresber.*, **1879**, 472; **1880**, 586; *Compt. rend.*, **89**, 104 (1879).

indigo white into indoxyl and isatinic acid followed by their condensation. Decarboxylation of the resulting quindolin carboxylic acid gives quindolin.

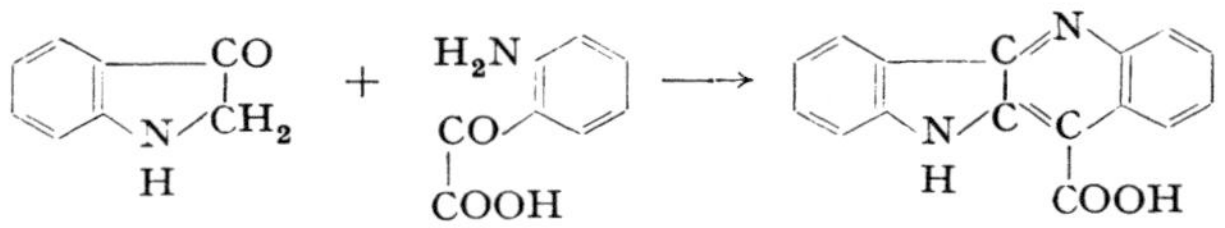

Acyl derivatives of indigo white are prepared from indigo as well as from indigo white itself.[85] Indigo on treatment with zinc dust and acetic anhydride in presence of sodium acetate gives diacetylindigo white (I), insoluble in cold alkali, which is oxidized by nitrous acid to diacetylindigo, a change evidently involving migration of the acetyl groups. The same diacetyl derivative can be prepared by the action of acetic anhydride on indigo white. An isomeric compound (II) soluble in cold alkali and not

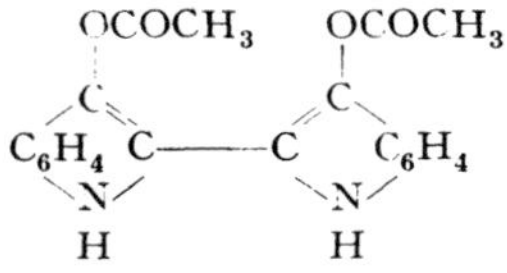

(I) 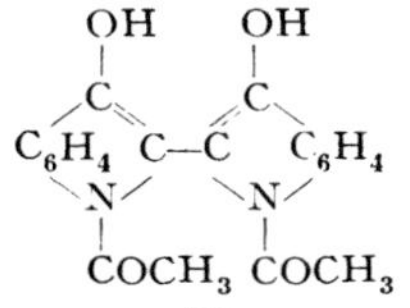 (II) 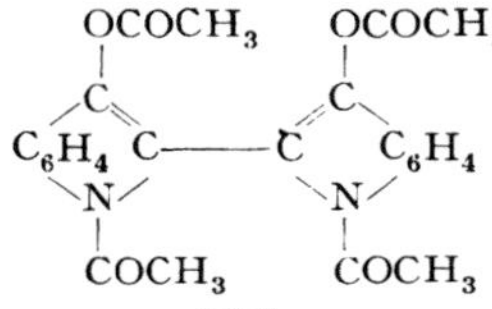 (III)

changed by nitrous acid is formed from diacetylindigo through reduction with phenylhydrazine in hot benzene. Both compounds on treatment with acetic anhydride and sodium acetate give the same tetraacetylindigo white (III), m.p. 258°.[86] Long heating of indigo with benzoyl chloride in pyridine gives tetrabenzoylindigo white (see p. 191).[87] Heating indigo with an alkaline alcoholic hydrazine solution yields desoxyindigo (IV).[87a]

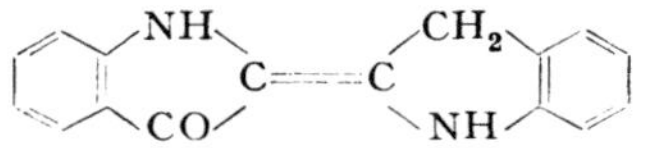

(IV)

Halogenated Indigos

Indigo derivatives substituted in the two benzene rings can be prepared both by direct substitution in indigo and by synthesis from substituted

[85] Liebermann, *Ber.*, **21**, 442 (1882). Liebermann and Dickhuth, *ibid.*, **24**, 4130 (1901). Vorländer and Drescher, *ibid.*, **34**, 1858 (1901). German patent 126,799; *Chem. Zentr.*, **1902**, I, 82. Heller, *Ber.*, **36**, 2765 (1903). Falk and Nelson, *J. Am. Chem. Soc.*, **29**, 1743 (1907).

[86] Vorländer and Pfeiffer, *Ber.*, **52**, 325 (1919). Posner and Aschermann, *ibid.*, **53**, 1925 (1920).

[87] Heller, *ibid.*, **36**, 2764 (1903). Posner, *ibid.*, **59**, 1799 (1926).

[87a] Borsche and Meyer, *ibid.*, **54**, 2854 (1921). Seidel, *ibid.*, **77**, 788 (1944).

intermediates. As examples, substituted indigos can be prepared through the Baeyer-Drewson synthesis from derivatives of *o*-nitrobenzaldehyde[88] or by the condensation of substituted isatin chlorides with indoxyl.

Cl CHO NO₂ —acetone, KOH→ Cl CO H N C=C N OC H Cl

Br CO N CCl Br + NH CH₂ CO → Br CO N H C=C N H OC Br

Direct substitution of indigo, indigo white, and dehydroindigo gives the corresponding sulfonic acid, halogen, and nitro derivatives. In these substitutions the entering groups enter first position 5, then position 7, and finally position 4.

The action of halogens on indigo in the presence of water gives principally halogen derivatives of isatin (see p. 110) but the action of halogens in glacial acetic acid or in the presence of mineral acids gives primarily the monobromo and dibromo or monochloro and dichloro derivatives of indigo.[89] In nitrobenzene and at higher temperatures the bromination of indigo yields a tetrabromoindigo which is also obtained when the bisulfite addition compound of dehydroindigo is brominated in aqueous solution at ordinary temperature and the resulting bromo derivative is decomposed by boiling aqueous hydrochloric acid:

$$C_{16}H_4O_2N_2Br_4 \cdot 2NaHSO_3 \rightarrow C_{16}H_6O_2N_2Br_4 + SO_2 + Na_2SO_4$$

More highly brominated indigos (pentabromo- and hexabromoindigo) can be prepared from indigo or from the lower bromo derivatives through the action of excess bromine in concentrated sulfuric acid or in chlorosulfonic acid. Similarly 5,7,5′,7′-tetrachloroindigo (Brilliant Indigo B) has been prepared from indigo by the action of chlorine on a suspension of indigo in sulfuryl chloride[90] as well as by the reduction of 5,7-dichloroisatin chloride.[91] The dye

[88] Schwalbe and Jochheim, *ibid.*, **41,** 3297 (1908). Gindraux, *Helv. Chim. Acta,* **12,** 921 (1929). German patent 112,400; *Chem. Zentr.*, **1900,** II, 700.

[89] Grandmougin, *Ber.*, **42,** 4408 (1909).

[90] German patent 226,319; *Chem. Abstr.*, **5,** 1196 (1911).

[91] Danaila, *Compt. rend.*, **149,** 1385 (1909). Kalb, *Ber.*, **42,** 3657 (1909). German patent 222,460; *Chem. Zentr.*, **1910,** II, 119.

has also been prepared from 5,7-dichloroindoxylic acid[92] and from 5,7,5′,7′-tetrachlorodehydroindigo acetate.[93] 5-Chloroindigo has been prepared by the direct chlorination of both indigo and indigo white[94] while 5,5′-dichloroindigo has been synthesized by a number of workers.[95] Both 4,4′-dichloroindigo[96] and 6,6′-dichloroindigo[97] have been prepared by conventional methods. 5,6,7,5′,6′,7′-Hexachloroindigo has been prepared from 2-nitro-3,4,5-trichlorobenzaldehyde, acetone, and alkali[98] while 4,5,6,7,4′,5′,6′,7′-octachloroindigo has been prepared by the oxidation of 1-acetyl-4,5,6,7-tetrachloroindoxylic acid.[99]

Prepared by direct bromination of indigo are: 5,5′-dibromoindigo,[100] 5,7,5′-tribromoindigo,[101] 5,7,5′,7′-tetrabromoindigo,[102] while 4,5,7,4′,5′,7′-hexabromoindigo has been prepared by the bromination of indigo in chlorosulfonic acid[103] and from 5,7,5′,7′-tetrabromoindigo by bromination in fuming sulfuric acid.[104] 4,4′-Dibromoindigo has also been prepared from

[92] German patent 226,689; *Chem. Zentr.*, **1910**, II, 1257.

[93] German patent 237,689; *Chem. Zentr.* **1911**, II, 579.

[94] German patent 163,280; *Chem. Zentr.*, **1935**, II, 1144.

[95] Eichengrün and Einhorn, *Ann.*, **262**, 138 (1891). Baeyer and Wirth, *ibid.*, **284,** 156 (1895). German patents 30,329, 33,064; *Friedl.*, **1**, 143, 146 (1888). German patent 120,321; *Chem. Zentr.*, **1901**, I, 1131.

[96] German patent 112,400; *Chem. Zentr.*, **1900**, II, 700. Schwalbe and Jochheim, *Ber.*, **41**, 3797 (1908). Gindraux, *Helv. Chim. Acta*, **12**, 921 (1929).

[97] Sachs and Kempf, *Ber.*, **36**, 3301 (1903). German patent 128,727; *Chem. Zentr.*, **1902**, I, 552. German patent 195,291; *Chem. Zentr.*, **1908**, I, 1230. Sachs and Sichel, *Ber.*, **37**, 1866 (1904).

[98] Van de Bunt, *Rec. trav. chim.*, **48**, 121 (1929).

[99] Orndorff and Nichols, *Am. Chem., J.*, **48**, 473 (1912). Grandmougin and Seyder, *Ber.*, **47**, 2365 (1914).

[100] German patent 128,575; *Chem. Zentr.*, **1902**, I, 551. German patent 149,940; *Chem. Zentr.*, **1904**, I, 1046. Fichter and Cuena, *Helv. Chim. Acta*, **14**, 651 (1931).

[101] British patent 4,423; *Chem. Abstr.*, **3**, 1930 (1909). German patent 193,438; *Chem. Abstr.*, **2**, 1894 (1908). German patent 208,471; *Chem. Abstr.*, **3**, 2058 (1909). German patent 228,137; *Chem. Abstr.*, **5**, 2185 (1911).

[102] German patent 209,078; *Chem. Abstr.*, **3**, 2238 (1909). German patent 224,204; *Chem. Abstr.*, **4**, 3305 (1910). German patent 205,699; *Chem. Abstr.*, **3**, 1933 (1909). German patent 224,460; *Chem. Zentr.*, **1910**, II, 119. Grandmougin, *Ber.*, **42**, 4408 (1909); **43**, 937 (1910). Engo, *Chem.-Ztg.*, **32**, 1178 (1908). Kalb, *Ber.*, **42**, 3657 (1909); **43**, 937 (1910).

[103] Grandmougin, *Ber.*, **43**, 937 (1910). German patent 231,407; *Chem. Abstr.*, **5**, 2736 (1911).

[104] German patent 229,351; *Chem. Abstr.*, **5**, 2436 (1911). German patent 229,352; *Chem. Abstr.*, **5**, 2436 (1911). German patent 228,960; *Chem. Abstr.*, **5**, 2337 (1911). German patent 236,902; *Chem. Abstr.*, **6**, 1535 (1912).

5-bromo-2-nitrobenzaldehyde, acetone, and alkali[105] as well as through the reduction of 5-bromoisatin chloride.[106] 5,7,5,'7'-tetrabromoindigo has been prepared by the reduction of 5,7-dibromoisatin chloride with hydriodic acid[107] as well as from 4,6-dibromophenylglycine-2-carboxylic acid,[108] from 3,5-dibromo-2-nitrobenzaldehyde,[109] and from 3,5-dibromo-3-methylanthranil.[110] 4,4'-Dibromoindigo,[111] 7,7'-dibromoindigo,[111] 5,6,5',6'-tetrabromoindigo,[112] and 4,5,6,7,4',5',6',7 -octabromoindigo[113] have also been synthesized by conventional methods.

Of historic and biological interest is the proof of structure and synthesis, by Friedländer, of "antique purple," the dye obtained from the purple snail, *Murex brandaris*, found in the Mediterranean Sea. A gland of this snail contains a colorless substance which in sunlight develops the dye. In 1879 Schunck[114] isolated a small quantity of the dye. Years later Friedländer[115] isolated 1.5 grams of the dye from 12,000 snails and established its constitution as 6,6'-dibromoindigo which was synthesized in the following manner from 1-bromoanthranilic acid.[116] The compound has also been pre-

Br–C$_6$H$_3$(COOH)(NH$_2$) $\xrightarrow{ClCH_2COOH}$ Br–C$_6$H$_3$(COOH)(NHCH$_2$COOH) $\longrightarrow$ Br–C$_6$H$_3$<(CO)(NH)>CHCOOH $\longrightarrow$ 6,6'-dibromoindigo (Br–C$_6$H$_3$<(CO)(NH)>C=C<(NH)(CO)>C$_6$H$_3$–Br)

pared from 2-nitro-4-bromobenzaldehyde, acetone, and alkali.[117]

[105] Einhorn and Gernsheim, *Ann.*, **284**, 144 (1895). German patent 33,064; *Friedl.*, **1**, 146 (1888). German patent 135,564; *Chem. Zentr.*, **1902**, II, 1234. German patent 222,460; *ibid.*, **1910**, II, 119. Baeyer and Bloem, *Ber.*, **17**, 968 (1884). Kalb, *ibid.*, **42**, 3663 (1909).

[106] Baeyer and Bloem, *Ber.*, **12**, 1315 (1879). For other syntheses see Friedländer, Bruckner, and Deutsch, *Ann.*, **388**, 32 (1912) and German patent 273,340; *Chem. Zentr.*, **1914**, I, 1793.

[107] Danaila, *Compt. rend.*, **149**, 1384 (1909).

[108] Ullmann, and Kopetschni, *Ber.*, **44**, 430 (1911).

[109] Janse, *Rec. trav. chim.*, **40**, 285 (1921). [110] Bruining, *ibid.*, **41**, 655 (1922).

[111] Friedländer, Bruckner and Deutsch, *Ann.*, **388**, 32 (1912).

[112] Majima and Kotake, *Ber.*, **63**, 2237 (1930).

[113] Grandmougin, *Compt. rend.*, **173**, 982 (1921). Van de Bunt, *Rec. trav. chim.*, **48**, 121 (1929). [114] Schunck, *J. Chem. Soc.*, **35**, 591 (1879).

[115] Friedländer, *Monatsh.*, **28**, 991 (1907); *Ber.*, **42**, 765 (1909); *Chem. Ztg.*, **35**, 640 (1911).

[116] Friedländer, *Monatsh.*, **30**, 247 (1909). Sachs and Sichel, *Ber.*, **37**, 1868 (1904).

[117] Sachs and Kampf, *Ber.*, **36**, 3302 (1903). Rottig, *J. prakt. Chem.*, **142**, 35 (1935).

5,5′-Diiodoindigo has been prepared by the reduction of 5-iodoisatin-2-anilide[118] while 5,7,5′,7′-tetraiodoindigo has been prepared by the action of iodine monochloride on the bisulfite addition compound of dehydroindigo.[119] Similarly 4,4′-dichloro-5,7,5′,7′-tetraiodoindigo and 5,6,7,5′,6′,7′-hexaiodoindigo were prepared through the action of iodine monochloride on the sodium bisulfite addition compounds of 4,4′-dichlorodehydroindigo and of 6,6′-diiododehydroindigo.[120] 6,6′-Diiodoindigo has been prepared by diazotization of 6,6′-diaminoindigo and subsequent treatment with potassium iodide.[121]

5,5′-Difluoroindigo and 7,7′-difluoroindigo have both been prepared from *p*-fluoroaniline through use of the Sandmeyer indigo synthesis.[122]

5,5′-Dinitroindigo has been prepared by the direct nitration of indigo[123] as well as by the reduction of 5-nitroisatin chloride[124] and from 3-amino-5-nitrophenacyl bromide and sodium carbonate.[125] 4,4′-Dinitroindigo has been prepared from 3-nitrophenylglycine-2-carboxylic acid[126] while 6,6′-dinitroindigo has also been synthesized by conventional methods.[127]

The sulfonation of indigo[128] can be controlled to give either a mono- or a disulfonic acid while with fuming sulfuric acid either three or four sulfonic acid groups can be introduced. The tetrasulfonic acid has been shown to be the 5,7,5′,7′-derivative[129] since oxidation to the isatindisulfonic

[118] Borsche, Weussmann, and Fritzsche, *Ber.*, **57**, 1770 (1924).

[119] Kalb and Berrer, *ibid.*, **57**, 2105 (1924).

[120] Kalb and Berrer, *ibid.*, **57**, 2117 (1924).

[121] Grandmougin and Seyder, *ibid.*, **47**, 2370 (1914).

[122] Roe and Teague, *J. Am. Chem. Soc.*, **71**, 4019 (1949).

[123] German patent 242,149; *Chem. Abstr.*, **6**, 2182 (1912). Van Alphen, *Rec. trav. chim.*, **57**, 837 (1938).

[124] Baeyer, *Ber.*, **12**, 1316 (1879).

[125] Borsche and Berbert, *Ann.*, **546**, 293 (1941).

[126] Grandmougin, *Chimie & Industrie*, **11**, 113 (1923); *Chem. Abstr.*, **18**, 1201 (1924).

[127] Friedländer and Cohn, *Monatsh.*, **23**, 1006 (1902). Schwarz, *ibid.*, **26**, 1261 (1905).

[128] Crum, *Berz. Jahresb.*, **4**, 189 (1825). Berzelius, *ibid.*, **7**, 262 (1828). Joss, *ibid.*, **14**, 316 (1835). Dumas, *Ann.*, **22**, 73 (1873); **48**, 338 (1843). Schlieper and Schlieper, *ibid.*, **120**, 1 (1861). Vogel, *Ber.*, **11**, 1365 (1878). Heymann, *ibid.*, **24**, 1476 (1891). Juillard, *Bull. soc. chim.*, [3] **7**, 619 (1892). Vorländer and Schubert, *Ber.*, **34**, 1860 (1901). Eder, *Monatsh.*, **24**, 15 (1903). German patent 143,141; *Chem. Zentr.*, **1903**, II, 272. German patent 168,302; *Chem. Zentr.*, **1906**, I, 1204. Heiduschka, *Arch. Pharm.*, **244**, 569 (1905). German patent 180,097; *Chem. Zentr.*, **1907**, I, 1371. Schwalbe and Jochheim, *Ber.*, **41**, 3798 (1908). Tauber, *Chem.-Ztg.*, **33**, 418 (1909). Hofmann, *Ber.*, **45**, 3332 (1912). Martell, *Z. angew. Chem.*, **26**, 198 (1913). Wagner, *J. prakt. Chem.*, [2] **89**, 383 (1914).

[129] Grandmougin, *Compt. rend.*, **173**, 586 (1921).

acid and replacement of the $-SO_3H$ groups by bromine gave 5,7-dibromoisatin. The disulfonic acid, indigo-5,5′-disulfonic acid, has long been used as a vat dye under the trade names "Indigo Carmine" and "Saxony Blue" (the compound is blue in solution). It is also easily prepared through the action of fuming sulfuric acid on phenylglycine as well as from phenylglycine-2-carboxylic-4-sulfonic acid through treatment with acetic anhydride and sodium acetate which yield the indoxyl derivative which is then oxidized to the dye.

Alkyl and Acyl Derivatives of Indigo

Indigo reacts with the Grignard reagent to give a complex which gives on decomposition in the usual way a 3-hydroxy-3′-keto-3-alkyl-2,2′-diindolinylidene, also prepared by the action of alkyl halides or alkyl sulfates on the sodium salt of leucoindigo.[130]

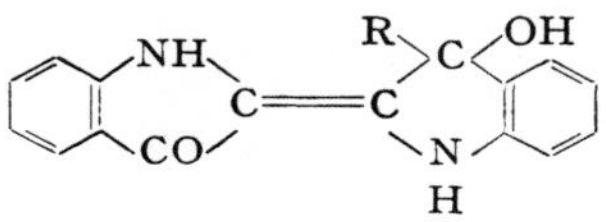

R C OH
NH
C=C
CO N
H

The *N*-alkyl derivatives of indigo[131] are not prepared by the alkylation of indigo but by synthesis from *N*-substituted intermediates. For example, *N*-methylisatin-2-anil yields 1,1′-dimethylindigo on reduction with ammonium sulfide. The substance is more soluble and more basic than indigo.

Many alkyl derivatives of indigo have been prepared through the application of conventional synthetic procedures. Thus 5,5′-dimethylindigo has been prepared from 6-nitro-3-methylbenzaldehyde through treatment with acetone and alkali,[132] by heating 4-acetamino-3-chloroacetyltoluene with alkali,[133, 133a] by the reduction of 5-methylisatin-2-*p*-tolylimide,[134] and by oxidation of 5-methylindoxylic acid.[135] 6,6′-Dimethylindigo has been prepared by heating chloroacetyl-*m*-toluidide with alkali and subsequent oxidation of the resulting product.[133a] 7,7′-Dimethylindigo has been prepared

[130] Sachs and Kantorowicz, *Ber.*, **42,** 1565 (1909). Chilkin, *J. Russ. Phys.-Chem. Soc.*, **48,** 1834 (1934); *Chem. Abstr.*, **8,** 912 (1914). Madelung, *Ber.*, **57,** 246 (1924).
[131] Baeyer, *Ber.*, **16,** 2201 (1883). Ettinger and Friedländer, *ibid.* **45,** 2074 (1912).
[132] German patent 113,604; *Chem. Zentr.*, **1900,** II, 751. Mayer, *Ber.*, **47,** 406 (1914).
[133] Kunckell, *Ber.*, **33,** 2648 (1909).
[133a] Kuhara and Chikashiga, *Am. Chem. J.*, **27,** 12 (1902).
[134] German patent 119,831; *Chem. Zentr.*, **1901,** I, 1075.
[135] Blank, *Ber.*, **31,** 1817 (1898).

similarly by several procedures.[136] 5,7,5′,7′-Tetramethylindigo has been prepared from 2-nitro-3,5-dimethylbenzaldehyde through the Baeyer-Drewson synthesis,[137] as well as by several other methods[133a, 138] while 4,6,7,4′,6′,7′-hexamethylindigo has been prepared in similar fashion from 6-nitro-2,4,5-trimethylbenzaldehyde.[139] α-Naphthindigo has been prepared from α-naphthylglycine by alkali fusion and subsequent oxidation[140, 140a] as well as from 6,7-benzoisatin-2-anil by reduction with ammonium sulfide,[141] while β-naphthindigo has been prepared similarly.[140, 142]

A number of dialkylindigos[143] have been prepared from the appropriate *N*-alkylanthranilic acids while 5,6,5′,6′-tetramethylindigo has been prepared by treating 2-chloroacetyl-4,5-dimethylacetanilide with alkali.[144]

1,1′-Diphenylindigo has been prepared through the oxidation of *N*-phenylindoxyl prepared in turn from *N*-phenyl-phenylglycine-*o*-carboxylic acid.

1,1′-Diacetylindigo is easily prepared from indigo by the combined action of acetyl chloride and acetic anhydride.[145] The compound has also been prepared through the oxidation of *N*-acetylindoxyl.[146] Oxalylindigo

[136] German patent 113,604; *Chem. Zentr.*, **1900**, II, 751. Mayer, *Ber.*, **47**, 406 (1914). Heumann, *ibid.*, **24**, 978 (1891). German patent 58,276; *Friedl.*, **3**, 276 (1896). German patent 63,310; *Friedl.*, **3**, 279 (1896). German patent 119,831; *Chem. Zentr.*, **1901**, I, 1075.

[137] Bamberger and Wiler, *J. prakt. Chem.*, [2] **57**, 361 (1898).

[138] German patent 61,711; *Friedl.*, **3**, 277 (1896).

[139] Gattermann, *Ann.*, **347**, 378 (1906).

[140] Wichelhaus, *Ber.*, **26**, 2547 (1893).

[140a] German patent 69,636; *Friedl.*, **3**, 286 (1896).

[141] German patent 153,418; *Chem. Zentr.*, **1904**, II, 679.

[142] Blank, *Ber.*, **31**, 1817 (1898).

[143] Van Alphen, *Rec. trav. chim.*, **61**, 201 (1942).

[144] Kranzlein, *Ber.*, **70**, 1776 (1937). For the synthesis of other homologs of indigo see: Grandmougin and Dessoulavy, *ibid.*, **42**, 3641, 4407 (1909). German patent 61,712; *Friedl.*, **3**, 278 (1896). Kunckell, *J. prakt. Chem.*, **89**, 324 (1914). Kunckell and Lillig, *ibid.*, **116**, 17 (1927). Rubenstein, *J. Chem. Soc.*, **127**, 1998 (1925). Tröger and Eicker, *J. prakt. Chem.*, **116**, 1295 (1933). Feist, Arve, and Volksen, *Ber.*, **69**, 2743 (1936). Chardonnens, *Helv. Chim. Acta*, **16**, 1295 (1933). Chardonnens and Venetz, *ibid.*, **22**, 853 (1939). Chardonnens and Heinrich, *ibid.*, **23**, 1399 (1940). Löw, *Ber.*, **18**, 950 (1885). Heumann, *ibid.*, **24**, 978 (1891). Gattermann, *Ann.*, **393**, 222 (1912). Grandmougin, *Ber.*, **42**, 4218 (1909). Kunckell and Schneider, *J. prakt. Chem.*, [2] **86**, 429 (1912). Ettinger and Friedländer, *Ber.*, **45**, 2078 (1912). Benedicenti, *Z. physiol. Chem.*, **53**, 188 (1907).

[145] Liebermann and Dickhuth, *Ber.*, **24**, 4131 (1891). Vorländer and Drescher, *ibid.*, **34**, 1859 (1901).

[146] Vorländer and Pfeiffer, *ibid.*, **52**, 325 (1919).

is obtained through the action of oxalyl chloride on a solution of indigo in nitrobenzene.[147]

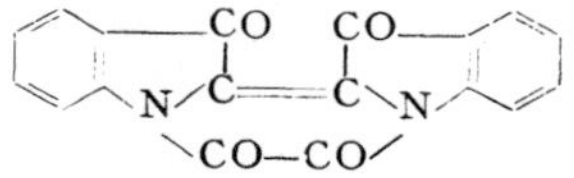

Indigo is converted into *N,N'*-dibenzoylindigo, m.p. 254° (286°)[150] by benzoylation in pyridine solution.[87, 148] Further heating with benzoyl chloride gives the "Dessoulavy" compound.[149] The exact structure of the Dessoulavy compound, $C_{30}H_{17}O_3N_2Cl$, colorless, m.p. 243°, as well as of the dye Höchst yellow, $C_{30}H_{18}O_4N_2$, m.p. 359°, formed from the compound by warming with sulfuric acid, Höchst yellow U, $C_{23}H_{12}O_2N_2$, orange-yellow crystals, m.p. 287°, obtained from the Dessoulavy compound by heating to 300–380° alone or in molten paraffin, and Ciba yellow 3 G, $C_{23}H_{12}O_2N_2$, greenish-yellow needles, m.p. 275–276°, prepared by heating indigo with benzoyl chloride and copper powder has been a matter of some controversy.[150, 150a]

Preparation of Indigoids

Indigoid derivatives can be obtained by condensing isatin chloride,

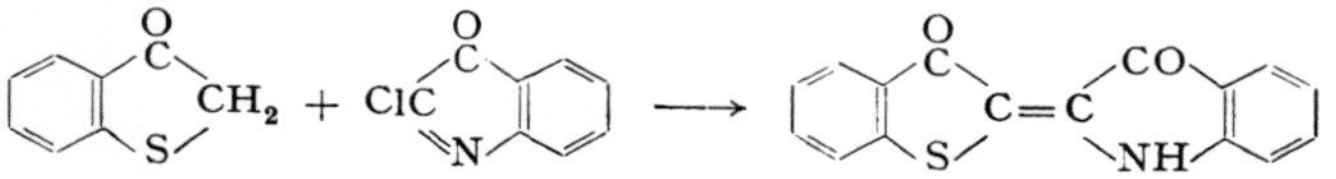

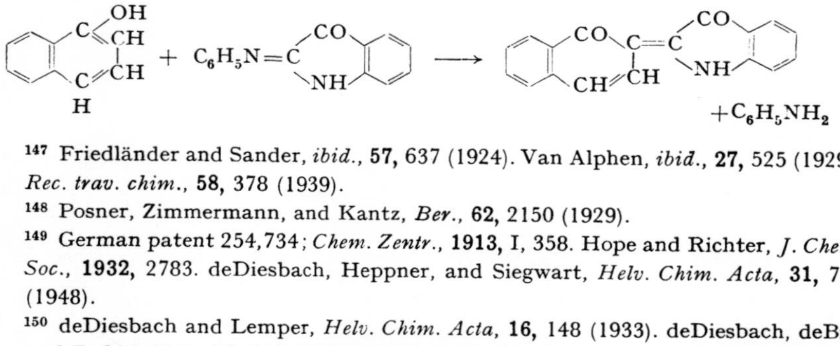

$+ C_6H_5NH_2$

[147] Friedländer and Sander, *ibid.*, **57**, 637 (1924). Van Alphen, *ibid.*, **27**, 525 (1929); *Rec. trav. chim.*, **58**, 378 (1939).

[148] Posner, Zimmermann, and Kantz, *Ber.*, **62**, 2150 (1929).

[149] German patent 254,734; *Chem. Zentr.*, **1913**, I, 358. Hope and Richter, *J. Chem. Soc.*, **1932**, 2783. deDiesbach, Heppner, and Siegwart, *Helv. Chim. Acta*, **31**, 724 (1948).

[150] deDiesbach and Lemper, *Helv. Chim. Acta*, **16**, 148 (1933). deDiesbach, deBie, and Rubki, *ibid.*, **17**, 113 (1934). deDiesbach, and Heppner, *ibid.*, **32**, 687 (1949). deDiesbach, Cupponi, and Farquet, *ibid.*, **32**, 1214 (1949). Hope, Kersey, and Richter, *J. Chem. Soc.*, **1933**, 1000. Hope and Anderson, *ibid.*, **1936**, 1474. Posner and Hofmeister, *Ber.*, **59**, 1827 (1926). Posner, Zimmermann, and Kantz, *ibid.*, **62**, 2153 (1929).

[150a] For a complete account of these compounds see Allen, *Six-Membered Heterocyclic Nitrogen Compounds with Four Condensed Rings* (this series, Vol. II) Interscience, New York-London, 1951, pp. 101–111.

isatin-2-anil, α-thioisatin, or *O*-alkyl ethers of isatin with compounds containing active methylene groups.

With isatin itself the β-carbonyl group reacts, isoindigoids being

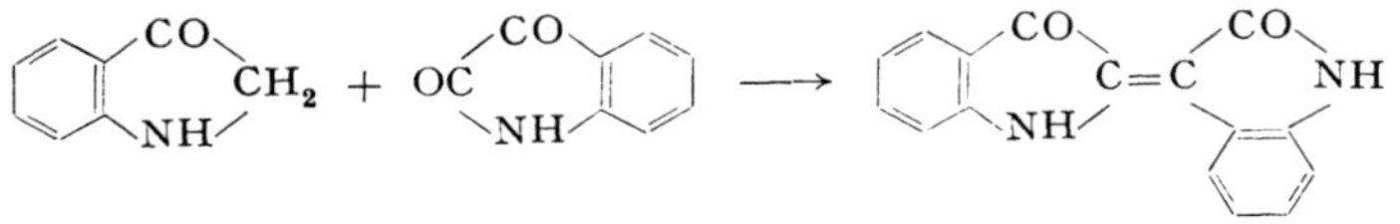

CO CH$_2$ NH + OC CO NH ⟶ CO C=C CO NH NH

obtained. Indigoids can also be prepared by the condensation of indoxyl with compounds containing active carbonyl groups. While the indigoids are

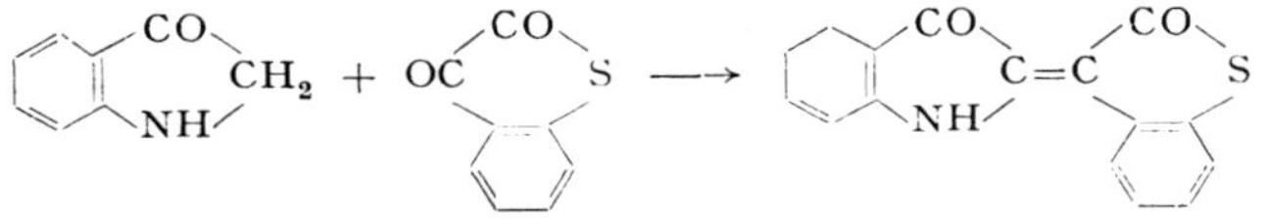

CO CH$_2$ NH + OC CO S ⟶ CO C=C CO S NH

are stable against acids they are split by alkali. With some of the derivatives the splitting takes place on warming with dilute alkali while with others long

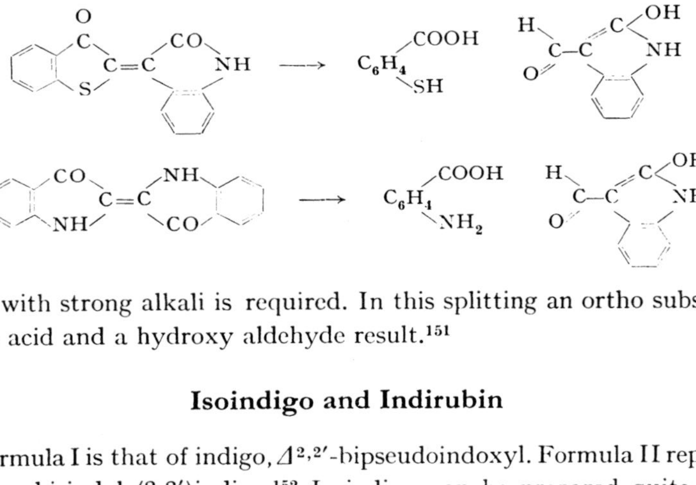

O C C=C CO NH S ⟶ C_6H_4 COOH SH H C OH C–C NH O

CO C=C NH NH CO ⟶ C_6H_4 COOH NH_2 H C OH C–C NH O

boiling with strong alkali is required. In this splitting an ortho substituted benzoic acid and a hydroxy aldehyde result.[151]

Isoindigo and Indirubin

Formula I is that of indigo, $\Delta^{2,2'}$-bipseudoindoxyl. Formula II represents isoindigo, bisindole(3,3')indigo.[152] Isoindigo can be prepared quite readily by condensing oxindole with isatin in acetic acid in the presence of hydrochloric acid. The substance crystallizes from methyl alcohol in garnet-red

[151] Friedländer, *Monatsh.*, **29,** 359 (1908); *Ber.*, **41,** 1035 (1908). Friedländer and Schwenck, *ibid.*, **43,** 1973 (1910).

[152] Dornier and Martinet, *Bull. soc. chim.*, [4] **33,** 786 (1923). Wahl and Hansen, *Compt. rend.*, **176,** 1071 (1923). Hansen, *Ann. chim. phys.*, [10] **1,** 100 (1924). Sander, *Ber.*, **58,** 820 (1925). Wahl and Bagard, *Compt. rend.*, **148,** 718 (1909); *Bull. soc. chim.*, [4] **5,** 1039 (1909); [4] **15,** 329 (1914). Stolle, *Ber.*, **47,** 2121 (1914).

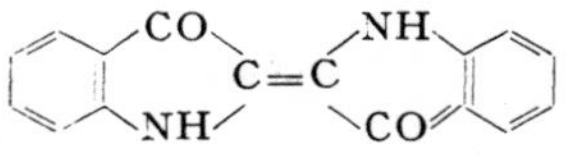

(I) 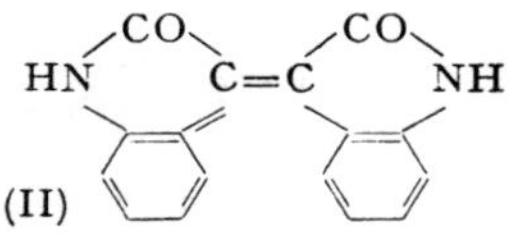 (II)

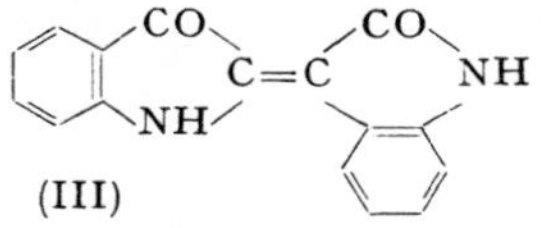

(III)

crystals. Unlike indigo and indirubin it is not reduced by sodium hydrosulfite solutions. Isoindigo has been prepared also by heating isatane[153] as well as from dithioisatide by the action of alcoholic alkali.

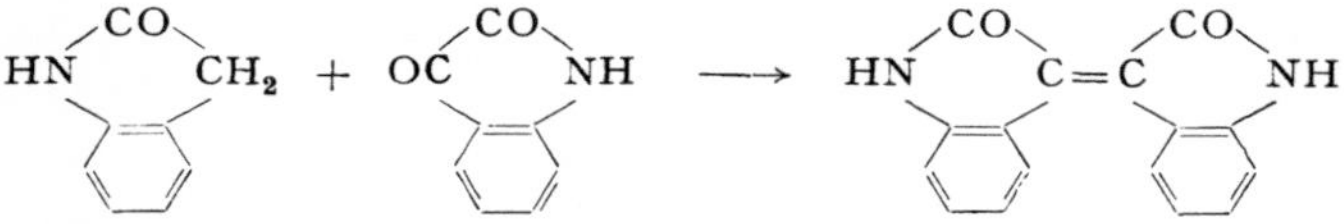

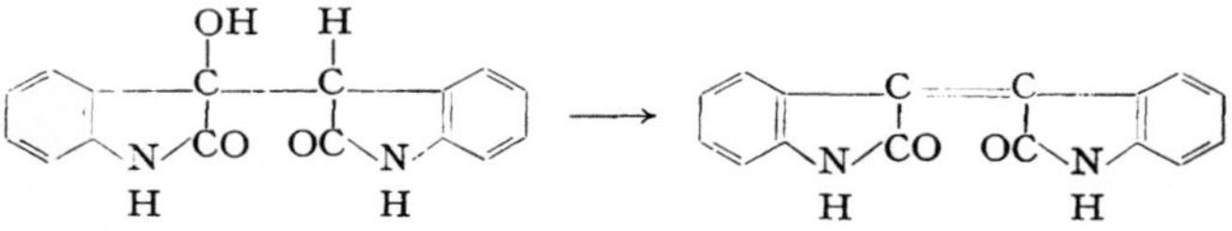

Formula III represents indirubin, bisindole(2,3′)indigo.[154] Indirubin was discovered in 1856 by Schunck as a by-product in the conversion of indican to indigo. It also results in small amounts when synthetic indoxyl is oxidized to indigo. It consequently is present as an impurity in most samples of indigo. Indirubin was first obtained artificially by Baeyer along with

[153] Laurent, *Ann. chim. phys.*, [3] **3**, 473 (1840).

[154] Schunck, *J. prakt. Chem.*, [1] **66**, 328 (1855); *Jahresber.*, **1855**, 666; **1858**, 468; *Ber.*, **12**, 1220 (1879). Baeyer and Emmerling, *Ber.*, **3**, 515 (1870). Baeyer, *ibid.*, **12**, 457 (1879); **14**, 1745 (1881); **16**, 2200 (1883). Forrer, *ibid.*, **17**, 975 (1884). Schunck and Marchlewski, *ibid.*, **28**, 539 ,2535 (1895). Kley, *Rec. trav. chim.*, **19**, 16 (1900). Maillard, *Compt. rend.*, **132**, 990 (1901); **134**, 470 (1902); *Bull. soc. chim.*, [3] **29**, 756 (1903); [4] **5**, 1153 (1909); [4] **9**, 202 (1911); *Z. physiol. Chem.*, **41**, 445 (1904). Vaubel, *Chem.-Ztg.*, **25**, 726 (1901). Ellinger, *Z. physiol. Chem.*, **41**, 29 (1904). Porcher, *Bull. soc. chim.*, [4] **5**, 531 (1909). Wahl and Bagard, *ibid.*, [4] **5**, 1043 (1909); *Compt. rend.*, **148**, 719 (1909); **156**, 898 (1913); *Bull. soc. chim.*, [4] **7**, 1090 (1910); [4] **9**, 56, 574 (1911); [4] **15**, 333 (1914). Thomas, Bloxam, and Perkin, *J. Chem. Soc.*, **95**, 824 (1909). Perkin and Thomas, *ibid.*, **95**, 801 (1909). Perkin, *ibid.*, **95**, 847 (1909). Bohn, *Ber.*, **43**, 997 (1910). Bloxam and Perkin, *J. Chem. Soc.*, **97**, 1460 (1910). Friedländer and Schwenck, *Ber.*, **43**, 1971 (1910). Ettinger and Friedländer, *ibid.*, **45**, 2081 (1912). Martinet, *Compt. rend.*, **169**, 183 (1919).

indigo in the reduction of isatin chloride and synthetically in pure form by condensing indoxyl with isatin. The compound can also be prepared by

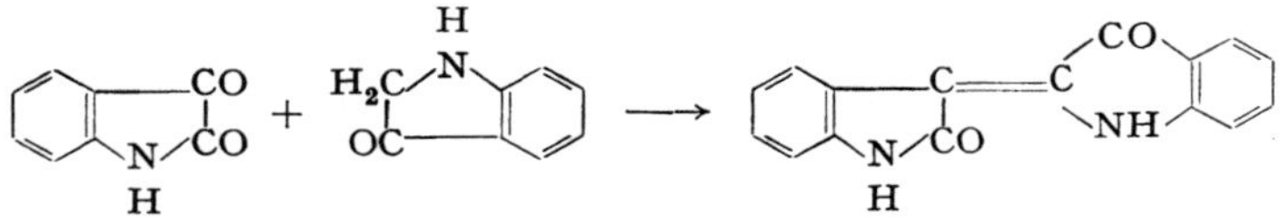

condensing oxindole with isatin chloride, isatin 2-anilide, or with *O*-methylisatin. Indirubin crystallizes from aniline in chocolate-brown needles and from alcohol and other solvents in cherry-red crystals. It is oxidized in alcoholic solution by air or by hydrogen peroxide to isatin.[155] Reduction of indirubin with zinc dust in boiling acetic acid gives indileucin.[155,156] Thio-

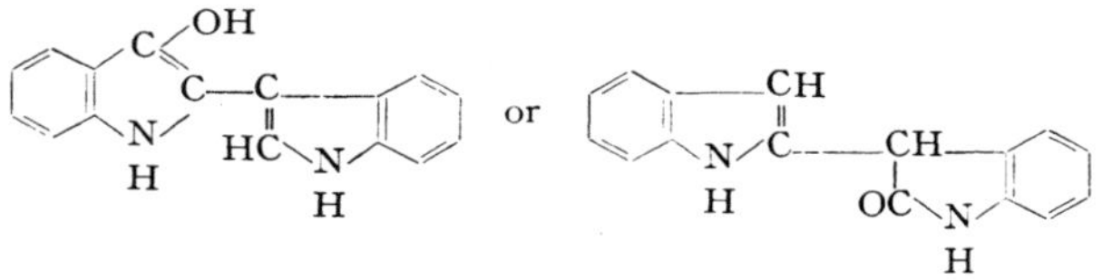

or

indigo scarlet-R is prepared by condensing thionaphthenequinone and oxindole (see page 144).[157]

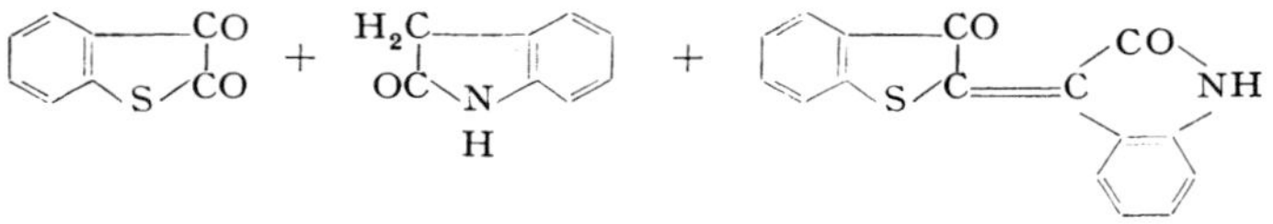

The condensation of 3-oxythionaphthene and isatin-2-anil gives thioindigo scarlet-2′-anil[158] which like isatin-anil occurs in tautomeric forms.

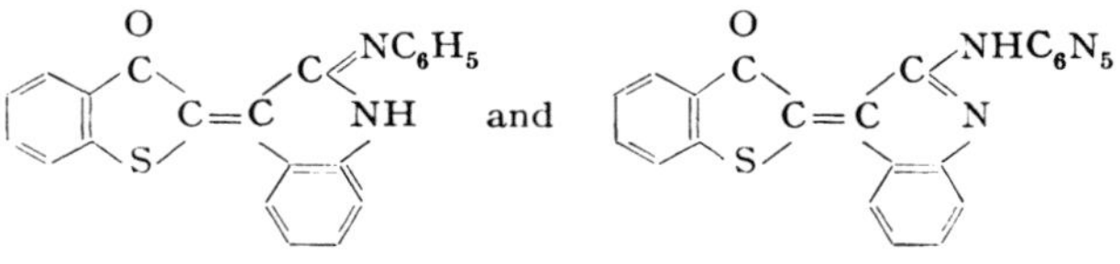

and

Indigoids in numbers far too great to permit enumeration have been prepared through the condensation of indoxyl with compounds containing

[155] Schunck and Marchlewski, *Ber.*, **28**, 541 (1895).

[156] Forrer, *ibid.*, **17**, 976 (1884).

[157] Bezdzik and Friedländer, *Monatsh.*, **29**, 375 (1908). Felix and Friedländer, *ibid.*, **31**, 55 (1910).

[158] Pummerer, *Ber.*, **44**, 341 (1911).

carbonyl groups and of isatin-2-anil, isatin chloride, and isatin-*O*-alkyl ethers with compounds containing active methylene groups while iso-indogenides in large numbers have been prepared from isatin and compounds containing active methylene groups. Many of these compounds have found application as vat dyes.

CHAPTER VIII

Natural Products Containing the Indole Nucleus

Of the large number of nitrogen bases occurring in nature, a considerable portion contain the indole nucleus. These bases, the majority of which are true alkaloids, that is, occur in seed-bearing plants, vary in complexity from the relatively simple gramine to the extremely intricate strychnine molecule.

The members of this class of alkaloids, with the single exception of gramine, have a common grouping other than the indole nucleus, in the form of the β-aminoethylindole skeleton (I). The presence of such a system

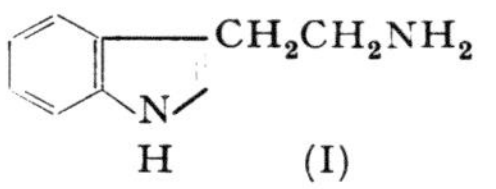

$CH_2CH_2NH_2$

H (I)

in the structures which have been established suggests a relationship to the amino acid tryptophan, and indicates that this substance, or its precursors or transformation products, may be the actual progenitors of these bases.

Simple Bases

Gramine

The alkaloid gramine, $C_{11}H_{14}N_2$, m.p. 133–136°, $[\alpha]_D \pm 0°$, was first observed in barley mutants,[1] and later shown[2] to be identical with donaxine, obtained from the Asiatic reed, *Arundo donax*.[3] Evidence for the indole nucleus came from the ultraviolet absorption spectrum and from a positive pine splinter test, and was confirmed by the isolation of skatole (II) from a

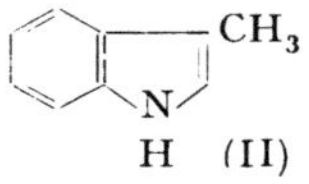

CH_3

H (II)

[1] von Euler and Helström, *Z. physiol. Chem.*, **208**, 43 (1932); **217**, 23 (1923). Von Euler and Erdtman, *Ann.*, **520**, 1 (1935).

zinc dust distillation of the alkaloid.[2] The original supposition that a substituent was present in the α-position of the indole nucleus was thus eliminated by this evidence and the demonstration[4] that gramine was identical with 3-(dimethylaminomethyl)indole (III), prepared from indolylmagnesium iodide and dimethylaminoacetonitrile. Gramine has since been

C_2H_5MgI → N-MgI; $NCCH_2N(CH_3)_2$ → $CH_2N(CH_3)_2$ (III)

+ CH_2O + $HN(CH_3)_2$

prepared by the Mannich reaction from indole, formaldehyde, and dimethylamine.[5]

Donaxarine

This base, $C_{13}H_{16}O_2N_2$, m.p. 217°, $[\alpha]_D \pm 0°$, was also isolated from *Arundo donax*.[6] Although it does not give the normal indole color reactions, the vapor produces a positive pine splinter test. It is reported to contain an active hydrogen atom and a methylamino group, but no methoxyl or *C*-methyl groups.

Tryptophan

This amino acid, $C_{11}H_{12}O_2N_2$, m.p. 289°, $[\alpha]_D$ —33°, was first detected and named[7] as the result of several color reactions given by proteins containing this substance. The crystalline material was first isolated in 1901.[8]

Tryptophan was found to give indole and skatole on heating, and to be oxidized by ferric chloride to indole-3-aldehyde (V), and by permanganate to indole-3-carboxylic acid.[9] This led to the postulation of its structure as 2-amino-3-(3-indolyl)propionic acid (IV). Confirmation of this structure came

[2] von Euler, Erdtman, and Helström, *Ber.*, **69**, 743 (1936). Orekhov and Norkina *J. Gen. Chem.* (*U. S. S. R.*), **7**, 673 (1937); *Chem. Abstr.*, **31**, 5801 (1937).

[3] Orekhov, Norkina, and Maximova, *Ber.*, **68**, 436 (1935).

[4] Wieland and Hsing, *Ann.*, **526**, 188 (1936).

[5] Kuhn and Stein, *Ber.*, **70**, 567 (1937). Snyder and Eliel, *J. Am. Chem. Soc.*, **70**, 107 (1948).

[6] Madinaveita, *Ber.*, **68**, 436 (1935); *J. Chem. Soc.*, **1937**, 1927.

[7] Neumeister, *Z. Biol.*, **26**, 329 (1890).

[8] Hopkins and Cole, *J. Physiol.*, **27**, 418 (1901); *J. Chem. Soc.*, **82**, I, 193 (1902).

[9] Ellinger, *Ber.*, **39**, 2515 (1906).

(IV) [indole-3-]$CH_2CHCOOH$ with NH_2 on the CH; indole N–H

as the result of the synthesis of the racemic amino acid from indole-3-aldehyde through the azlactone (VI).[10]

[indole-3-]CHO + $C_6H_5CONHCH_2COOH$ (V) ⟶

(VI) [indole-3-]$CH{=}C$–CO (azlactone ring: N=C(–C_6H_5)–O) ⟶ [indole-3-]$CH{=}C$–COOH with NH–CO–C_6H_5 on the C; $\xrightarrow[C_2H_5OH]{Na}$ (IV)

Several recent syntheses of tryptophan have been based on the discovery that gramine and its methiodide will condense with compounds having active methylene groups.[11] Thus, the methiodide (VII) will alkylate

[indole-3-]$CH_2N(CH_3)_3I$ (VII) + $CH_3CONHCH(COOC_2H_5)_2$ → [indole-3-]$CH_2C(COOC_2H_5)_2$ with $NHCOCH_3$ on the C (VIII) → (IV)

acetylaminomalonic ester to give VIII, which may be hydrolyzed to tryptophan.

A similar preparation using gramine (IX) has been developed[12] and applied to the synthesis of trytophan derivatives.[13]

[indole-3-]$CH_2N(CH_3)_2$ (XI) + $O_2NCH_2COOC_2H_5$ → [indole-3-]$CH_2CHCO_2C_2H_5$ with NO_2 on the CH $\xrightarrow[(2)\ H_2O]{(1)\ H_2}$ (IV)

[10] Ellinger and Flamand, *Ber.*, **40**, 3029 (1907); *Z. physiol. Chem.*, **55**, 8 (1908). Boyd and Robson, *Biochem. J.*, **29**, 2256 (1935). Robson, *J. Biol. Chem.*, **62**, 405 (1924). Majima, *Ber.*, **55**, 3859 (1922).

[11] Snyder and Smith, *J. Am. Chem. Soc.*, **66**, 350 (1944). Albertson and Tuller, *ibid.*, **67**, 502 (1945). Albertson, Archer, and Suter, *ibid.*, **66**, 500 (1944); **67**, 36 (1945). Howe, Zambito, Snyder, and Tishler, *ibid.*, **67**, 38 (1945). Jackson and Archer, *ibid.*, **68**, 2105 (1946). Albertson, Tuller, King, Fishburn, and Archer, *ibid.*, **70**, 1150 (1948). Hegedus, *Helv. Chim. Acta*, **29**, 1499 (1946). Vejdelek, *Chem. Listy*, **44**, 73 (1950); *Chem. Abstr.*, **45**, 8004 (1951). Snyder, Beilfuss, and Williams, *J. A. C. S.*, **75**, 1873 (1953).

[12] Lyttle and Weisblatt, *J. A. C. S.*, **69**, 2118 (1947). U.S. pat. 2, 557, 041.

[13] Snyder and Katz, *ibid.*, **69**, 3140 (1947).

In addition to the resolution of the synthetic material, L-tryptophan can be prepared by the tryptic digestion of casein.[14] Hydrolysis with acid or base results in destruction of the amino acid.

L-Tryptophan is one of the essential amino acids,[15] normal growth being impossible on diets deficient in it. Nevertheless, it may be replaced by the D-isomer, as well as a number of derivatives and related products.[16]

The amino acid is converted by the dog, rabbit, and rat into kynurenic acid (X). Microorganisms effect several types of modifications of the

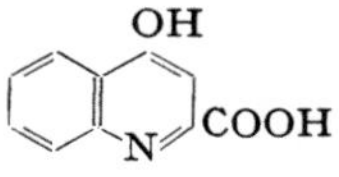

OH
COOH
N
(X)

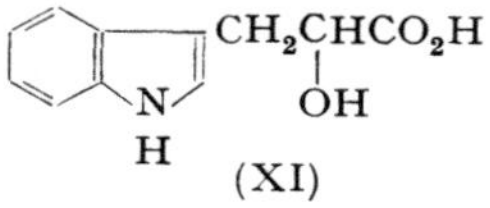

CH_2CHCO_2H
OH
N
H
(XI)

tryptophan molecule, producing such products as indolyllactic acid (XI),[17] β-aminoethylindole (XII),[18] and indole-3-propionic acid.[19]

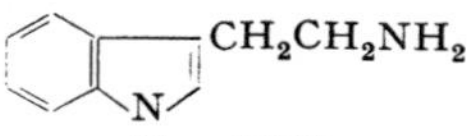

$CH_2CH_2NH_2$
N
H (XII)

Serotonin

The vasoconstrictor substance, $C_{14}H_{21}O_3N_5 \cdot H_2SO_4$, m.p. 214–215°, $[\alpha]_D \pm 0°$, was isolated from beef serum.[20] It contains an *N*-methyl group, and was found by hydrolysis to be composed of equimolecular portions of sulfuric acid, creatinine (XIII), and an indole base, $C_{10}H_{12}ON_2$, named serotonin.[21] On the basis of its absorption spectrum, color tests, and a

[14] Hopkins and Cole, *Zentr.*, **1903,** II, 1011. Onslow, *Biochem. J.*, **15,** 383 (1931). Cox and King, *Organic Synthesis*, Vol. 10, Wiley, New York, 1930, p. 100.

[15] Willock and Hopkins, *J. Physiol.*, **35,** 88 (1906–1907). Fasal, *Bio. Zeit.*, **44,** 400 (1912). Abderhalden, *Z. physiol. Chem.*, **83,** 452 (1913).

[16] du Vigneaud, Sealock, and Van Etten, *J. Biol. Chem.*, **98,** 565 (1932). Berg, *ibid.*, **91,** 513 (1931); **104,** 373 (1934). Bauguess and Berg, *ibid.*, **104,** 675 (1934); **106,** 615 (1934). Berg, Rose, and Marvel, *ibid.*, **85,** 207 (1929).

[17] Ehrlich and Jacobson, *Ber.*, **44,** 890 (1911).

[18] Bertholet and Bertrand, *Compt. rend.*, **154,** 1826 (1912).

[19] Woods, *Biochem. J.*, **29,** 640 (1935).

[20] Rapport, Green, and Page, *J. Biol. Chem.*, **174,** 735 (1948); **176, 1243** (1948).

[21] Rapport, *ibid.*, **180, 961** (1949).

possible biological relationship to bufotenine and physostigmine, the structure assigned was that of 5-hydroxytryptamine (XIV).[21] This structure for

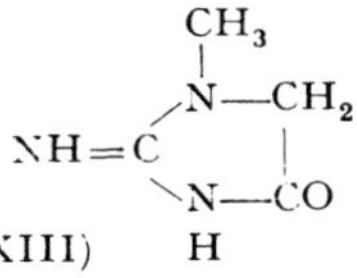

CH_3 / N—CH_2 / NH=C / N—CO / H (XIII)

HO / $CH_2CH_2NH_2$ / N / H (XIV)

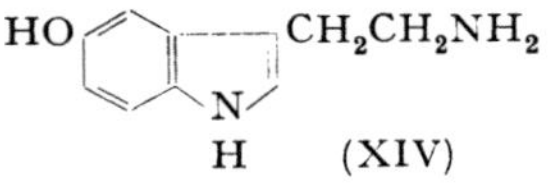

serotonin has been confirmed by synthesis,[22,23] and the synthetic material converted into serotonin creatinine sulfate, identical with the natural product.[23]

Abrine

This tryptophan derivative, $C_{12}H_{14}O_2N_2$, m.p. 295°, $[\alpha]_D$ +44°, was isolated from the seeds of *Abrus precatorus*.[24] Its nature was determined[25] as *N*-methyltryptophan (XV) by decarboxylation to *N*-methyltryptamine (XVI), and the racemic material (XV) has been synthesized.[26] The con-

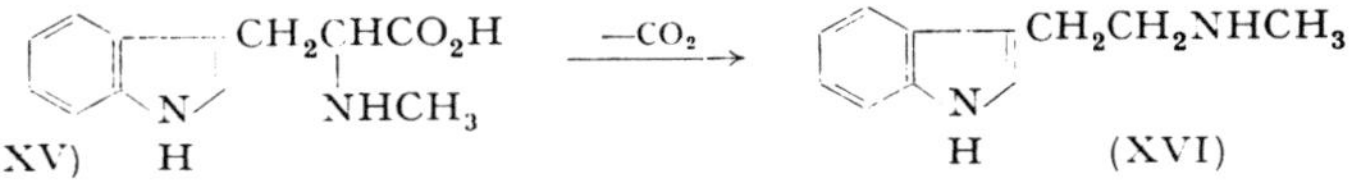

CH_2CHCO_2H / $NHCH_3$ / N / H (XV) $\xrightarrow{-CO_2}$ $CH_2CH_2NHCH_3$ / N / H (XVI)

figuration of the asymmetric carbon atom has been shown to be the same as in tryptophan, and the alkaloid is then L-(–)-abrine.[27]

Hypaphorine

Several species of *Erythrina*,[28] originally *E. hypaphorus*,[29] have been found to contain hypaphorine, $C_{14}H_{18}O_2N_2$, m.p. 255°, $[\alpha]_D$ +91–93°. The indole nature of the molecule was shown by its cleavage to indole and trimethylamine in the presence of hot aqueous potassium hydroxide,[30] and the

[22] Hamlin and Fischer, *J. Am. Chem. Soc.*, **73**, 5007 (1951).

[23] Speeter, Heinzelmann, and Weisblat, *ibid.*, **73**, 5515 (1951).

[24] Ghatak, *Bull. Acad. Sci. United Provinces Agra and Oudh India*, **3**, 295 (1934); *Chem. Abstr.*, **29**, 3344 (1935). Ghatak and Kaul, *J. Indian Chem. Soc.*, **9**, 383 (1932)

[25] Hoshino, *Proc. Imp. Acad. (Tokyo)*, **11**, 227 (1935); *Chem. Abstr.*, **29**, 6596 (1935); *Ann.* **520**, 31 (1935).

[26] Miller and Robson, *J. Chem. Soc.*, **1938**, 1910.

[27] Cahill and Jackson, *J. Biol. Chem.*, **126**, 29 (1938).

[28] Maranon and Santos, *Philippine J. Sci.*, **48**, 563 (1932); *Chem. Abstr.*, **26**, 5609 (1932). Rao, Rao, and Seshadre, *Proc. Indian Acad. Sci.*, **7**, 179 (1938). Deulofeu, Hug, and Mazzocco, *J. Chem. Soc.*, **1939**, 1841.

[29] Greshoff, *Ber*, **23**, 3537 (1890).

[30] van Romburgh, *Koninkl. Akad. Wetenschap. Amsterdam, wisk. en Natk. Afd.*, **19**, 1250 (1912); *Chem. Abstr.*, **5**, 3417 (1912).

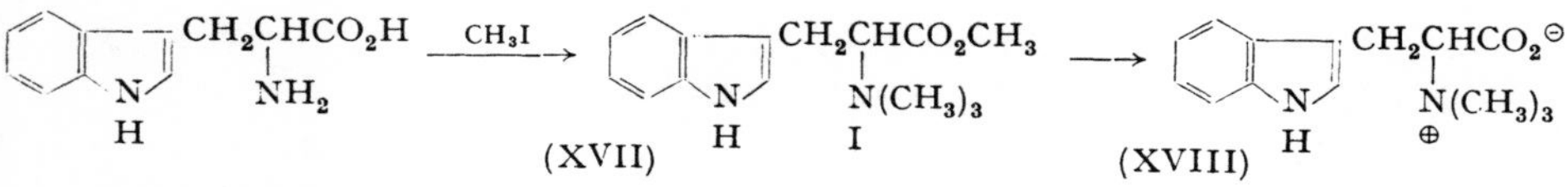

(XVII) (XVIII)

proposed betaine structure (XVIII) was confirmed by synthesis[31] from L-tryptophan through the ester-quaternary iodide (XVII).

Bufotenines

Along with steroidal substances, and sometimes adrenaline, the skin secretions of toads contain several alkaloid-like bases. The presence of these animal alkaloids was established by Handrovsky,[32] although the name bufotenine had previously been applied to a crude base.[33]

Bufotenine, $C_{12}H_{16}ON_2$, m.p. 146–147°, $[\alpha]_D \pm 0°$, from the European toad *Bufo vulgaris*,[34] was found to contain one basic nitrogen atom and two active hydrogen atoms.[35] It gives a blue color with ferric chloride and a positive test with the Hopkins–Cole reagent, indicating an indole derivative unsubstituted in the 2-position, possibly containing a phenolic hydroxyl group. Bufotenine reacts with one mole of methyl iodide to produce a quaternary base, the picrate of which is identical with the picrate of bufotenidine, $C_{15}H_{18}ON_2$, a similar product obtained from the Japanese toad *Senso*.[35] The quaternary base, on heating, splits into trimethylamine and a substance giving indole color tests. That bufotenine and bufotenidine are derivatives of tryptophan was proposed[36] and the structures of the bases were established by synthesis. The methoxytryptamine (XIX), prepared from 5-methoxyindole, on methylation yields *O*-methylbufotenine methi-

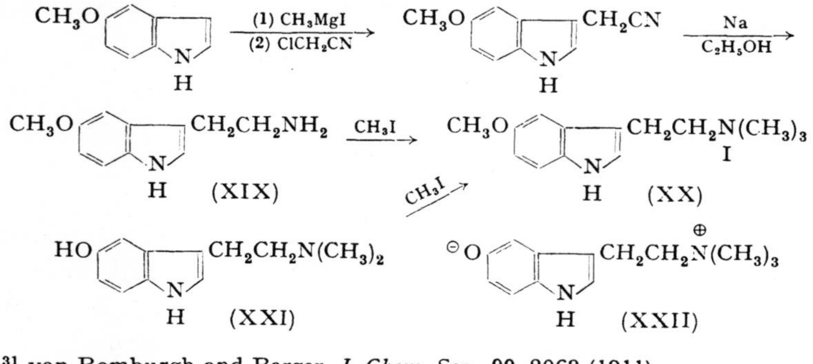

(XIX) (XX) (XXI) (XXII)

[31] van Romburgh and Barger, *J. Chem. Soc.*, **99**, 2068 (1911).

[32] Handrovsky, *Arch. exptl. Pathol. Pharmakol.*, **86**, 138 (1920).

[33] Phisalix and Bertrand, *Compt. rend. soc. biol.*, **45**, 477 (1893).

[34] Wieland, Hesse, and Mittasch, *Ber.*, **64**, 2099 (1931).

[35] Wieland, Konz, and Mittasch, *Ann.*, **513**, 1 (1934).

[36] Jensen and Chen, *Ber.*, **65**, 1310 (1932).

odide (XX), identical with the material obtained from bufotenine on methylation. Thus bufotenine is XXI and bufotenidine is XXII.

The synthesis of bufotenine has been accomplished[37] from 3-ethoxytryptophol (XXIII) through the bromide by amination and subsequent ether cleavage.

C_2H_5O–indole–CH_2CH_2OH (XXIII) ⟶ C_2H_5O–indole–CH_2CH_2Br $\xrightarrow{(CH_3)_2NH}$

C_2H_5O–indole–$CH_2CH_2N(CH_3)_2$ $\xrightarrow[C_6H_6]{AlCl_3}$ HO–indole–$CH_2CH_2N(CH_3)_2$ (XXI)

Accompanying bufotenine and bufotenidine in toad venom are the related bufothionene, $C_{12}H_{14}N_2SO_4$, m.p. 250°,[35,38] and dehydrobufotenine, $C_{12}H_{14}ON_2$, m.p. 199°.[39] Bufothionine is hydrolysed to dehydrobufotenine by dilute acid. Hydrogenation of the latter yields bufotenine. The unsaturation was located by oxidation, which gave formic acid and dimethylamine, establishing the structure XXIV and XXV for bufothionine and dehydrobufotenine, respectively.[40]

$^{\ominus}O_2SO$–indole–$CH{=}CH\overset{\oplus}{N}H(CH_3)_2$ (XXIV) ⟶ HO–indole–$CH{=}CHN(CH_3)_2$ (XXV)

(XXIV) $\xrightarrow{(O)}$ $HCO_2H + (CH_3)_2NH$; (XXV) $\xrightarrow{H_2}$ (XXI)

It is interesting to note that the 5-hydroxy group in the bufotenines is known to occur in only two other natural products, physostigmine and serotonin. The unsaturated side chain in the dehydro compounds bears a formal resemblance to neurine, $CH_2{:}CHN(CH_3)_2$, a substance found in the adrenal glands.

Dipterine

This alkaloid, optically inactive, $C_{11}H_{14}N_2$, m.p. 87–88°, is found in *Girgensohnia dipteria*.[41] It contains a secondary amino group, gives indole

[37] Hoshino and Shimodaira, *Ann.*, **520,** 19 (1935). See also Hoshino and Shimodaira *Bull. Chem. Soc. Japan,* **11,** 221 (1936); *Chem. Abstr.*, **30,** 5892 (1936).

[38] Wieland and Vocke, *Ann.*, **481,** 215 (1930). Jensen, *J. Am. Chem. Soc.*, **57,** 1765 (1935).

[39] Chen and Chen, *J. Pharmacol.*, **49,** 1, 514 (1933).

[40] Wieland and Wieland, *Ann.*, **528,** 324 (1937).

[41] Yurashevskii and Stepanov, *J. Gen. Chem.* (*U. S. S. R.*), **9,** 2203 (1939); *Chem. Abstr.*, **34,** 4071 (1940).

color reactions, and furnishes skatole on zinc dust distillation. Exhaustive methylation yielded trimethylamine, and comparison with the previously

$CH_2CH_2NHCH_3$

N
H (XXVI)

prepared[42] *N*-methyltryptamine (XXVI) showed the two to be identical.[43]

Calabar Alkaloids

The seeds of *Physostigma venenosum*, long used for the administration of "divine justice," contain several alkaloids, the most prominent of which is physostigmine.

Physostigmine

This base, also known as eserine, $C_{15}H_{21}O_2N_3$, m.p. 105–106°, $[\alpha]_D$ —76°, was first isolated in 1864[44] and obtained crystalline the following year.[45] It behaves as a monoacidic tertiary base, containing two methylamino groups.[46] The rather unusual presence of a urethan grouping was indicated by the basic hydrolysis of the alkaloid to eseroline, methylamine, and carbon dioxide.[47] If sodium ethoxide is used for the hydrolysis, the products are eseroline and methyl urethan. Heating physostigmine at the melting point

$$\underset{\text{Physostigmine}}{C_{15}H_{21}O_2N_3} \xrightarrow{H_2O} \underset{\text{Eseroline}}{C_{13}H_{18}ON_2} + CH_3NH_2 + CO_2$$

$$C_{15}H_{21}O_2N_3 \underset{CH_3NCO}{\overset{NaOC_2H_5}{\rightleftarrows}} C_{13}H_{18}ON_2 + CH_3NHCO_2C_2H_5$$

evolves methyl isocyanate. Physostigmine may be reformed by treating eseroline with methyl isocyanate.[48]

An indole nucleus was suggested by the isolation of 1-methylindole (I) from the zinc dust distillation of physostigmine.[49] Confirmation of the indole grouping was obtained by the characterization of a second degradation

[42] Hoshino and Kobayashi, *Ann.*, **520**, 31 (1935).
[43] Yurashevskii, *J. Gen. Chem.* (*U. S. S. R.*), **10**, 1781 (1940); *Chem. Abstr.*, **35**, 4016 (1941).
[44] Jobst and Hesse, *Ann.*, **129**, 115 (1864).
[45] Vee, *Jahresber.*, **1865**, 456.
[46] Straus, *Ann.*, **406**, 332 (1914).
[47] Polonovski, *Bull. soc. chim.*, **17**, 235 (1915).
[48] Polonovski and Nitzberg, *ibid.*, **19**, 270 (1916).
[49] Salway, *J. Chem. Soc.*, **101**, 978 (1912).

(I)

product. Eseroline methiodide, on heating to 200°, furnishes the base $C_{10}H_{11}ON$, physostigmol, which exhibits typical indole color reactions.[46] This substance contains a methylamino group and a phenolic hydroxyl function. Eseroline ethyl ether (eserethole) may be similarly degraded, giving physostigmol ethyl ether, $C_{12}H_{15}ON$.

The constitution of physostigmol (II, H for C_2H_5), as proposed by Straus,[46] has been confirmed by synthesis.[50] Condensation of *p*-ethoxy-*unsym*-phenylmethylhydrazine with α-ketoglutaric acid, followed by decarboxylation, yielded 1,3-dimethyl-5-ethoxyindole, identical with physostigmol ethyl ether (II). This product has been obtained by other methods.[51]

(II)

The reduction of eserethole[52,53] adds two hydrogen atoms, producing a secondary base, indicating that a ring has been opened. Expansion of the C_3H_7N residue necessary to develop the physostigmol formula into that of eserethole to $-CH_2CH_2N(CH_3)-CH_3$, remembering that the nitrogen atom is

(III) (IV)

tertiary, suggests a pyrrolidine ring as in the representation (III) for eserethole.[53] Dihydroeserethole would then be IV.

This formulation for eserethole is further substantiated by the reactions of the quaternary salts derived from the molecule. Eserethole

[50] Stedman, *ibid.*, **125**, 1373 (1924).

[51] Spath and Brunner, *Ber.*, **58**, 518 (1925). Keimatsu and Sugasawa, *J. Pharm. Soc. Japan*, **48**, 348 (1928); *Chem. Abstr.*, **22**, 3163 (1928).

[52] Polonovski, *Bull. soc. chim.*, **23**, 217 (1918).

[53] Stedman and Barger, *J. Chem. Soc.*, **127**, 274 (1925).

methiodide (V) is transformed into eserethole methine by treatment with base,[54] a reaction process which is reversed by the action of hydriodic acid.[55] This transformation has its counterpart in the case of the alkylene indolines.[56] It was found[57] that the methine is in reality a pseudo base (VI) formed by isomerization, and not a des-base produced by loss of water, as originally assumed. Oxidation of the methine gives dehydroeserethole methine (VII), which may be degraded to 1,3-dimethyl-3-ethyl-5-ethoxyoxindole (VIII).[53]

Eserethole methine may be further characterized as an indolinol by the formation of a methiodide (IX), which with picric acid yields a diquaternary picrate.[53]

The structure of dehydroesermethole methine has been confirmed by synthesis.[58] 2-Methyl-3-(2-phenoxyethyl)-5-methoxyindole (XI), prepared from the *p*-methoxyphenylhydrazone of 1-phenoxy-4-pentanone, was alkylated with methyl iodide under pressure to give the indolinium iodide

[54] Polonovski and Polonovski, *Bull. soc. chim.*, **23**, 357 (1918).

[55] Polonovski and Polonovski, *ibid.*, **23**, 335 (1918); *ibid.*, **33**, 1126 (1923).

[56] Brunner, *Ber.*, **38**, 1359 (1905).

[57] Polonovski and Polonovski, *Compt. rend.*, **177**, 127 (1923).

[58] King and Robinson, *J. Chem. Soc.*, **1932**, 326.

(XII). Treatment of the quaternary base with cold alkali produces the methylene indoline (XIII), which on oxidation with permanganate is converted to the oxindole (XIV). Cleavage of the ether with hydrobromic acid and replacement of the bromine with a dimethylamino group yielded *dl*-dehydroesermethole methine (XV), resolvable into its antipodes, the l-isomer proving identical with the corresponding base from the degradation of physostigmine.

The first synthesis of the complete ring system of physostigmine was achieved by the preparation of *dl*-noreserethole,[59] which could be methylated to give *dl*-eserethole. 3-Methyl-3-(2-phthalimidoethyl)-5-ethoxy indoline (XVI), produced by the Fischer synthesis, was methylated, and the phthalimido group removed. Ring closure occurred upon acidification to form *dl*-

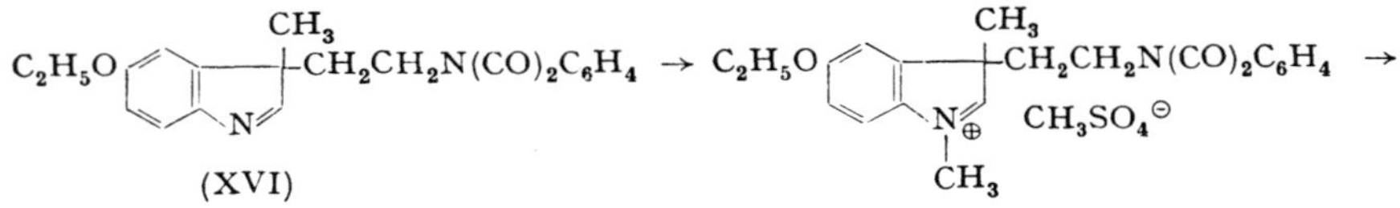

C_2H_5O, CH_3, $-CH_2CH_2N(CO)_2C_6H_4$ (XVI) → C_2H_5O, CH_3, $-CH_2CH_2N(CO)_2C_6H_4$, $N^{\oplus}$–CH_3, $CH_3SO_4^{\ominus}$ →

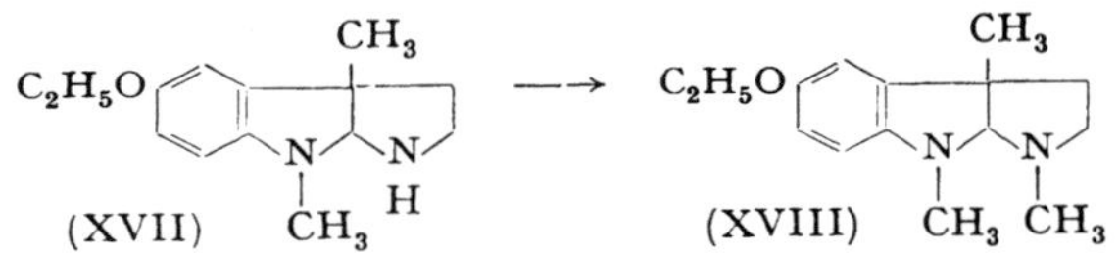

C_2H_5O, CH_3, N–CH_3, N–H (XVII) ⟶ C_2H_5O, CH_3, N–CH_3, N–CH_3 (XVIII)

noreserethole (XVII). Methylation to *dl*-eserethole (XVIII) was effected with methyl *p*-toluenesulfonate.

Other forms of the physostigmine system have been prepared in an extensive series of investigations by Hoshino and collaborators[60] and by Julian and Pikl.[61]

The synthesis of l-eseroline,[62] and hence the total synthesis of physostigmine, in view of the previous conversion of l-eseroline into the parent alkaloid, was completed by Julian and Pikl. Alkylation of 1,3-dimethyl-5-ethoxyoxindole (XIX) with chloroacetonitrile yielded XX, which was catalytically hydrogenated to the amine (XXI). The monomethyl derivative (XXIII) was formed through the anil (XXII). Resolution of XXIII and reductive cyclization by sodium in alcohol of the l-isomer furnished

[59] Robinson and Suginome, *ibid.*, **1932**, 298, 304. King, Liquori, and Robinson, *ibid.*, **1933**, 1475. See also: *ibid.*, **1932**, 1433; **1933**, 270, 1472.

[60] Hoshino and co-workers, *Proc. Imp. Acad.* (*Tokyo*), **8**, 171 (1932); *Chem. Abstr.*, **26**, 4814 (1932); *ibid.*, **10**, 99 (1934); *Chem. Abstr.*, **28**, 3411 (1934); *ibid.*, **10**, 564 (1934); *Chem. Abstr.*, **29**, 1829 (1935); *Ann.*, **516**, 76, 81 (1935); **520**, 11 (1935).

[61] Julian and Pikl, *J. Am. Chem. Soc.*, **56**, 1797 (1934); **57**, 539 (1935).

[62] Julian and Pikl, *ibid.*, **57**, 563, 755 (1935).

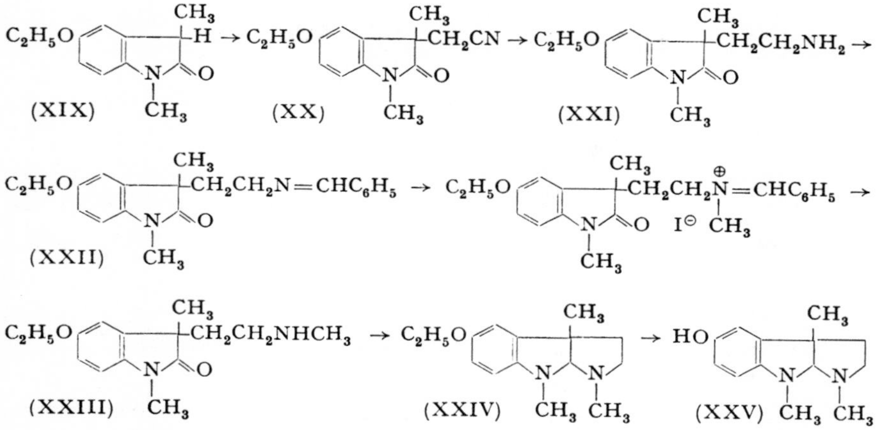

(XIX) → (XX) → (XXI) → (XXII) → C_2H_5O–oxindole–$CH_2CH_2\overset{\oplus}{N}(CH_3)=CHC_6H_5$ $I^{\ominus}$ → (XXIII) → (XXIV) → (XXV)

l-eserethole (XXIV). Cleavage of the ether with aluminum chloride yielded l-eseroline (XXV), identical with the natural material from the hydrolysis of physostigmine.

Geneserine

This base, $C_{15}H_{21}O_3N_3$,[63] m.p. 128–129°, $[\alpha]_D$ −175°, differs from physostigmine by one oxygen atom, and exhibits completely parallel reactions in being hydrolyzed to geneseroline, $C_{13}H_{18}O_2N_2$, analogous to eseroline, and from which geneserethole may be obtained. Geneserine is reduced by zinc and acetic acid to physostigmine, and may be reformed by oxidizing physostigmine with hydrogen peroxide. On this basis geneserine is

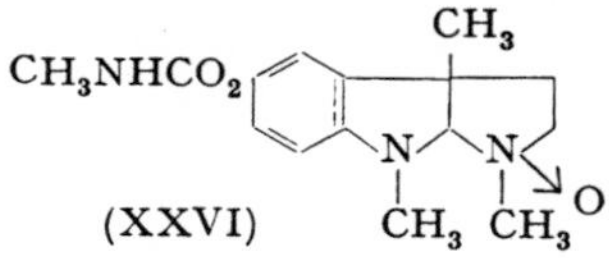

(XXVI)

regarded as the amine oxide of physostigmine and represented by structure XXVI.

Other bases which have been reported in the calabar bean are eseramine, $C_{16}H_{25}O_3N_4$, m.p. 245°,[64] isophysostigmine, $C_{15}H_{21}O_2N_3$, m.p. 200–202°[65] (unconfirmed), physovenine, $C_{14}H_{18}O_3N_2$, m.p. 123°,[66] and eseridine,

[63] Polonovski and Nitzberg, *Bull. soc. chim.*, **17**, 244 (1915); **21**, 191 (1917); **23**, 335 (1918); **23**, 356 (1918).

[64] Ehrenberg, quoted by Salway, *J. Chem. Soc.*, **99**, 2148 (1911).

[65] Ogui, quoted by Salway, *loc. cit.*

[66] Salway, *loc. cit.*

$C_{15}H_{23}O_3N_3$, m.p. 132°,[67] convertible to physostigmine by mineral acids and suggested to be identical with geneserine.[68]

Harmala Alkaloids

The seeds and roots of *Peganum harmala* contain three bases having an indole nucleus and which are of considerable interest from a structural aspect, since the harman nucleus is to be found in several more complex alkaloids.

Harmaline, Harmine, and Harmalol

The *harmala* alkaloids, all optically inactive, are harmaline, $C_{13}H_{14}ON_2$, m.p. 239–240° (dec.),[69] harmine, $C_{13}H_{12}ON_2$, m.p. 260–261.5°,[70] and harmalol, $C_{12}H_{12}ON_2$, m.p. 212°.[71] That the three are related is shown by the following series of transformations: harmaline contains a methoxyl group, but no methylamino group; cleavage of this ether linkage produces harmalol,[72] and gentle oxidation yields harmine.[73] Hydrogenation of both harmaline and harmine gives tetrahydroharmine.[74]

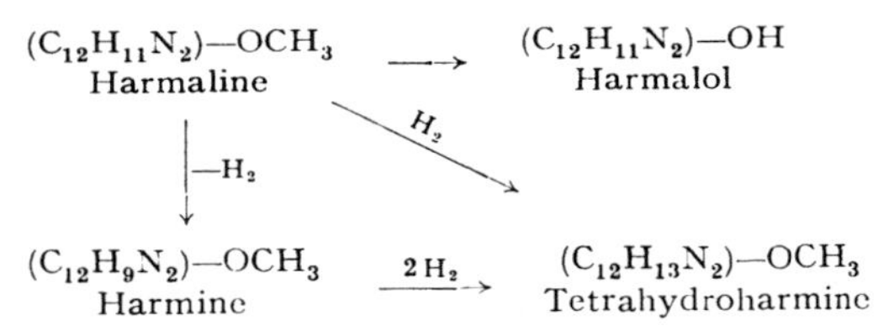

Chromic acid oxidation of both harmine and harmaline yields harminic acid, $C_8H_6N_2(COOH)_2$, an orthodicarboxylic acid which undergoes stepwise decarboxylation to apoharminic acid and apoharmine, $C_8H_8N_2$, a secondary base.[75] Further oxidation of harminic acid furnishes isonicotinic acid (I).[76] Harmine and harmaline, on oxidation with nitric acid, give harminic acid and *m*-nitroanisic acid (II).[77]

[67] Böhringer and Söhne, and Eber, quoted by Salway, *loc. cit.*

[68] *Merck's Ber.*, **40**, 37 (1926).

[69] Goebel, *Ann.*, **38**, 363 (1841).

[70] Fritsche, *ibid.*, **64**, 365 (1847).

[71] O. Fischer, *Ber.*, **18**, 400 (1885).

[72] O. Fischer, *Chem. Soc. Abstr.*, **1**, 405 (1901).

[73] O. Fischer, *Ber.*, **22**, 640 (1889); **30**, 2482 (1897).

[74] O. Fischer, *ibid.*, **22**, 637 (1889).

[75] O. Fischer and Täuber, *ibid.*, **18**, 403 (1885).

[76] O. Fischer, *ibid.*, **47**, 99 (1914).

[77] O. Fischer and Boesler, *ibid.*, **45**, 1930 (1912).

Harmine, Harmaline $\xrightarrow{CrO_3}$ Harminic acid $\xrightarrow{-2CO_2}$ Apoharmine

Harmine, Harmaline $\xrightarrow{HNO_3}$ (II) + (I)

(II): COOH, O_2N, OCH_3 (benzene ring)

(I): COOH (pyridine ring, N)

Harmine condenses with benzaldehyde to produce a benzylideneharmine,[78] indicative of an α-methylpyridine structure. The benzylideneharmine is oxidized to norharmine carboxylic acid, which may be decarboxylated to norharmine.

$$(C_{12}H_9ON_2)CH_3 \rightarrow (C_{12}H_9ON_2)CH{=}CHC_6H_5 \rightarrow (C_{12}H_9ON_2)COOH \rightarrow C_{12}H_{10}ON_2$$

Norharmine

Twelve of the thirteen carbon atoms are accounted for in the known degradation products I and II, and the remaining is located by the benzaldehyde condensation reaction. Perkin and Robinson[78] suggested that the *harmala* nucleus is a tricylic system of pyridine, benzene, and pyrrole rings. Further evidence for the pyrrole ring was afforded by the observation that harmaline couples with diazonium salts to yield dyes.[77]

Definite proof of the ring system present was obtained when it was found[79] that harman (III), produced from harmine by ether cleavage, treatment of the harmalol with zinc chloride-ammonia, and replacement of the amino group by hydrogen,[80] was identical with a substance

$$(C_{12}H_9N_2){-}OCH_3 \longrightarrow (C_{12}H_9N_2){-}OH \longrightarrow (C_{12}H_9N_2){-}NH_2$$

Tryptophan (indole N–H, $CH_2CHCOOH$, NH_2) $\xrightarrow{CH_3CHO}$ [tetrahydro-β-carboline: COOH, NH, CH_3, N–H] $\longrightarrow$ (III) (β-carboline: N, CH_3, N–H)

obtained from tryptophan by treatment with acetaldehyde followed by potassium dichromate oxidation.[81]

The position of the methoxyl group is located by consideration of the *m*-nitroanisic acid. Thus harmine is IV,[79] and harminic acid and apoharmine

[78] Perkin and Robinson, *J. Chem. Soc.*, **101,** 1775 (1912).

[79] Perkin and Robinson, *ibid.*, **115,** 933, 967 (1919).

[80] O. Fischer and Täuber, *Ber.*, **30,** 2482 (1897).

[81] Hopkins and Cole, *J. Physiol.*, **29,** 451 (1903).

are represented by V and VI, respectively. Norharmine would then have structure VII. The methyl group must be located in the position indicated, since attachment to the alternate carbon atom adjacent to the nitrogen atom would leave it unreactive toward benzaldehyde, as is the case with 3-methyl-

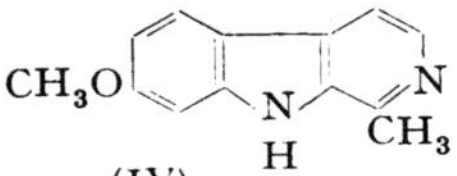

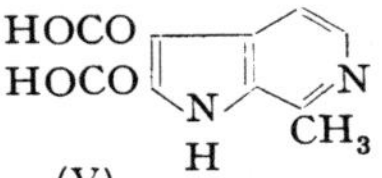

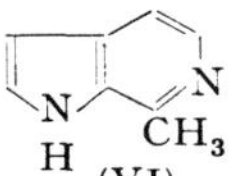

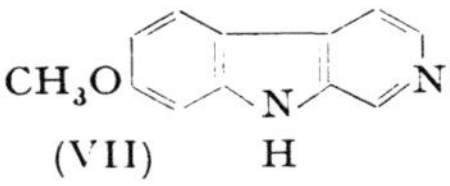

isoquinoline (VIII). The formation of harman from tryptophan and acetaldehyde supports this conclusion.

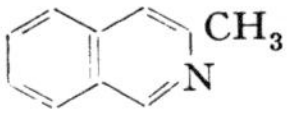

The unsubstituted ring system found in the *harmala* alkaloids has been designated the carboline system (IX)[82] with the position of the pyridine nitrogen indicated by the two systems shown. The less correct notation 4-carboline[79] still occurs in the literature, however, along with 3-(or β)-

8-Methyl-7-pyrindole

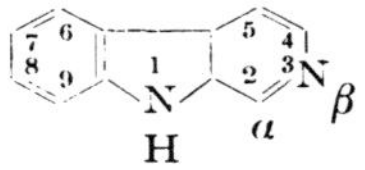

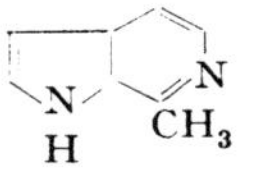

carboline. The system is named more systematically as pyridindole, not to be confused with the pyrindole nomenclature applied to apoharmine (X).[79]

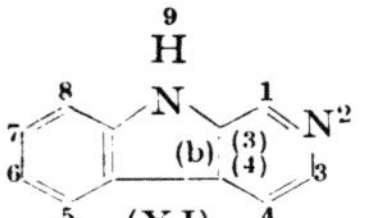

Thus norharman, according to *The Ring Index*,[83] is 9-pyrid[3,4-*b*]indole (XI) shown with proper orientation.

Harmaline could be either XII or XIII. Structure XIII would explain readily the formation of an acetyl derivative,[84] and of *N*-methylharmaline,

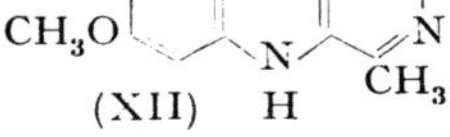

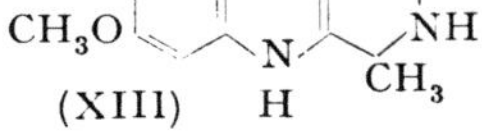

[82] Gulland, Robinson, Scott, and Thornley, *J. Chem. Soc.*, **1929**, 2924.
[83] Patterson and Capell, *The Ring Index*, Reinhold, New York, 1940, No. 1646
[84] O. Fischer and Täuber, *Ber.*, **18**, 405 (1885).

produced when harmaline methiodide is treated with barium hydroxide.[80] However, XII would be preferred on the basis of the optical inactivity of harmaline, its addition of hydroxylamine,[84] and the fact that it forms quaternary salts with only one mole of methyl iodide.

The apparent difficulty is resolved on further examination of the properties of *N*-methylharmaline. Oxidation[79] with permanganate converts this substance into a neutral compound, $C_{13}H_{14}O_2N_2$, which may be reduced with sodium in alcohol to *N*-methyltetrahydronorharmine (XVI), synthesized[85] by an unambiguous method. Thus the methyl group was lost in the oxidation, and since it was adjacent to this nitrogen atom, the neutral oxidation product is the amide (XV). *N*-Methylharmaline is then rep-

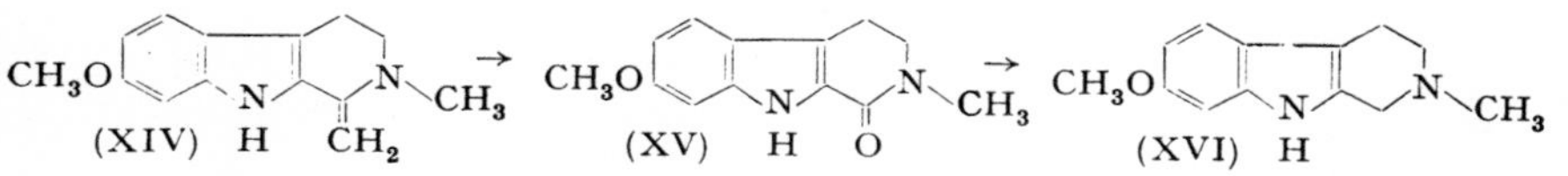

CH_3O N N CH_3 (XIV) H CH_2 → CH_3O N N CH_3 (XV) H O → CH_3O N N CH_3 (XVI) H

resented by XIV.[85] The acetyl derivative has been formulated similarly.[86]

These facts indicate that harmaline is 4,5-dihydroharmine and this has been confirmed by synthesis. The tautomerism exhibited in the formation of the *N*-acetyl- and *N*-methylharmaline is characteristic of α-methylpyridine derivatives.[87]

Since the discovery of the carboline nucleus, numerous derivatives of the basic structure have been synthesized. The synthesis of harman from tryptophan and acetaldehyde[81] has been mentioned. This has been extended to norharman[88] by substituting formaldehyde for acetaldehyde, and to many other examples.[89]

Synthesis of the amide (XVIII), obtained from *N*-acetylharmaline by oxidation and hydrolysis, was effected from ethyl 3-(6-methoxy-3-indolyl)-propionate (XVII) through the azide. Treatment of the azide with hydrochloric acid effects rearrangement to the isocyanate, which in turn cyclizes to the amide (XVIII).[86]

[85] Kermack, Perkin, and Robinson, *J. Chem. Soc.*, **121**, 1872 (1922). Nishikawa, Perkin, and Robinson, *ibid.*, **119**, 657 (1924).

[86] Barrett, Perkin, and Robinson, *ibid.*, **1929**, 2942.

[87] Taylor and Baker, *Sidgwick's Organic Chemistry of Nitrogen*, Oxford, London, 1945, Ch. XVIII.

[88] Kermack, Perkin, and Robinson, *J. Chem. Soc.*, **119**, 1617 (1921).

[89] Snyder, Hansch, Katz, Parmerter, and Spaeth, *J. Am. Chem. Soc.*, **70**, 219 (1948).

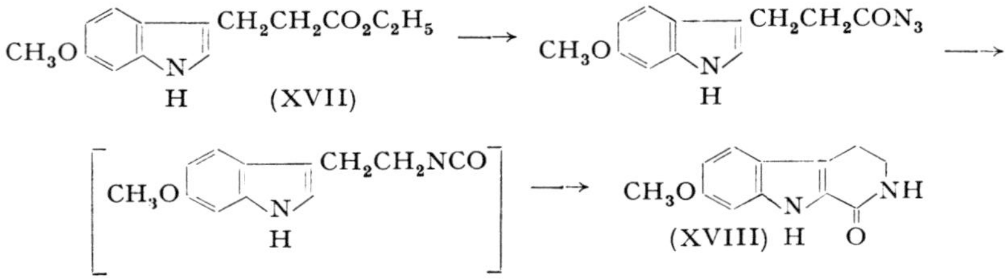

CH_3O–indole–$CH_2CH_2CO_2C_2H_5$ (XVII) → CH_3O–indole–$CH_2CH_2CON_3$ →

[CH_3O–indole–CH_2CH_2NCO] → CH_3O–(lactam, NH, O) (XVIII)

The amide reduction product (XVI) was prepared from 6-methoxyindole-2-carboxylic acid, by conversion to the amide acetal (XIX), followed by treatment with alcoholic hydrochloric acid to give the unsaturated amide (XX). Reduction with sodium and butanol furnished *N*-methyltetrahydronorharmine (XVI).[85]

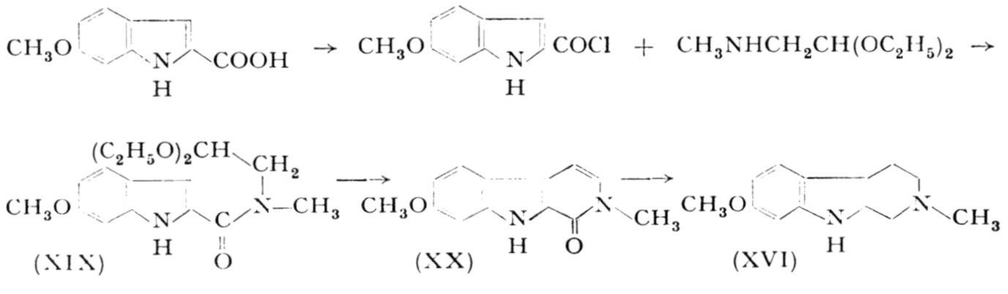

CH_3O–indole–COOH → CH_3O–indole–COCl + $CH_3NHCH_2CH(OC_2H_5)_2$ →

CH_3O–indole–C(=O)–N(CH_3)–CH_2–$CH(OC_2H_5)_2$ (XIX) → CH_3O–(N–CH_3, O) (XX) → CH_3O–(N–CH_3) (XVI)

The total synthesis of harmine and harmaline has been effected. The phthalimidoacetylvaleric acid (XXI) was saponified and treated with *m*-methoxybenzene diazonium chloride in the Japp-Klingemann synthesis. Ring closure of the phenylhydrazone (XXII) gave the indole derivative (XXIII), and subsequent treatment with hydrazine to remove the phthalic acid group yielded harmaline (XII).[90]

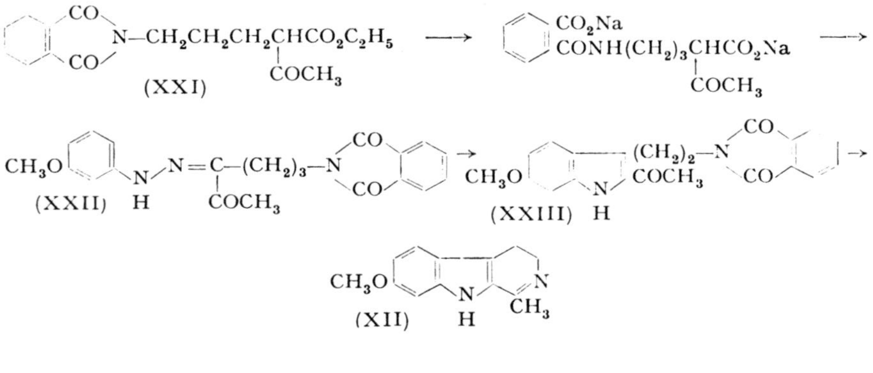

$C_6H_4(CO)_2N–CH_2CH_2CH_2CH(COCH_3)CO_2C_2H_5$ (XXI) → $C_6H_4(CO_2Na)CONH(CH_2)_3CH(COCH_3)CO_2Na$ →

CH_3O–C_6H_4–NH–N=C(COCH_3)–$(CH_2)_3$–N$(CO)_2C_6H_4$ (XXII) → CH_3O–indole–$(CH_2)_2$–N$(CO)_2C_6H_4$, $COCH_3$ (XXIII) →

CH_3O–(dihydro-β-carboline, CH_3) (XII)

[90] Manske, Perkin, and Robinson, *J. Chem. Soc.*, **1927**, 1.

A simpler synthesis of harmaline has been effected by acetylating 6-methoxytryptamine (XXIV) and cyclizing the acetyl derivative with phosphorus pentoxide. Dehydrogenation of the harmaline furnished harmine.[90,91]

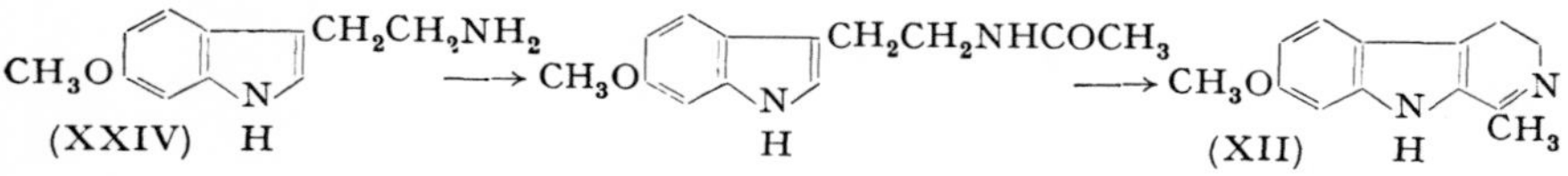

CH_3O–indole–$CH_2CH_2NH_2$ (XXIV) $\longrightarrow$ CH_3O–indole–$CH_2CH_2NHCOCH_3$ $\longrightarrow$ (XII)

Harman

Although harman (III) has not yet been found in any of the species yielding harmine or harmaline, it has been observed[92] that loturine from

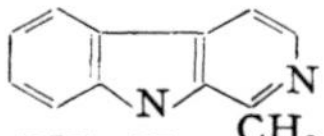

(III)

Symplocos racemosa[93] and aribine from *Ariba rubra*[94] are identical with harman.

Eleaginine

This optically inactive base, $C_{12}H_{14}N_2$, m.p. 180°,[95] obtained from *Eleagnus angustifolia*, has been shown[96] to be identical with tetrahydronorharman (XXV) on the basis of its dehydrogenation to norharman and re-formation from the latter by reduction with sodium and alcohol.

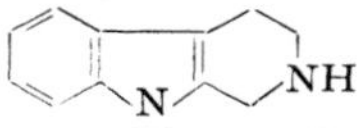

(XXV)

Leptocladine

A further alkaloid having the carboline nucleus, leptocladine, has been isolated from *Anthrophytum leptocladium*, along with dipterine.[97] This base,

[91] Späth and Lederer, *Ber.*, **63**, 122 (1930). Akabori and Kojiro, *Ber.*, **63**, 2245 (1930).

[92] Späth, *Monatsh.*, **40**, 351 (1919); *ibid.*, **41**, 401 (1920).

[93] Hesse, *Ber.*, **11**, 1542 (1879).

[94] Rieth and Wöhler, *Ann.*, **120**, 247 (1861).

[95] Massagetov, *J. Gen. Chem.* (*U. S. S. R.*), **16**, 139 (1946); *Chem. Abstr.*, **40**, 6754 (1946).

[96] Men'shikov, Gurevich, and Samsonova, *J. Gen. Chem.* (*U. S. S. R.*) **20**, 1927 (1950); *Chem. Abstr.*, **45**, 2490 (1951).

[97] Yurashevskii, *J. Gen. Chem.* (*U. S. S. R.*), **9**, 595 (1939); *Chem. Abstr.*, **33**, 7800 (1939).

$C_{13}H_{16}N_2$, m.p. 109–110°, $[\alpha]_D$ $\pm 0°$, gives the Ehrlich reaction, indicating an indole nucleus. It contains a basic nitrogen atom, not capable of nitrosation, but yielding a methiodide. A Hofmann degradation resulted in the isolation of trimethylamine, and a fecal odor was observed on distillation of the base hydrochloride. The structure has been determined by synthesis.[98] *N*-Methyltryptamine (XXVI), on treatment with acetaldehyde, followed by heating at 110° with sulfuric acid led to the isolation of a picrate of a base identical with the picrate of the alkaloid. The structure must then be XXVII.

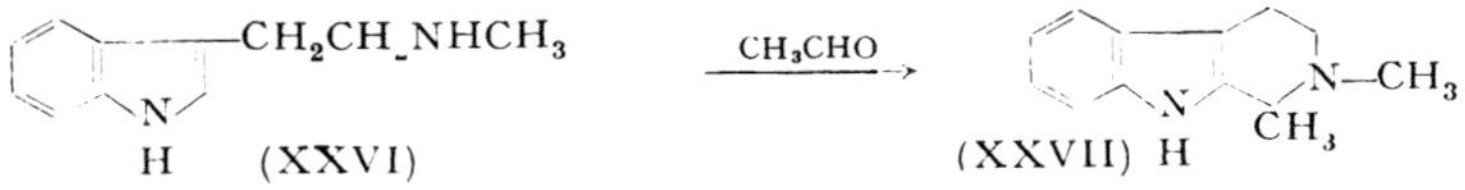

$CH_2CH_2NHCH_3$ (XXVI) $\xrightarrow{CH_3CHO}$ (XXVII) N—CH_3, CH_3

Calycanthine

This substance was first isolated from the seeds of *Calycanthus glaucus*,[99] but has since been found in several other species: *C. floridus*,[100] *C. occidentalis*,[101] and *Meraita praecox*,[102] An isocalycanthine has been reported,[103] but later evidence indicates its identity with calycanthine.[101]

Calycanthine, $C_{22}H_{26}N_4$, m.p. 245°, $[\alpha]_D$ $+684°$, contains one methylamino group, yields a nitrosoamine,[104] and exhibits two active hydrogen atoms at 22° and four at 95°.[105] Its behaviour with methyl iodide is not decisive, although two basic nitrogen atoms are indicated.[100,103,105]

Treatment of calycanthine with benzoyl chloride yields an amorphous material.[104] The latter, on treatment with alkali or on permanganate oxidation, gives benzoyl-*N*-methyltryptamine (XXVIII). This material has

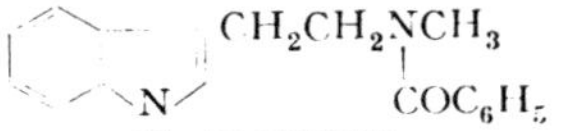

$CH_2CH_2NCH_3$ / COC_6H_5 (XXVIII)

also been obtained by heating the amorphous benzoylation product with calcium oxide.[105] *N*-Methyltryptamine is produced by heating cyalcanthine with soda lime.[105]

[98] Yurashevskii, *J. Gen. Chem.* (*U. S. S. R.*), **11**, 157 (1941); *Chem. Abstr.*, **35**, 5503 (1941).
[99] Eccles, *Proc. Am. Pharm. Assoc.*, **84**, 383 (1888).
[100] Späth and Stroh, *Ber.*, **58**, 2131 (1925).
[101] Manske and Marion, *Can. J. Research*, **17**, 293 (1939).
[102] Manske, *J. Am. Chem. Soc.*, **51**, 1836 (1929).
[103] Gordin, *ibid.*, **27**, 144, 1418 (1905); **31**, 1305 (1909); **33**, 1626 (1911).
[104] Manske, *Can. J. Research*, **4**, 275 (1931).
[105] Barger, Madinaveita, and Streuli, *J. Chem. Soc.*, **1939**, 510.

Oxidation of the alkaloid with lead tetraacetate results in the loss of two hydrogen atoms. The natural material may be reformed by reduction with zinc and acetic acid.[106] Several other degradation products have been characterized, but without lending definite support to any one formulation for calycanthine. Selenium dehydrogenation.[106] furnished skatol, 3-ethylindole (XXIX), norharman (XXX), 4-methylquinoline (XXXI), and a base

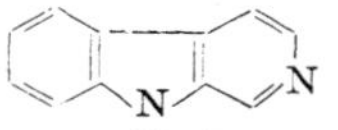

(XXX) (XXIX) 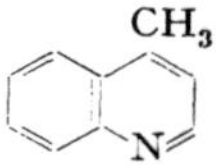 (XXXI)

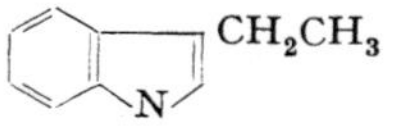

named calycanine. Heating the alkaloid with calcium oxide[105] gives, along with alkylindoles, calycanine, *N*-methyltryptamine, and a base, $C_{12}H_{10}N_2$, appearing to be a methyl carboline. Similar treatment of benzoyl calycanthine yielded a neutral substance, $C_{11}H_{11}N$, quinoline, and 2-phenylindole (XXXII). Silver acetate hydrolysis yields two products,[107] one of

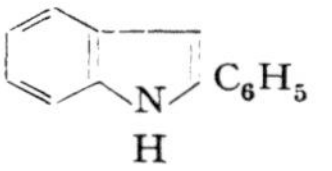

(XXXII) 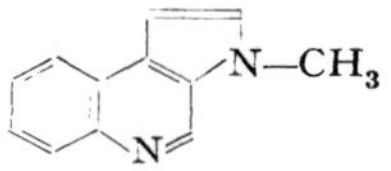 (XXXIII)

which, $C_{12}H_{10}N_2$, has been shown to have structure (XXXIII) by synthesis.[108]

Calycanine has been obtained from the alkaloid in a number of degradative procedures other than those mentioned. It is formed by heat alone,[105] by zinc dust distillation,[105, 106] and by oxidation with chromic acid in acetic acid.[105] That such variant procedures all yield this product has been taken to indicate that it constitutes an integral portion of the nucleus.

Calycanine contains one active hydrogen atom.[105] It is an extremely weak base, and since it does not give the Ehrlich color reaction, the α-position of the indole ring is presumed to be blocked. Molecular weight determinations furnish spurious results, and the various compositions $C_{16}H_{10}N_2$,[105, 106] and $C_{21}H_{13}N_3$ or $C_{21}H_{15}N_3$[101] have been proposed. X-ray crystallographic studies[109] favor the smaller formula, and in view of this $C_{15}H_{10}N_2$, in agreement with combustion results, has recently been forwarded.

[106] Marion and Manske, *Can. J. Research,* **16,** 432 (1938).

[107] Späth, Stroh, Lederer, and Eiter, *Monatsh.,* **79,** 11 (1948). Eiter, *ibid.,* **79,** 17 (1948).

[108] Eiter and Nagy, *ibid.,* **80,** 607 (1949).

[109] Hargreaves, *Nature,* **152,** 600 (1943).

Structures XXXIV[105] and XXXV or XXXVI[100] have been suggested for calycanine, but the substance, represented by XXXV has been synthe-iszed and found to be different.[110] This along with the X-ray data prompted the proposal of XXXVII.[110]

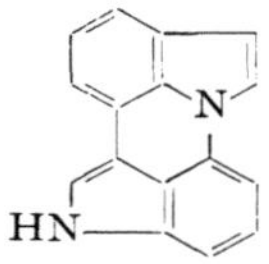

HN N (XXXIV)

N H CH3 N N (XXXV)

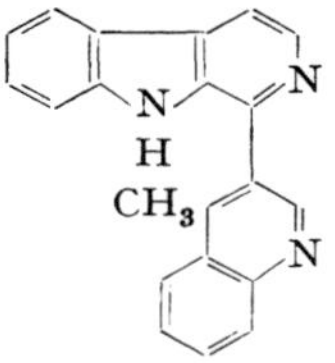

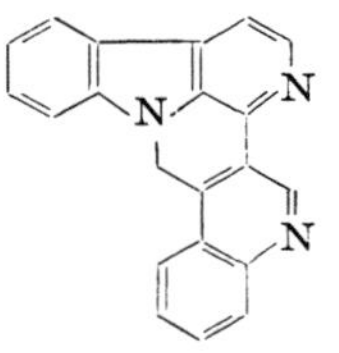

N N N (XXXVI)

N H N (XXXVII)

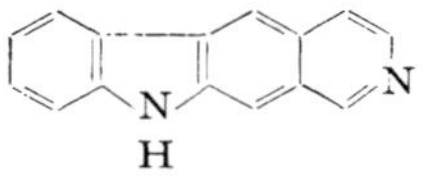

Two groups of workers have suggested structures for calycanthine. On the assumption that two tryptamine residues are present, one isolated in the form of *N*-methyltryptamine and the other as the methyl carboline, XXXVIII was offered.[105] With the idea that tryptamine and 3-carboline represent one portion of the molecule and 4-methylquinoline the other, XXXIX has been suggested.[100]

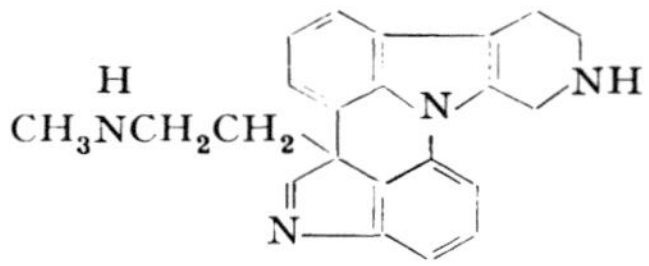

H $CH_3NCH_2CH_2$ N NH N XXXVIII)

N N CH_3 HN NH (XXXIX)

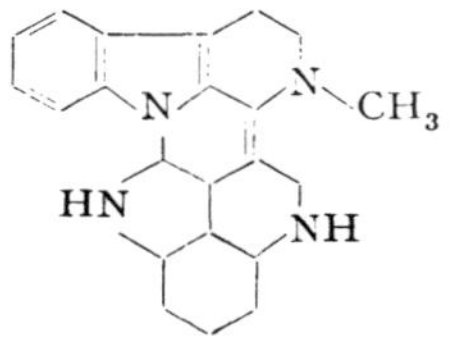

Calycanthidine

This alkaloid, $C_{13}H_{16}N_2$, m.p. 142°, $[\alpha]_D$ —285°, was found in *C. floridus*.[111] It contains one *N*-methyl group, one active hydrogen atom and perhaps one *C*-methyl group. Although color reactions indicate an indole nucleus, the Hopkins–Cole test is negative, showing a blocked α-position in the indole nucleus.

Reactions with methyl iodide in ether gives ill-defined products, but in methanol containing potassium carbonate, methyl iodide converts caly-

[110] Marion, Manske, and Kulka, *Can. J. Research*, **24**, 22 (1948).

[111] Barger, Jacob, and Madinaveita, *Rec. trav. chim.*, **57**, 548 (1938).

canthidine to a methiodide, $C_{16}H_{25}N_2OI$, in which the oxygen atom is introduced as a methoxyl group. This bit of evidence, together with the composition of the alkaloid, led to the postulation[111] of *N*-methyltetrahydroharman (XL) as the structure. Treatment of the methiodide with base might be expected to produce the pseudo base (XLI) which would be methylated to give XLII, having the composition observed. Such a structure for XL was

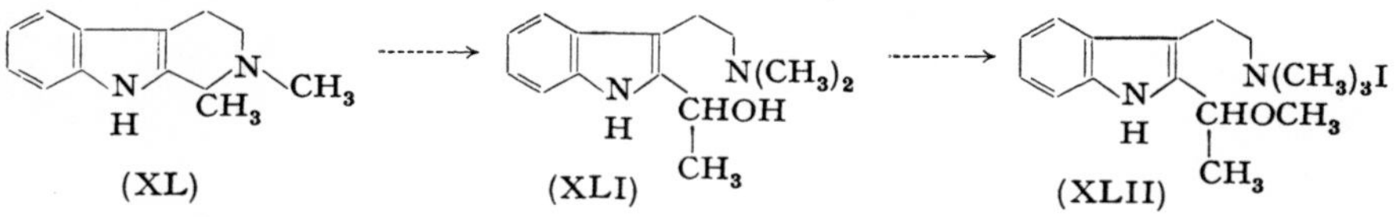

N H CH_3 N CH_3 (XL) ------→ N H $N(CH_3)_2$ CHOH CH_3 (XLI) ------→ N H $N(CH_3)_3I$ $CHOCH_3$ CH_3 (XLII)

partially supported by the observation that XLII on treatment with silver oxide, followed by heating, causes the loss of the methylamine to yield a neutral oil.

N-Methyltetrahydroharman (XL) was synthesized, from *N*-methyltryptamine and acetaldehyde, but the synthetic base could not be resolved, nor could calycanthidine be racemized, preventing direct comparison. However, identity of the two does not seem likely, since the synthetic product yields a normal methiodide readily, reacts differently with nitrous acid, and forms a full mole of acetic acid in the *C*-methyl determination, the natural base giving consistently low results. On the other hand, coupling products of both synthetic XL and calycanthidine with diazotized *p*-nitroaniline are similar.

It should be noted that structure XL has reportedly been established for leptocladine, an optically inactive alkaloid, and that the properties of XL and those reported for both natural and synthetic leptocladine are in agreement.

Folicanthine

A third alkaloid has recently been obtained from *C. floridus* and named folicanthine.[112] This material, $C_{18}H_{23}N_3$, m.p. 118–119°, $[\alpha]_D$ −364°, contains two *N*-methyl groups, but no *C*-methyl or active hydrogen atoms. Treatment with dilute hydrochloride acid at room temperature yields 1,7-dimethyltryptamine (XLIII) along with a second product $C_{12}H_{16}O_2N_2$, for which structure XLIV has been proposed. Zinc dust distillation forms an oily base, $C_{12}H_{16}N_2$, containing one methylamino and one *C*-methyl group and giving typical indole color reactions.

[112] Eiter and Svierak, *Monatsh.*, **82,** 186 (1951).

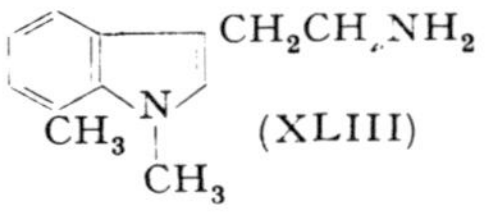

$CH_2CH_2NH_2$ (XLIII) CH_3 CH_3 $CH_2CH_2NH_2$ OH (XLIV) CH_2 OH CH_3

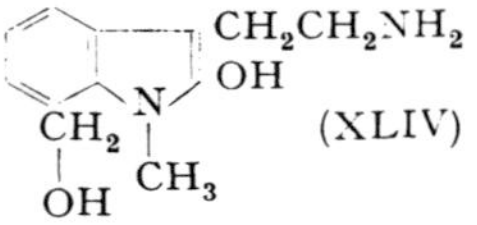

Evodia Alkaloids

The fruit of *Evodia rutecarpa,* known in China as the drug "wou chou yu," contains three bases, rutecarpine, evodiamine,[113] and wuchuyine,[114] $C_{15}H_{13}O_2N$, m.p. 237°, $[\alpha]_D$ −60.8°.

Evodiamine

This alkaloid, $C_{19}H_{17}ON_3$, m.p. 272°, $[\alpha]_D$ +352°, is cleaved by alcoholic potassium hydroxide into *N*-methylanthranilic acid (I) and a base, $C_{11}H_{10}N_2$, which on oxidation with potassium dichromate yields norharman (II).[115] The base must then be dihydronorharman (III).

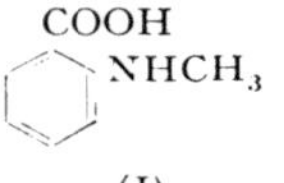

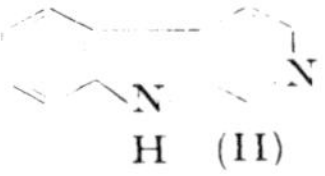

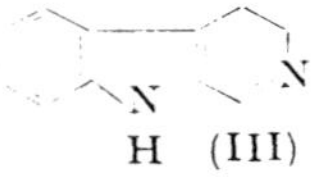

COOH NHCH$_3$ (I) N N H (II) N N H (III) $CH_2CH_2NH_2$ N H (IV)

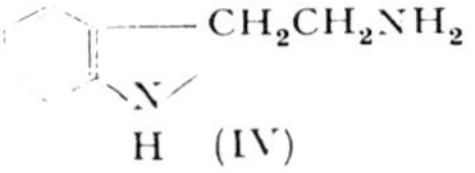

When evodiamine is heated with alcoholic hydrochloric acid, one molecule of water is added, forming isoevodiamine (evodiamine hydrate). Alcoholic potassium hydroxide ruptures the isoevodiamine molecule, yielding *N*-methylanthranilic acid (I) and tryptamine (IV).[115,116] Isoevodiamine is converted into racemic evodiamine by treatment with acetic anhydride.[117] Pyrolysis of dry isoevodiamine hydrochloride eliminates water and methyl chloride to form rutecarpine.[117]

Rutecarpine

The reactions of rutecarpine, $C_{18}H_{13}ON_3$, m.p. 258°, $[\alpha]_D$ ±0°, parallel

[113] Asahina and Kashiwagi, *J. Pharm. Soc. Japan,* **No. 405,** 1293 (1913); *Chem. Abstr.,* **10,** 607 (1916). Asahina and Mayeda, *J. Pharm. Soc. Japan,* **No. 416,** 871 (1916); *Chem. Abstr.,* **11,** 332 (1917).

[114] Chen and Chen, *J. Am. Pharm. Assoc.,* **22,** 716 (1933).

[115] Asahina, *J. Pharm. Soc. Japan,* **No. 503,** 1 (1924); *Chem. Abstr.,* **18,** 1667 (1925).

[116] Kermack, Perkin, and Robinson, *J. Chem. Soc.,* **119,** 1615 (1921).

[117] Asahina and Ohta, *J. Pharm. Soc. Japan,* **No. 530,** 293 (1926); *Chem. Abstr.,* **21,** 2134 (1927). Ohta, *J. Pharm. Soc. Japan,* **65,** No. 2A, 15 (1945); *Chem. Abstr.,* **45,** 5697 (1951).

those of evodiamine in many ways.[113] Amyl alcoholic potassium hydroxide converts it into anthanilic acid, rather than the methyl derivatives, and 3-(2-aminoethyl)indole-2-carboxylic acid (V), which easily loses carbon dioxide to form tryptamine (IV).[116,118]

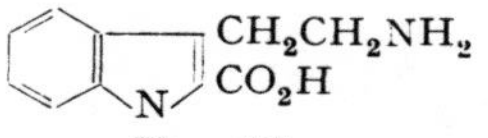

$CH_2CH_2NH_2$ / CO_2H (V)

On the basis of these reactions, evodiamine may be represented as VI and rutecarpine as VII,[115,116] structures which readily explain the transformation of evodiamine to rutecarpine via the hydrated intermediate, isoevodiamine (VIII).

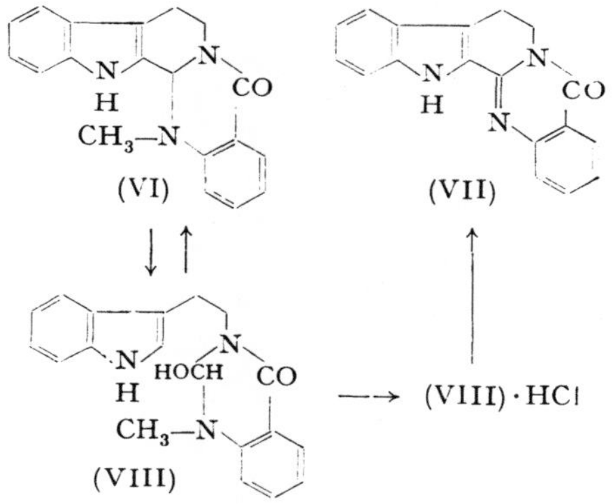

(VI) ⇅ (VIII) → (VIII)·HCl → (VII)

Several syntheses of rutecarpine have been effected. A partial synthesis was accomplished from the tryptamine carboxylic acid (V) by forming the *o*-nitrobenzoyl derivative (IX) and reducing this to the amine (X), which

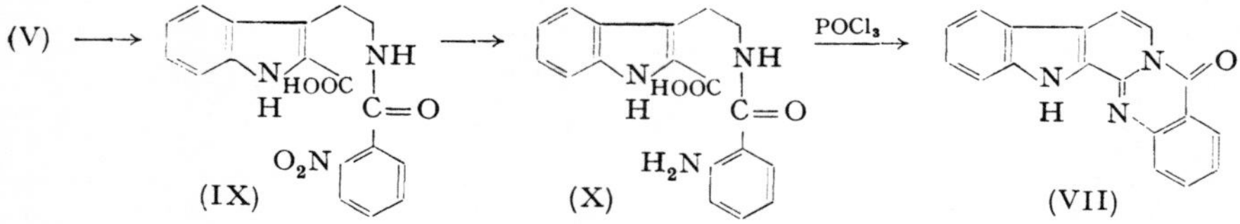

(V) → (IX) → (X) $\xrightarrow{POCl_3}$ (VII)

on treatment with phosphorus oxychloride in carbon tetrachloride yielded rutecarpine.[119] The preparation of V by the action of alcoholic potassium

[118] Asahina and Fujita, *J. Pharm. Soc. Japan,* **No. 476,** 863 (1921); *Chem. Abstr.,* **16,** 1584 (1922).

[119] Asahina, Irie, and Ohta, *J. Pharm. Soc. Japan,* **No. 543,** 51 (1927); *Chem. Abstr.,* **21,** 3054 (1927).

hydroxide on 2-keto-2,3,4,5-tetrahydro-3-carboline (XI) by the same authors[120] completed the synthesis.

An extremely simple synthesis was achieved by condensing XI with methyl anthranilate in the presence of phosphorus trichloride.[121]

(XI) + H_2N–C_6H_4–CO_2CH_3 $\xrightarrow{PCl_3}$ (VII)

A more recent preparation of rutecarpine involves the condensation of dihydronorharman (III) with *o*-aminobenzaldehyde, followed by oxidation with chromic acid in acetic acid.[122]

(III) + H_2N–C_6H_4–CHO ⟶ [OH intermediate] $\xrightarrow{CrO_3}$ (VII)

dl-Evodiamine has also been synthesized.[123] *N*-Methylisatoic anhydride (XII), prepared from *N*-methylanthranilic acid and ethyl chloroformate, reacts with tryptamine to yield 3-[2-(*N*-methylanthranylamino)ethyl]indole

$C_6H_4(CO_2H)NHCH_3$ + $ClCO_2C_2H_5$ → (XII) $\xrightarrow{(IV)}$ (XIII) $\xrightarrow{HC(OC_2H_5)_3}$ (VI)

(XIII), and this with ethyl orthoformate furnishes *dl*-evodiamine (VI). Isoevodiamine was produced by treating the evodiamine with alcoholic hydrochloric acid.

Cinchona Indole Alkaloids

Although the majority of the *cinchona* alkaloids of known constitution, such as quinine (I), contain a quinoline nucleus,[124] it has recently been

[120] Asahina, Irie, and Ohta, *J. Pharm. Soc. Japan*, **48,** 313 (1928); *Chem. Abstr.*, **22,** 3393 (1928).

[121] Asahina, Manske, and Robinson, *J. Chem. Soc.*, **1927,** 1708.

[122] Schöpf and Steuer, *Ann.*, **558,** 109 (1947).

[123] Asahina and Ohta, *Ber.*, **61,** 319 (1928).

[124] Henry, *The Plant Alkaloids*, 4th ed. Blakiston, New York, 1949, p. 412.

observed that two, and possibly four of the *cinchona* bases contain an indole

$CH=CH_2$
CHOH
CH_3O
N
(I)

moiety. The first evidence for the existence of this type of alkaloid in the *cinchona* or plants related botanically was in the discovery that cinchonamine gives color reactions much like yohimbine.[125] This was supported by a comparison of the ultraviolet absorption spectrum of cinchonamine with those of alkaloids of the indole group,[126,127] and further substantiated by degradative experiments on both cinchonamine and quinamine.

Cinchonamine

This base, $C_{19}H_{24}ON_2$, m.p. 194°, $[\alpha]_D + 121°$,[128] is diacidic and contains no methoxyl group.[129] It yields a methiodide and a monoacetyl derivative, absorbs one mole of hydrogen,[130] and gives typical indole color reactions. Its absorption spectrum resembles those of the other indole alkaloids.

The action of chromic acid on cinchonamine yields an acid, $C_{10}H_{15}O_2N$, identified in the form of its copper salt as 3-vinylquinuclidine-6-carboxylic acid (II).[131] Treatment of the alkaloid with acetic anhydride furnishes diacetyl allocinchonamine, oxidation of which with permanganate furnishes

$CH=CH_2$
HOCO
N
(II)

an aldehyde, characterized through its phenylhydrazone and the corresponding primary alcohol as III.[131] These two fragments permit an unequivocal representation of cinchonamine as IV.[131] Confirmation of this structure by synthesis has been reported without detail.[132]

CH_2CH_2OH
CHO
N
H (III)

CH_2CH_2OH
N
N
(IV) H
$CH=CH_2$

[125] Raymond-Hamet, *Compt. rend.*, **212,** 135 (1941).

[126] Janot and Berton, *ibid.*, **216,** 564 (1943).

[127] Raymond-Hamet, *ibid.*, **220,** 670 (1945); **221,** 307 (1945); **227,** 1182 (1948).

[128] Arnaud, *ibid.*, **93,** 593 (1881); **97,** 174 (1883).

[129] Howard and Chick, *J. Soc. Chem. Ind.*, **24,** 53 (1909).

[130] Hesse, *Ber.*, **10,** 2158 (1877); *Ann.*, **209,** 621 (1881).

[131] Goutarel, Janot, Prelog, and Taylor, *Helv. Chim. Acta*, **33,** 150 (1950). Taylor, *ibid.*, **33,** 164 (1950).

[132] Culvenor, Goldsworthy, Kirby, and Robinson, *Nature*, **166,** 105 (1950).

Quinamine

Quinamine, $C_{19}H_{24}O_2N_2$, m.p. 185–186°,[133] $[\alpha]_D$ +104°, is found in the bark of *Cinchona succirubra*, *C. ledgeriana*, and others of the *Cinchona* genus. It yields a methiodide and a nitrosoamine, and a single double bond is indicated by catalytic hydrogenation to dihydroquinamine. Neither a methoxyl, *C*-methyl, nor methylamino group is present in the molecule. A Zerewitinoff determination shows two active hydrogen atoms.

Treatment of the base with acetyl chloride yields acetyl apoquinamine,[134] hydrolyzable to apoquinamine, $C_{19}H_{22}ON_2$. Hydrogenation of the apo-base results in the absorption of one mole of hydrogen rapidly, then more slowly the tetrahydro derivative is formed.

Quinamine, when treated with amyl alcoholic potassium hydroxide, is converted into isoquinamine.[132,135] This isomer yields a methiodide and a nitroso derivative, and on hydrogenation is transformed into dihydroisoquinamine, obtainable directly from dihydroquinamine. That a modification of the carbon skeleton has occurred is shown by the presence of a *C*-methyl group not found in the parent alkaloid. This is supported by the pyrolysis of isoquinamine, which yields acetaldehyde as compared with the formaldehyde obtained from the same treatment of quinamine.[136] Both isoquinamine and its dihydro derivative are dehydrated on treatment with acetic anhydride.

The action of chromic acid on quinamine[134] yields the same vinylquinuclidine carboxylic acid (II) obtained from cinchonamine. The other portion of the alkaloid is isolated in the form of 2,3-dimethylindole (V) by distilling dihydroquinamine from zinc dust.[136] That this indole nucleus is

CH_3
CH_3
N
(V) H

present in a hydrogenated form is shown by the absorption spectrum of quinamine,[127] which does not resemble that of cinchonamine or other indole alkaloids with a true indole nucleus, and substantiation of this idea is found

[133] Hesse, *Ber.*, **5**, 265 (1872); **10**, 2152 (1877); *Ann.*, **166**, 217 (1873); **199**, 333 (1879); **207**, 788 (1881).

[134] Henry, Kirby, and Shaw, *J. Chem. Soc.*, **1945**, 524.

[135] Kirby, *ibid.*, **1949**, 735. Culvenor, Goldsworthy, Kirby, and Robinson, *ibid.*, **1950**, 1478.

[136] Kirby, *ibid.*, **1945**, 528.

[137] Witkop, *J. Am. Chem. Soc.*, **72**, 2311 (1950).

in the coupling reaction which quinamine undergoes with diazobenzene sulfonic acid.

That quinamine and cinchonamine are closely related is indicated by the facile reduction of quinamine to cinchonamine through the agency of lithium aluminium hydride.[131]

Several structures have been proposed[131, 136] for the quinamine molecule on the basis of this evidence, the most notable being the epoxide (VI)[131] to

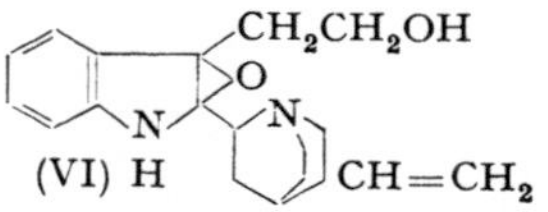

CH_2CH_2OH O N N (VI) H $CH{=}CH_2$

account for the above reduction to cinchonamine. The discovery that the reverse transformation, *i.e.*, oxidation of cinchonamine to quinamine by peracetic acid, was possible, has led to a more adequate formulation (VII)[137] for quinamine, in view of the fact that peracetic acid does not produce epoxides in the indole series. The oxidation results first in a 3-hydroxy-

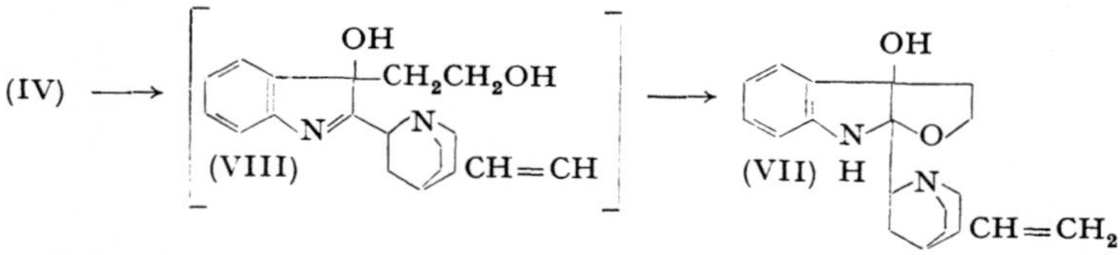

(IV) ⟶ [OH CH_2CH_2OH N N (VIII) $CH{=}CH$] ⟶ OH N O (VII) H N $CH{=}CH_2$

indolenine (VIII) which undergoes internal cyclization by addition of the alcoholic hydroxyl across the aldimine system.

Two other bases of *Cinchona* appear to be related to the indole series. Conquinamine, $C_{19}H_{24}O_2N_2$, m.p. 121°, $[\alpha]_D$ +204°, is reported to be convertible to quinamicine (a rearrangement product of quinamine) and apoquinamine,[130] and aricine[138] also give indole color reactions.

Yohimbe Alkaloids

The bases which may be considered as members of the yohimbine group are so classified on the basis of structure rather than on a strong relationship of the species in which they occur. Yohimbehe bark,[139] the original source of the parent alkaloid, yohimbine, has been the subject of some discussion as regards botanical origin,[140] but is now generally believed to come from

[138] Pelletier and Carriol, *J. Pharm.*, **15,** 563 (1829); Hesse, *Ann.*, **166,** 254 (1873); **181,** 58 (1876).
[139] Spiegel, *Chem. Ztg.*, **20,** 970 (1896); **21,** 833 (1897); **23,** 59 (1899).
[140] Schomer, *Arch. Pharm.*, **265,** 509 (1927). Brandt, *ibid.*, **260,** 49 (1922).

Pausinystalia yohimbe.[141] The confusion has been deepened by the occurrence of identical bases in distinctly different plants.

Also occurring in *P. yohimbe*, which has the synonym *Corynanthe yohimbe*, and in *Pseudocinchona africana*, are twelve other alkaloids, ten of which are isomeric with yohimbine. These isomerides are listed in Table I for convenience, and their known relationships to the parent base will be considered subsequently, along with the remaining two alkaloids of this group. Table I is on page 231.

Yohimbine

This alkaloid, $C_{21}H_{26}O_3N_2$, m.p. 234°, $[\alpha]_D$ +62°, contains one alcoholic hydroxyl group, yielding an acetyl derivative, and forms a methodide, indicating a tertiary nitrogen atom.[139] An *O,N*-diacetyl derivative may also be produced,[140] the nitrogen atom in this case being that of the indole nucleus. Treatment with base, followed by acidification, yields yohimbic acid,[142] which is reconverted to yohimbine by esterification,[143,144] locating the single methoxyl function as a carbomethoxyl group.

The nature of the nucleus to which these functional groups are attached is known from the study of a number of degradation products. Indications of an indole nucleus were provided by the isolation of an unidentified alkylindole by soda lime distillation of the alkaloid.[145] Permanganate oxidation of yohimbine yields *N*-oxalylanthranilic acid (I),[146] *o*-hydroxyphenylisocyanate, isolated as the benzoxazolone (II), and an indole carboxylic acid.[147] Oxidation with nitric acid resulted in the formation of 6-nitroindazole-3-carboxylic acid (III),[148] but it was early recognized that this product was formed, probably from *o*-aminophenylacetic acid, by the action of nitrous acid in the oxidizing agent.

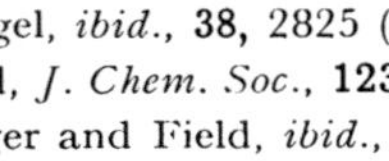

CO_2H / $NHCOCO_2H$ (I) — O / C=O / N H (II) — O_2N / CO_2H / N / N H (III)

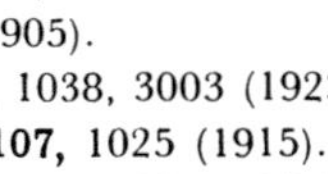

Potassium hydroxide fusion of yohimbine gave 3-ethylindole (IV)[149]

[141] Hahn and Schuch, *Ber.*, **63**, 1638 (1930).
[142] Spiegel, *ibid.*, **36**, 169 (1903).
[143] Spiegel, *ibid.*, **38**, 2825 (1905).
[144] Field, *J. Chem. Soc.*, **123**, 1038, 3003 (1923).
[145] Barger and Field, *ibid.*, **107**, 1025 (1915).
[146] Späth and Bretschneider, *Ber.*, **63**, 2997 (1930).
[147] Warnat, *ibid.*, **59**, 2388 (1926).
[148] Hahn and Just, *ibid.*, **65**, 717 (1932).
[149] Barger and Scholz, *J. Chem. Soc.*, **1933**, 614; *Helv. Chim. Acta*, **16**, 1343 (1933).

and indole-2-carboxylic acid.[149,150] The distillation of yohimbic acid[150] gave, among other substances, including quinoline bases, a compound identified as harman (V).[149] Distillation of yohimbine from zinc dust gave a mixture of bases, from which isoquinoline (VI) was identified.[151] Tetrahydroyobyrine,

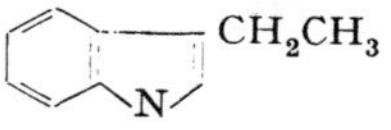

CH_2CH_3 N H (IV) — N N CH_3 H 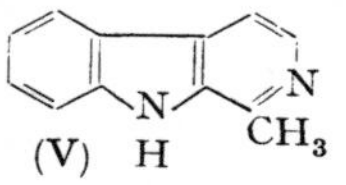 (V) — N (VI)

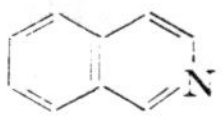

a dehydrogenation product of yohimbine, on oxidation with nitric acid gave berberonic acid (VII).[149] Alkaline cleavage of ketoyobyrine, another dehydrogenation product, yielded 2,3-dimethylbenzoic acid (VIII),[149,152] and similar fusion of yohimbic acid furnished *m*-toluic acid (IX).

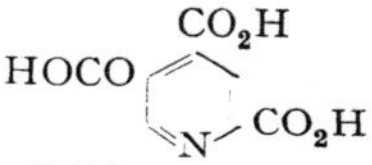

CO_2H HOCO N CO_2H (VII) — CO_2H CH_3 CH_3 (VIII — CO_2H CH_3 (IX)

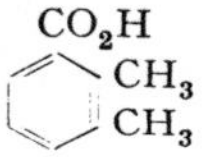

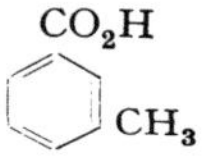

Correlation of these degradation products results in a skeletal structure such as X for yohimbine.[149] Rings A, B, and C are found in harman, and D and E in isoquinoline. Berberonic acid is derived from ring D, and ring E is represented by the alkylbenzoic acids. That rings C, D, and E are fully

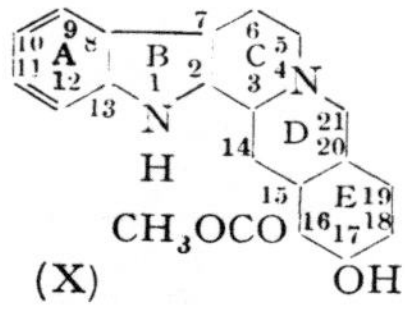

(X) CH_3OCO OH

hydrogenated is suggested by the oxidations of yohimbine or yohimbic acid; other than the indole derivatives, only succinic acid[151] has been obtained from such reactions, although oxidation of the dehydrogenation products results in fragments arising from these rings.

The position of the carbomethoxyl group is indicated by the isolation of 2,3-dimethylbenzoic acid, in which the orientation is established by the two methyl groups, arising from carbon atoms 14 and 21.

The original structural proposal[153] for yohimbine carried the hydroxyl group at position 14. That this possibility was unlikely was pointed out,[154]

[150] Warnat, *Ber.*, **60**, 1118 (1927).

[151] Winterstein and Walter, *Helv. Chim. Acta*, **10**, 777 (1927).

[152] Mendlik and Wibaut, *Rec. trav. chim.*, **50**, 91 (1931).

[153] Scholz, *Helv. Chim. Acta*, **18**, 423 (1935).

[154] Hahn and Werner, *Ann.*, **520**, 123 (1935).

since this carbon appears in the methyl group of harman, produced along with *m*-toluic acid by the hydrolysis of tetrahydroyohimbic acid. Position 17, or possibly 19 or 18, was suggested. Cited as evidence for the former was the synthesis of the yohimbine skeleton "under physiological conditions,"[154] which failed if the hydroxyl group were located other than at position 17.

More positive evidence for this position is furnished by the isolation of harman and *p*-cresol by the zinc dust distillation of yohimbine,[155] products formed by the smooth rupture of ring D in the process of aromatization.

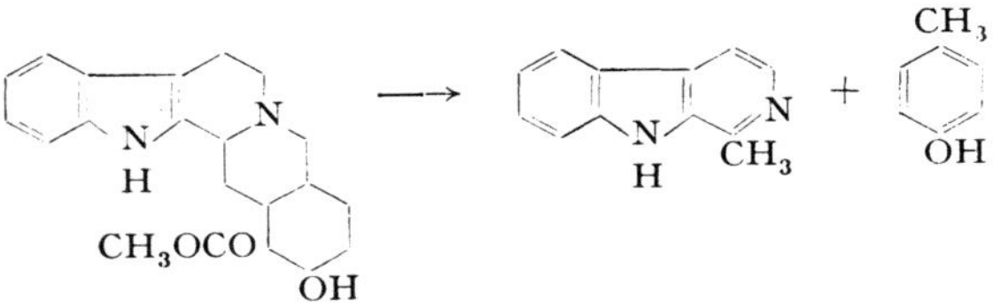

N N H CH_3OCO OH ⟶ N N H CH_3 + CH_3 OH

It has also been shown[154] that the decarboxylation of yohimbic acid proceeds with extreme ease under oxidizing conditions, yielding yohimbone (XI). This behavior would be expected on the basis of an intermediate *β*-keto acid.

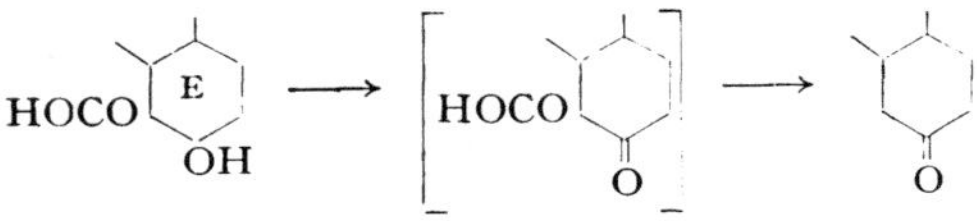

HOCO E OH ⟶ [HOCO O] ⟶ O (XI)

Other important transformation products of yohimbine are known, and the structures of these may be rationalized with the established structure for the alkaloid. Treatment of yohimbine with sulfuric acid furnishes a product, first taken to be a sulfonic acid,[145] but later shown[144] to be the sulfuric acid ester (XII), since sulfuric acid was eliminated on treatment with alkali. The products of this reaction are yohimbic acid hydrogen sulfate and apoyohimbine, $C_{21}H_{14}O_2N_2$ (XIII). Each may be hydrolyzed to apoyohimbic acid, reconvertible to apoyohimbine by esterification. Hydrogenation of apoyohimbine gives desoxyyohimbine (XIV).

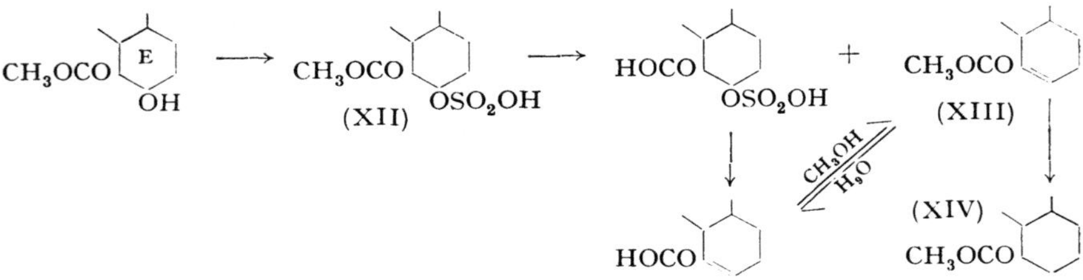

CH_3OCO E OH ⟶ CH_3OCO OSO_2OH (XII) ⟶ HOCO OSO_2OH + CH_3OCO (XIII)

HOCO ⇌ (CH_3OH / H_2O) (XIII); (XIV) CH_3OCO

The dehydrogenation of yohimbine with selenium[152,156] furnishes three products, yobyrine, $C_{19}H_{16}N_2$, tetrahydroyobyrine, $C_{19}H_{20}N_2$, also formed by soda lime distillation of yohimbic acid,[155] and ketoyobyrine, $C_{20}H_{16}ON_2$. The names of these derivatives are based on the hypothetical structure yobine, $C_{19}H_{24}N_2$ (XV), for which the more rational name yohimbane has been suggested.[157]

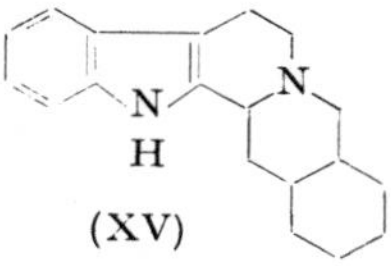

(XV)

Oxidation of tetrahydroyobyrine with ozone or chromic acid[153] yields a compound, $C_{19}H_{20}O_2N_2$, which is regarded as XVII, since acid hydrolysis yields *o*-aminopropiophenone and 5,6,7,8-tetrahydroisoquinoline-3-carboxylic acid (XVIII). This oxidation product may be further degraded by nitric acid to berberonic acid, also obtained directly from tetrahydroyobyrine, which must then be XVI.[154,155] Dehydrogenation of tetrahydro-

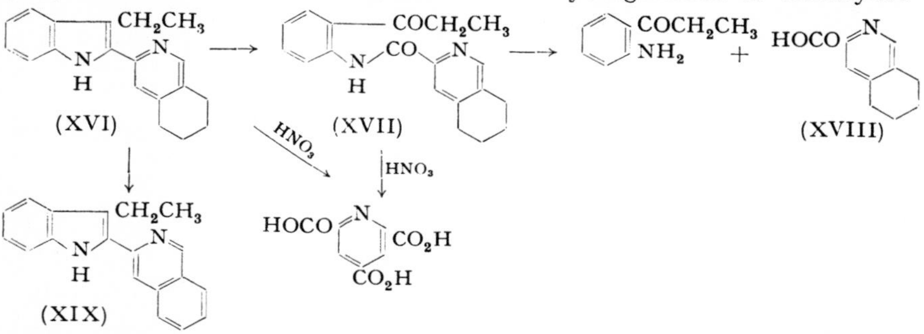

(XVI) (XVII) (XVIII)

(XIX)

yobyrine with platinum black produces 2-(3′-isoquinolinyl)-3-ethylindole (XIX).[154]

Tetrahydroyobyrine and its dehydrogenation product (XIX) have been synthesized.[158] The tetrahydroisoquinoline carboxylic acid (XVIII) was condensed with propyl lithium to give 3-butyrylisoquinoline (XX), from

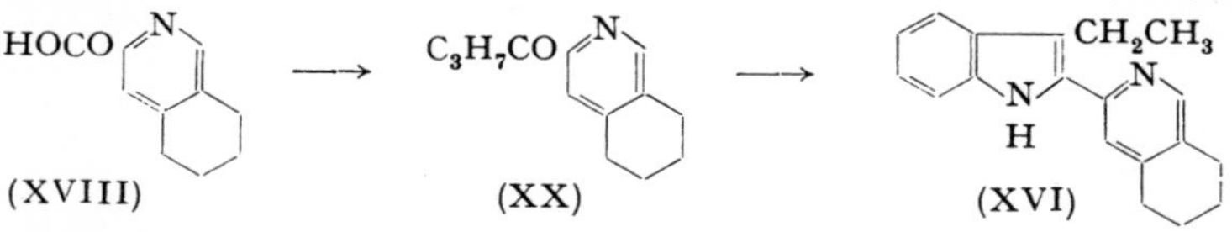

(XVIII) (XX) (XVI)

[155] Witkop, *ibid.*, **554**, 83 (1943). Pruckner and Witkop, *ibid.*, **554**, 127 (1943).

[156] Mendlik and Wibaut, *Rec. trav. chim.*, **48**, 191 (1929).

[157] Jost, *Helv. Chim. Acta*, **32**, 1247 (1949).

[158] Julian, Karpel, Magnani, and Meyer, *J. Am. Chem. Soc.*, **70**, 180 (1948).

[159] Wibaut, Wibaut-van Gastel, and Breizer, *Rec. trav. chem.*, **68**, 497 (1949).

which the Fischer indole synthesis yielded a compound identical with the yohimbine derivative (XVI). Dehydrogenation of synthetic XVI gave XIX.

The second dehydrogenation product of yohimbine, yobyrine, may be hydrogenated over platinum oxide to a hexahydro derivative, as well as tetra- and decahydro derivatives.[159] Oxidation with sodium chromate in acetic acid gives phthalic acid, *o*-toluic acid, and a further transformation product, yobyrone, also produced by action of selenium dioxide or ozone on yobyrine. An active methylene group is also indicated by the formation of alkylidene derivatives with aldehydes. On the basis of the degradation products, XXI was proposed as a formulation for this compound.[149] This was modified to XXII,[155] which accounts for the degradation products as well as the active methylene group, and also explains the failure of yobyrine to be dehydrogenated in the presence of platinum black, a reaction which would be expected to occur with a structure such as XXI. Yobyrone would then be represented as XXIII. The blocked position of the carbonyl group accounts for the lack of reactivity exhibited by this function.

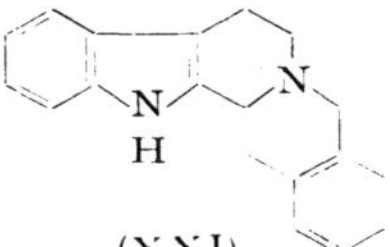

(XXI) 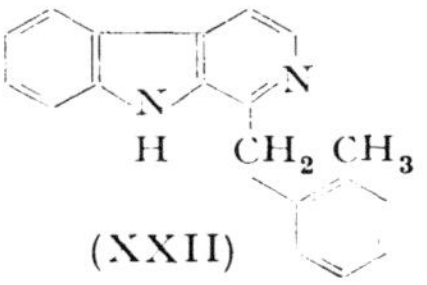 (XXII) CH_2 CH_3 (XXIII) CO CH_3

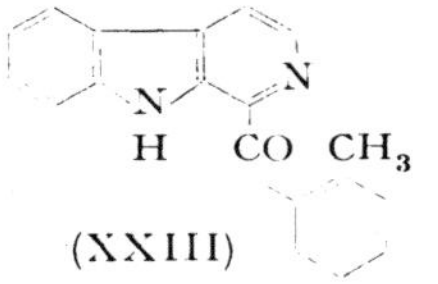

Structure XXII for yobyrine was readily confirmed by synthesis.[158,160,161] Tryptamine was condensed with *o*-tolylacetic acid to yield the amide (XXIV). Ring closure with phosphorus oxychloride, and dehydrogenation with palladium black gave 2-(2-methylbenzyl)-3-carboline (XXII), identical with yobyrine. Oxidation of the synthetic base gave yobyrone (XXIII).

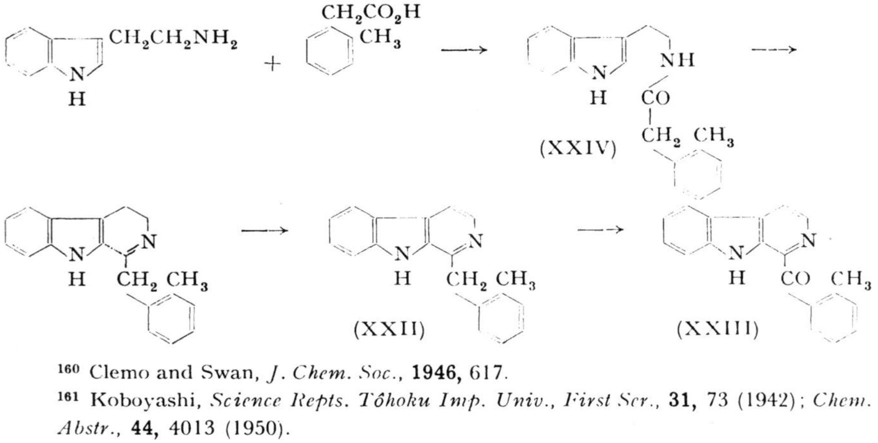

$CH_2CH_2NH_2$ + CH_2CO_2H, CH_3 ⟶ (XXIV) NH, CO, CH_2 CH_3 ⟶

CH_2 CH_3 ⟶ (XXII) CH_2 CH_3 ⟶ (XXIII) CO CH_3

[160] Clemo and Swan, *J. Chem. Soc.*, **1946,** 617.

[161] Koboyashi, *Science Repts. Tôhoku Imp. Univ., First Ser.*, **31,** 73 (1942); *Chem. Abstr.*, **44,** 4013 (1950).

The third product, ketoyobyrine, is cleaved by potassium hydroxide in amyl alcohol to norharman (XXV) and 2,3-dimethylbenzoic acid (VIII).[152] On this evidence, considering that no carbonyl derivatives are formed by this compound, XXVI was postulated.[155] This structure, however, does not

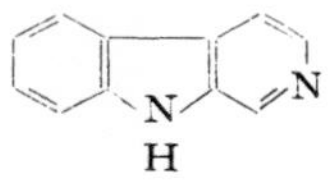

(XXV) 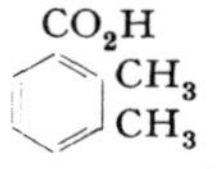 (VIII) 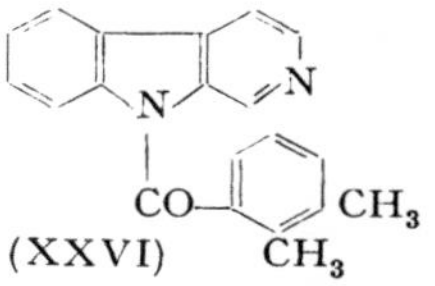 (XXVI)

account for the lack of basicity of ketoyobyrine. The properties of the substance are all reconciled with the structure (XXVII) recently proposed.[162,163] Further evidence is afforded[162] by the dehydrogenation of ketoyobyrine by palladium black to the dehydro derivative (XXVIII). This substance, on treatment with amyl alcoholic potassium hydroxide, is cleaved,

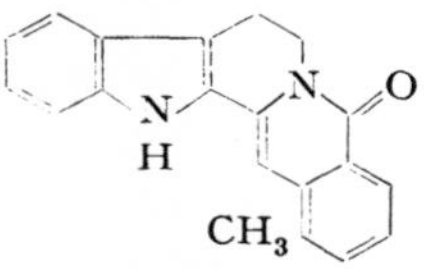

(XXVII) 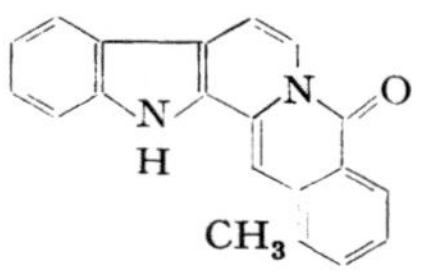 (XXVIII) 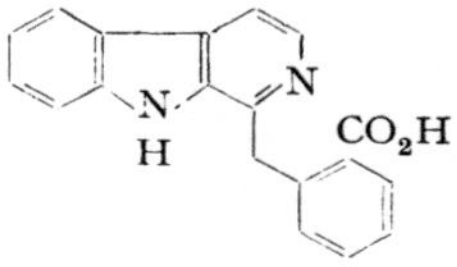 (XXIX)

not as is ketoyobyrine, but into an amphoteric compound (XXIX), whose absorption spectrum indicates its close relationship to yobyrine. Compound XXIX could not be decarboxylated to yobyrine, however, because of its facile reconversion to XXVIII, a process which occurs even on attempted recrystallization.

Direct establishment of the structure (XXVII) for ketoyobyrine has been accomplished by synthesis.[163] Acylation of tryptamine with 6-methylhomophthalic anhydride gives the amide (XXX) which may be converted to the imide through the ester. Ring closure with phosphorus oxychloride yields XXVII, identical with authentic ketoyobyrine.

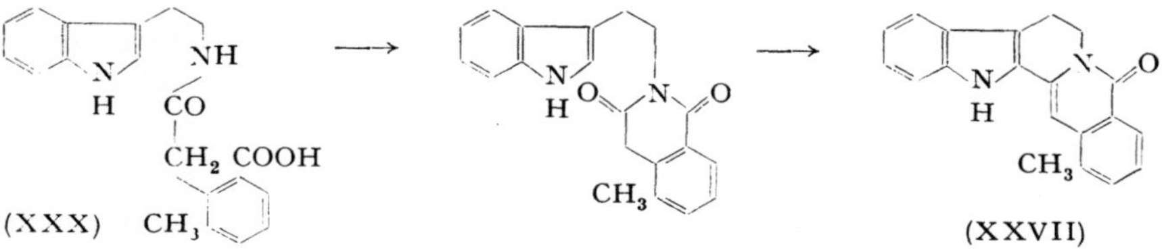

(XXX) → → (XXVII)

[162] Woodward and Witkop, *J. Am. Chem. Soc.*, **70**, 2409 (1948).
[163] Schlittler and Speitel, *Helv. Chim. Acta*, **31**, 1199 (1948).

Establishment of the structure of ketoyobyrine as XXVII provides direct proof[162] that the carboxyl group is yohimbine is located as proposed (X).

An attempt to prepare XXVI by acylation of norharman with 2,3-dimethylbenzoyl chloride resulted in acylation of the pyridine nitrogen atom, rather than that in the indole nucleus.[164]

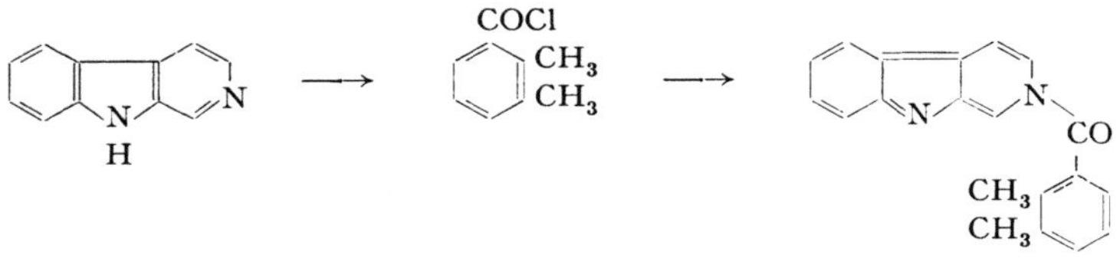

The mechanisms of the dehydrogenation of yohimbine have been considered by several workers.[162, 163, 165]

Dehydrogenation of yohimbine has also been accomplished through the use of lead tetraacetate.[166] The product, "tetradehydroyohimbine," is hydrolyzable to the corresponding acid. This material was assigned structure XXXI,[166] since the acid yielded harman and *m*-toluic acid on boiling with potassium hydroxide in amyl alcohol. However, examination of the absorption spectrum of the acid, that of the corresponding product from apoyohimbine and that of yobyrine, showed the three substances are closely related, and "tetradehydroyohimbic acid" has been renamed hexahydrohydroxyyohimbine carboxylic acid and given the structure represented by XXXII.[155]

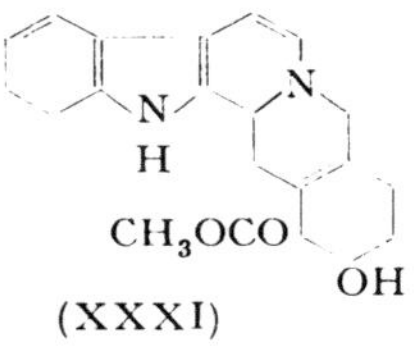

(XXXI) 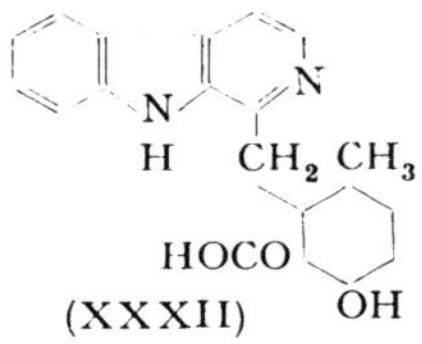 (XXXII)

The stereochemistry of the D/E ring fusion has recently been investigated.[167] It was found that yohimbic acid, on heating with thallous hydroxide, cleaved in ring C, between carbon atoms 2 and 3, to give *chano*desoxyyohimbol (XXXIII). Reduction and subsequent Hofmann degradation

[164] Speitel and Schlittler, *ibid.*, **32**, 860 (1949). Clemo and Swan, *J. Chem. Soc.*, **1949**, 487.

[165] Julian, Magnani, Pikl, and Kerpel, *J. Am. Chem. Soc.*, **70**, 174 (1948). Goutarel and Janot, *Int. Cong. Chem.*, London, 1947.

[166] Hahn, Kappes, and Ludewig, *Ber.*, **67**, 686 (1934).

[167] Witkop, *J. Am. Chem. Soc.*, **71**, 2559 (1949).

resulted in the isolation of optically active *trans-N*-methyldecahydroiso-quinoline (XXXIV), demonstrating that rings D and E are *trans* locked.

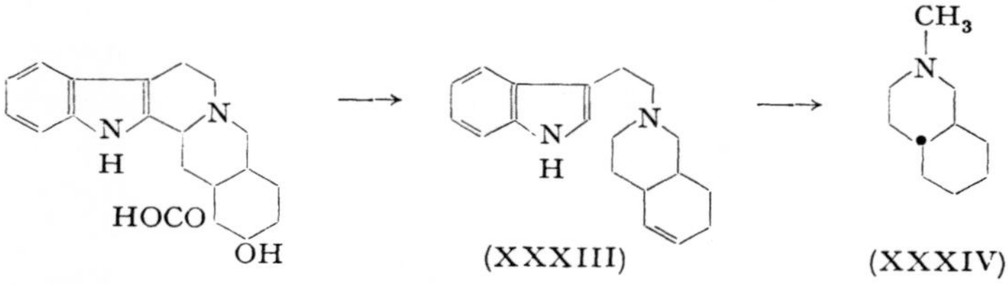

(XXXIII) (XXXIV)

Little is known regarding the constitution of the isomerides of yohimbine (Table I). Some yield the same ketone on oxidative decar-

TABLE I

Isomer	M.p. (°C.)	$[\alpha]_D$	Acid produced on hydrolysis	Ketone from decarboxylation of oxidized acid
Yohimbene[168]	276	+ 44	Yohimbenic[168]	Yohimbenone[155]
Alloyohimbine[147, 169]	105	— 74	Alloyohimbic[150, 169]	Alloyohimbone[170]
Isoyohimbine[147]	238	+ 57	Isoyohimbic[147, 150]	Yohimbone[170, 171]
Corynanthine[172]	241	—125	Corynanthic[172]	Yohimbone[172]
Corynanthidine[173]	243	— 11	Corynanthidic[173]	(Not yohimbone)[173]
ψ-Yohimbine[174]	264	+ 26	ψ-Yohimbic[175]	ψ-Yohimbone[175]
α-Yohimbine[176]	234	— 9	α-Yohimbic[176]	(Not yohimbone)[177]
β-Yohimbine[177]	235	— 54	β-Yohimbic[177]	
γ-Yohimbine[177]	240	— 28	γ-Yohimbic[177]	Yohimbone[177]
δ-Yohimbine[178]	254	— 50	δ-Yohimbic[178]	
Paniculatine[179]		— 42		

boxylation as does yohimbine, and those investigated yield identical dehydrogenation products.[175,177] It is probable that most, if not all, have a

[168] Hahn and Brandenberg, *Ber.*, **59,** 2189 (1926).
[169] Hahn and Brandenberg, *ibid.*, **60,** 669 (1927).
[170] Hahn and Stenner, *ibid.*, **61,** 278 (1928).
[171] Hahn and Schuch, *ibid.*, **62,** 2953 (1929).
[172] Perrot, *Compt. rend.*, **148,** 1465 (1909). Fourneau, *ibid.*, **148,** 1770 (1909); **150,** 976 (1910). Fourneau, *Bull. Soc. chim.*, **12,** 934 (1945).
[173] Janot and Goutarel, *Compt. rend.*, **220,** 617 (1945); *Bull. soc. chim.*, **13,** 535 (1946); **1949,** 509.
[174] Karrer and Salamon, *Helv. Chim. Acta*, **9,** 1059 (1926).
[175] Janot, Goutarel, and Amin, *Compt. rend.*, **230,** 2041 (1950).
[176] Lillig, *Merck's Jahresber.*, **42,** 20 (1929); *Chem. Abstr.*, **24,** 4517 (1930).
[177] Hahn and Schuch, *Ber.*, **63,** 1638 (1930).

common nucleus, identical with that of yohimbine, and that the isomerism is due to either epimerization at positions 16 and/or 17 or geometrical isomerism at the D/E ring fusion, or both. These ideas are supported by the fact that corynanthine is isomerized to yohimbine by treatment with base and that corynanthic acid produced from corynanthine is accompanied by yohimbic acid.

Other than those results previously mentioned, two synthetic efforts toward the yohimbine nucleus are notable. The natural unsubstituted nucleus, yohimbane (XXXV), may be prepared by the Wolf–Kishner reduction of yohimbone (XXXVI). An attempt has been made to prepare this compound from tryptamine and *trans*-hexahydrohomophthalic anhydride.[157] The product of this reaction (XXXVII) was cyclized in a manner employed previously[180] and in the synthesis of ketoyobyrine.[163] Electrolytic reduction

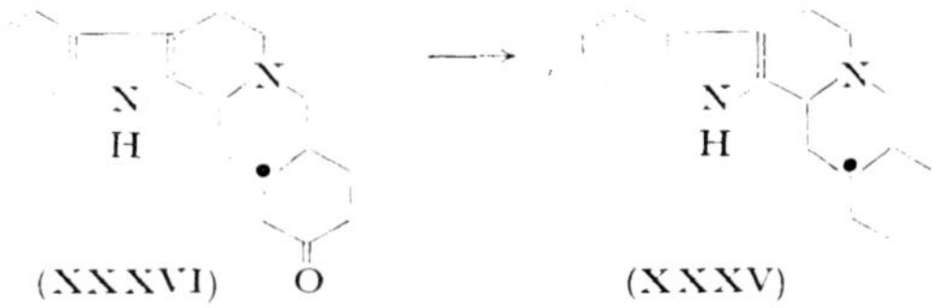

(XXXVI) → (XXXV)

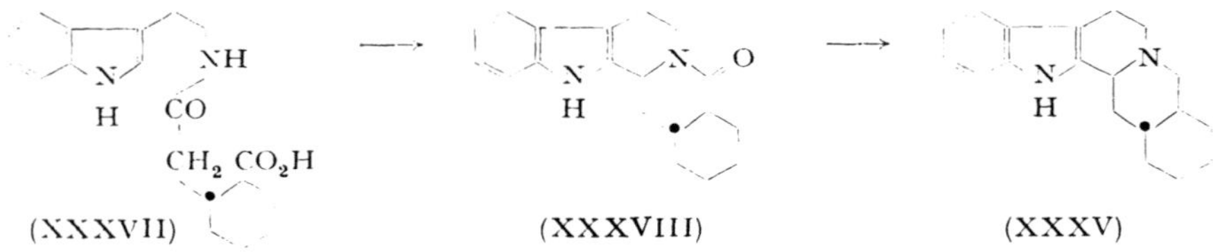

(XXXVII) → (XXXVIII) → (XXXV)

of XXXVIII furnished a product having the correct composition, but a direct comparison with natural yohimbane was not possible since no attempt was reported to resolve the synthetic material.

A more recent, and more successful, effort has resulted in the synthesis of yohimbone.[181] The methoxyphenylalanine (XXXIX) was converted to the carboline (XL) by condensation with formaldehyde. Esterification, followed by alkylation, gave the diester (XLI), which in a Dieckmann condensation yielded the ketoquinazoline (XLII). A Fischer indole synthesis with

[178] Heinemann, *ibid.*, **67,** 15 (1934).
[179] Raymond–Hamet, *Bull. sci. pharmacol.*, **44,** 54 (1937).
[180] Schlittler and Alleman, *Helv. Chim. Acta*, **31,** 128 (1948).
[181] Swan, *J. Chem. Soc.*, **1950,** 1534.

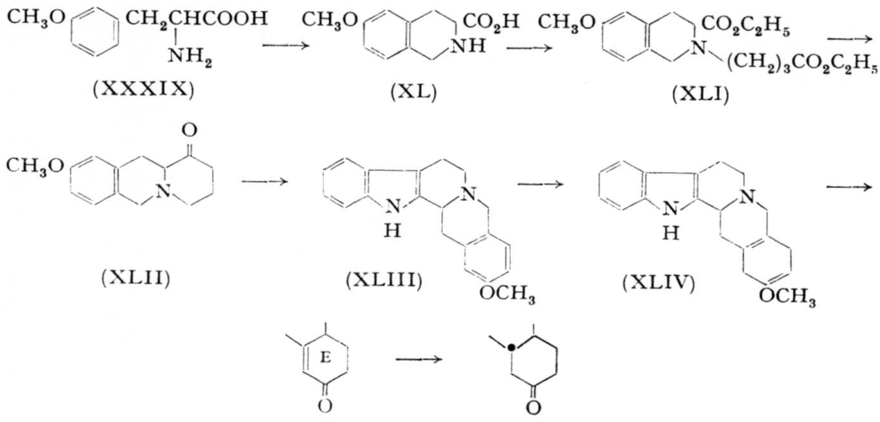

(XXXIX) (XL) (XLI)

(XLII) (XLIII) (XLIV)

(XXXVI)

the phenylhydrazone of XLII produced the yohimbine derivative (XLIII). This was reduced with sodium and the dihydro derivative (presumably XLIV) on demethylation and reduction over platinum gave racemic yohimbone (XXXVI). Resolution resulted in a material identical with the natural material.

Corynantheine

This base, $C_{22}H_{26}O_3N_2$,[182] m.p. 115–116°, $[\alpha]_D$ +28°,[183] was found in yohimbine residues.[174] It contains two methoxyl groups and may be hydrolyzed to corynantheic acid, containing only one methoxyl group.[184] Decarboxylation and hydrolysis of the acid furnish the ketone corresponding to yohimbone, named corynantheone, results indicating that the second methoxyl group is present as an enol ether.[182]

Selenium dehydrogenation of corynantheine or corynantheone[182] yields a base, corynanthyrene, $C_{19}H_{22}N_2$,[185] found[186] to be identical with a product, alstyrine, obtained similarly from alstonine (p. 238). Corynanthyrene may be ozonized to a ketone, $C_{19}H_{22}O_2N_2$, similar to the product (XVII) produced in the same manner from tetrahydroyobyrine. Acid hydrolysis of the ketone gives *o*-aminopropiophenone (XLV) and 3,4-diethylpyridine-6-carboxylic

[182] Janot and Goutarel, *Compt. rend.*, **231**, 152 (1950).
[183] Janot and Goutarel, *ibid.*, **206**, 1183 (1938).
[184] Raymond-Hamet, *J. pharm. chim.*, **22**, 306 (1935); Janot and Goutarel, *Ann. pharm. franc.*, **7**, 648 (1949).
[185] Karrer and Enslin, *Helv. Chim. Acta*, **32**, 1390 (1949).
[186] Karrer and Enslin, *ibid.*, **33**, 100 (1950).

acid (XLVI). On this basis, the structure of corynanthyrene, analogous to that proved for tetrahydroyobyrine, would be XLVII.[184]

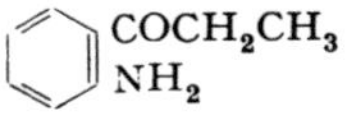

$COCH_2CH_3$, NH_2 (XLV) $HOCO$, CH_2CH_3, CH_2CH_3 (XLVI) CH_2CH_3, CH_2CH_3, CH_2CH_3 (XLVII)

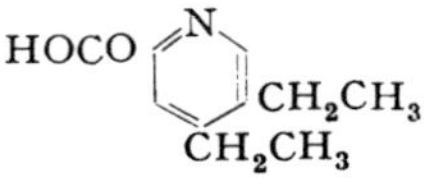

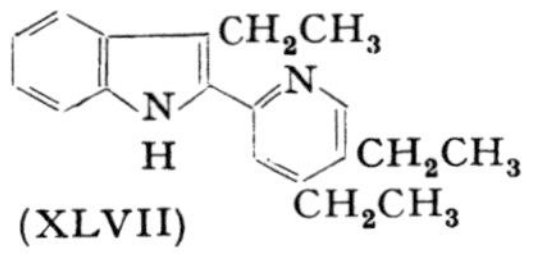

Corynantheidine

This alkaloid, $C_{22}H_{28}O_3N_2$, m.p. 117°, $[\alpha]_D$ —142°, was obtained from *P. africana*.[187] Its spectrum relates it to corynantheine.

Quebracho Alkaloids

From "quebracho bark," generally derived from *Aspidosperma quebracho*, four alkaloids have been isolated, and their presence in the plant confirmed. One of these, quebrachine,[188] has been shown to be identical with yohimbine.[189]

Aspidospermine

This alkaloid, $C_{22}H_{30}O_2N_2$, m.p. 208°, $[\alpha]_D$ —93°, was obtained from quebracho bark,[190] and since has been identified in *Vallesia glabra*[191] and *V. dichotoma*.[192] It does not give an Ehrlich color reaction. The base contains one methoxyl group and one *C*-methyl group, but no methylamino grouping,[193, 194] and is too feebly basic to yield distinct salts. Chromic acid oxidation gives a base, probably having the composition $C_{15}H_{24}O_2N_2$.[193]

Aspidospermine, on acid hydrolysis, loses an *N*-acetyl group to furnish deacetylaspidospermine, $C_{20}H_{28}ON_2$. Treatment with acetic anhydride reforms the alkaloid. The deacetyl derivative yields a benzoyl derivative, a dimethiodide, and with nitrous acid, apparently a nitro-nitroso derivative, $C_{20}H_{28}O_4N_4$, is produced. Boiling the deacetyl compound with hydriodic

[187] Janot and Goutarel, *Compt. rend.*, **218**, 852 (1944).
[188] Hesse, *Ann.*, **211**, 249 (1882).
[189] Fourneau and Page, *Bull. sci. pharmacol.*, **21**, 7 (1914). Spiegel, *Ber.*, **48**, 2077, 2084 (1915). Hahn, *ibid.*, **60**, 1681 (1927).
[190] Fraude, *Ber.*, **11**, 2189 (1878); *ibid.*, **12**, 1560 (1879).
[191] Hartmann and Schlittler, *Helv. Chim. Acta*, **22**, 547 (1939).
[192] Deulofeu and Carcamo, *J. Chem. Soc.*, **1940**, 1051.
[193] Ewings, *ibid.*, **105**, 2738 (1914).

acid converts it into aspidosine, $C_{19}H_{26}ON_2$, a phenolic substance in which an ether linkage and an amide grouping have been cleaved.

A comparison[194] of deacetylaspidospermine and aspidospermine with aromatic amines indicates that the acetylated nitrogen is attached to an aromatic system. This has been taken as evidence for a dihydroindole nucleus, and is supported by the similarity of its spectrum with that of strychnine.[195]

Vallesine

This base, $C_{20}H_{26}O_2N_2$, m.p. 154–156°, $[\alpha]_D$ —91°, was obtained from aspidospermine mother liquors.[196] It is a monoacidic base, containing one methoxyl group and one *C*-methyl group, but no *N*-methyl group or active hydrogen atom. Like aspidospermine, no Ehrlich color reaction is given, and hydrolysis furnishes deformylvallesine, $C_{19}H_{26}ON_2$, containing one active hydrogen atom and reconvertible to vallesine with formyl acetic anhydride. The deformyl compound is diacidic, and resembles deacetylaspidospermine. The physical constants of reformed vallesine and formyldeacetylaspidospermine are much the same, as are those of acetyldeformylvallesine and aspidospermine, although satisfactory analytical data could not be obtained. Further correlation is exhibited in the melting points of the benzoylated deacyl compounds, and both the infrared spectra and the x-ray powder diagrams of the natural products are indistinguishable.

Quebrachamine

Also obtained from the quebracho bark is quebrachamine, $C_{19}H_{26}N_2$, m.p. 147°,[193] $[\alpha]_D$ —109°. This base gives the Ehrlich color reaction, and yields a methiodide. It is saturated to hydrogen, and oxidation with nitric acid yields picric acid.[197] It has recently been found that the alkaloid kamassine is identical with quebrachamine (see p. 280).

Two further alkaloids have been reported to occur in "payta bark,"[198] also derived from species of *Aspidosperma*. These are paytamine, $C_{21}H_{24}ON_2$ and paytine, $C_{21}H_{24}ON_2$, m.p. 156°, $[\alpha]_D$ —150°.

Calabash Curare Alkaloids

Curare consists of plant extracts used by the Amazon Indians as arrow poisons. The recent interest in curare in controlling muscular rigidity in

[194] Openshaw and Smith, *Experientia*, **4,** 478 (1948).
[195] Raymond-Hamet, *Compt. rend.*, **226,** 2154 (1948).
[196] Schlittler and Rottenberg, *Helv. Chim. Acta*, **31,** 446 (1948).
[197] Field, *J. Chem. Soc.*, **125,** 1444 (1924).
[198] Hesse, *Ann.*, **154,** 287 (1870); **166,** 272 (1873); **211,** 280 (1882).

surgical operations, shock therapy of psychiatric cases, and in muscular spasms, such as poliomyelitis, has initiated a study into the origin of the poison, as well as the constitution of the active principles.

The material is of three types, differentiated by the character of the container in which it is packed. Tube or bamboo curare has been examined most extensively,[199] and is found to contain alkaloids having an isoquinoline nucleus, as does pot curare. The third form, gourd or calabash curare,[200] differs from the first two in that it is apparently derived, at least in part, from *Strychnos toxifera*[201] and contains alkaloids of the indole type, although the botanical origin of all three types is somewhat vague.

Calabash curare contains a mixture of quaternary alkaloids which are isolated from the accompanying material as the reineckate salts, and then separated by chromatography. For purposes of summation and comparison, these bases are listed in Table II. The full name, *e.g.*, Calabash-curarine I, is shortened to C-curarine I.

Some insight into the structure of these alkaloids is given by the degradative results thus far obtained. C-dihydrotoxiferine I, on heating with sulfur at 260°, yielded isoquinoline, and tetrahydro-C-curarine I gave 3-ethylindole, skatole, and isoquinoline when distilled from zinc dust. Similar treatment of C-toxiferine II furnished 3-ethylpyridine,[210] and of C-curarine I furnished 3-ethylpyridine, 3-ethylindole, skatole, carbazole, and *N*-methylcarbazole.[211] Indole color tests are given by calebassine chloride, and color reactions similar to those given by tetrahydrocarboline bases are exhibited by others of the group.

These results have been interpreted to indicate a relationship to either yohimbine or strychnine,[210, 211] in which one of the nitrogen atoms is quaternary (one of the alkyl groups being methyl) and the other secondary and

[199] Henry, *The Plant Alkaloids*. 4th ed., Blakiston, New York, 1949, p. 373.
[200] Boehm, *Arch. Pharm.*, **235**, 660 (1897).
[201] King, *J. Chem. Soc.*, **1932,** 1472.
[202] King, *ibid.*, **1949,** 3263.
[203] Wieland, Konz, and Sonderhoff, *Ann.*, **527,** 160 (1937).
[204] Wieland and Pistor, *ibid.*, **536,** 68 (1938).
[205] Wieland, Pistor, and Bähr, *ibid.*, **547,** 140 (1941).
[206] Wieland, Witkop, and Bähr, *ibid.*, **547,** 156 (1941).
[207] Karrer and Schmid, *Helv. Chim. Acta*, **29,** 1853 (1946).
[208] Schmid and Karrer, *ibid.*, **30,** 1162 (1947).
[209] Schmid and Karrer, *ibid.*, **30,** 2081 (1947).
[210] Wieland, Witkop, and Bähr, *Ann.*, **558,** 144 (1947).
[211] Schmid, Ebnöther, and Karrer, *Helv. Chim. Acta*, **33,** 1486 (1950).

nonbasic, incorporated in the indole nucleus. This has prompted the preparation of a series of quaternary salts of yobyrine and tetrahydroyobyrine which were tested for curare activity.[212] The results favor the apparent parallelism, as do the similarities in the ultraviolet absorption spectra of the synthetic bases and the natural alkaloids.

TABLE II

Alkaloid	Source	Composition of cation	M.p. (°C.) picrate	$[\alpha]_D$ chloride	Functions	Ref.
C-curarine I	Gourd	$C_{21}H_{21}N_2^+$	300	+70–73	One C–CH_3, no CH_3O or phenolic OH	203, 204, 205
C-curarine II	Gourd	$C_{20}H_{23}N_2^+$	203	+74	—	204, 205
C-curarine III	Gourd	$C_{21}H_{21}N_2^+$	189	—937	—	205
Toxiferine I	Gourd and *S. toxifera*	$C_{20}H_{22}ON_2^+$	270	—610	—	200, 202, 208, 211
Toxiferine II	Gourd and *S. toxifera*	$C_{20}H_{28}ON_2^+$	216	—	—	205, 202
Toxiferine IIa	—	$C_{20}H_{23}ON_2^+$	210	+67	—	205
Toxiferine IIb	—	$C_{20}H_{23}ON_2^+$	215	+78	—	205
C-toxiferine II	Gourd	$C_{20}H_{25}ON_2^+$	215	+72	One C–CH_3	206
C-dihydro-toxiferine I	Gourd	$C_{20}H_{23}N_2^+$	184	—610	One C–CH_3	206
C-isodihydro-toxiferine I	Gourd	$C_{20}H_{23}N_2^+$	242	—566	—	206
Calebassine	Gourd	$C_{20}H_{23}N_2^+$	192	—	—	207
Alkaloid A	Gourd	$C_{20}H_{23}ON_2^+$	260	—	—	207
Alkaloid B	Gourd	$C_{20}H_{25}ON_2^+$	196	—	—	208
Fluorocurine	Gourd	$C_{20}H_{21}ON_2^+$	178	+326 (I^-)	One N–CH_3, no CH_3O	209
Calebassinine	Gourd	$C_{19}H_{23}O_2N_2^+$	260	+63	No CH_3O	209
C-alkaloid UB	Gourd	$C_{19}H_{23}O_3N_2^+$	240	—	—	209
C-alkaloid X	Gourd	—	—	—	—	209
Toxiferine III	*S. toxifera*	$C_{20}H_{27}ON_2^+$	285	—	—	202
Toxiferine IV	*S. toxifera*	$C_{21}H_{26}O_4N_2^+$	238	—	—	202
Toxiferine V	*S. toxifera*	$C_{21}H_{26}O_3N_2^+$	270	—	—	202
Toxiferine VI	*S. toxifera*	$C_{21}H_{24}O_5N_2^+$	300	—	—	202
Toxiferine VII	*S. toxifera*	$C_{22}H_{24}O_3N_2^+$	300	—	—	202
Toxiferine VIII	*S. toxifera*	$C_{22}H_{24}O_3N_2^+$	300	—	—	202
Toxiferine IX	*S. toxifera*	$C_{23}H_{26}O_3N_2^+$	300	—	—	202
Toxiferine X	*S. toxifera*	$C_{19}H_{27}N_2^+$	268	—	—	202
Toxiferine XI	*S. toxifera*	$C_{21}H_{26}ON_2^+$	277	—	—	202
Toxiferine XII	*S. toxifera*	$C_{39}H_{46}ON_4^+$	333	—	—	202

[212] Karrer and Waser, *ibid.*, **32,** 409 (1949).

Alstonia Alkaloids

From the barks of the genus *Alstonia* a sizable number of alkaloids have been obtained, and there are indications of yet uncharacterized bases.

Alstonine

This unstable base, $C_{21}H_{20}O_3N_2$, sinters 77°,[213] sulfate ($B_2H_2SO_4$), m.p. 195–196°, $[\alpha]_D$ +127°,[214] is obtained from *A. constricta*.[215] It is a monoacidic base, containing a methoxyl but no methylamino group, and yielding a methiodide.[213] Hydrogenation gives tetrahydroalstonine, which contains a *C*-methyl group, a carbomethoxyl group, but no hydroxyl group.[216]

Permanganate oxidation of alstonine yields *N*-oxalyanthranilic acid (I). Both alstonine and the tetrahydro derivative on selenium dehydrogenation yield alstyrine, $C_{19}H_{22}N_2$ (II), identical with corynanthyrene, obtained

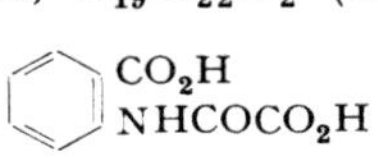

CO_2H
$NHCOCO_2H$
(I)

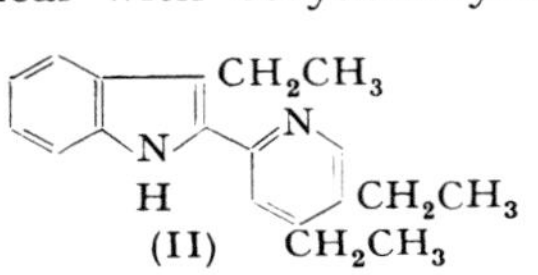

CH_2CH_3 N H CH_2CH_3 CH_2CH_3
(II)

from corynantheine in the same manner (see p. 233).[217] KOH fusion of alstonine produces harman (III), and tetrahydroalstonine in the same reaction yields, among other products, harman, nonharman, and indole-2-carboxylic acid (IV).

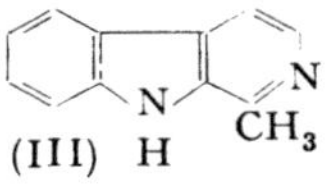

N N H CH_3
(III)

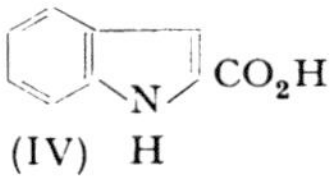

N H CO_2H
(IV)

Reduction of tetrahydroalstonine with lithium aluminium hydride yields tetrahydroalstonol, which readily forms ethers with alcohols, indicating an allyl type grouping.[216] This is confirmed by the ultraviolet spectrum, which indicates that the double bond still present in tetrahydroalstonine is conjugated with the ester carbonyl group, but not the indole nucleus.[216]

On the basis of the structure of alstyrine, tetrahydroalstonine has been

[213] Sharp, *J. Chem. Soc.*, **1934**, 287.
[214] Leonard and Elderfield, *J. Org. Chem.*, **7**, 556 (1942).
[215] Hesse, *Ann.*, **205**, 360 (1880).
[216] Elderfield and Gray, *J. Org. Chem.*, **16**, 506 (1951).
[217] Karrer and Enslin, *Helv. Chim. Acta*, **33**, 100 (1950).

represented as V. Tetrahydroalstonol would then be VI and alstonine would be formulated, similarly to sempervirine (see p. 242), as VII.[216]

N H N CH_3 CH_3OCO O
(V)

N CH_3 $HOCH_2$ O
(VI)

N N CH_3 CH_3OCO O
(VII)

Alstonilidine

This base was found in *A. constricta* bark,[218] and has the composition $C_{22}H_{18}O_3N_2$, m.p. 372°. Like alstonine, it is reduced to a tetrahydro derivative. Alstonilidine is susceptible to oxidation by air, and forms an oxide, $C_{22}H_{18}O_4N_2$.

The other known and well-characterized alstonia bases are listed in Table III.

TABLE III

Alkaloid	Composition	M.p. (°C.)	$[\alpha]_D$	Functions	Ref.
Villalstonine	$C_{40}H_{50}O_4N_4$	218	+56	Two N–CH_3, two CH_3O (one as ester), dimethiodide	219
Macralstonine	$C_{44}H_{54}O_5N_4$	293	+28	One CH_3O, three N–CH_3	219, 220
Macralstonidine	$C_{41}H_{50}O_3N_4$	270 (dec)	+174	Two N–CH_3, no CH_3O	219
Echitamine	$C_{22}H_{28}O_4N_2$	295	—58	One CH_3O (ester), one N–CH_3, diacetyl and nitroso derivatives	221
Echitamidine	$C_{20}H_{26}O_3N_2$	135	—515	Monoacidic, probably N–CH_3, no CH_3O	222

[218] Hawkins and Elderfield, *J. Org. Chem.*, **7**, 573 (1942).

[219] Sharp, *J. Chem. Soc.*, **1934**, 1227.

[220] Santos and Santos, *Bull. Univ. Phil. Nat. Appl. Sci.*, **5**, 153 (1936); *Chem. Abstr.*, **31**, 6243 (1938).

[221] Harnack, *Ber.*, **11**, 2004 (1878). Hesse, *Ann.*, **176**, 326 (1875). Goodson and Henry, *J. Chem. Soc.*, **127**, 1640 (1925).

[222] Goodson, *J. Chem. Soc.*, **1932**, 2626.

Gelsemium Alkaloids

Gelsemine

This base, $C_{20}H_{22}O_2N_2$,[223] m.p. 178°, $[\alpha]_D$ +16; was found[224] in *Gelsemium sempervirens*. The presence of one methylamino group, but no methoxyl or *C*-methyl groups has been established,[225,226] and a single olefinic linkage is indicated by the formation of a dihydro derivative.[227,228] Gelsemine may be isomerized to isogelsemine by treating with zinc and hydrochloric acid in the presence of palladium chloride.[227] Hydrogenation of isogelsemine yields the same dihydro derivative as the natural base, indicating a labile double bond, which may be hydrated to give either apogelsemine or isoapogelsemine, $C_{20}H_{24}O_3N_2$.[223] That this double bond in gelsemine is terminal ($=C=CH_2$) is shown by the presence of one $C—CH_3$ group in dihydrogelsemine,[226] and by the ozonolysis of gelsemine to furnish formaldehyde.[229]

That a benzene ring is present is demonstrated by the bromination of gelsemine and the nitration of dihydrogelsemine.[227] Gelsemine contains one active hydrogen atom,[230] and an *N*-acetyl derivative is known.[223,229,231] The alkaloid behaves as a monoacidic base,[228] yielding a methiodide. In the absence of a methoxyl, the second oxygen is presumed to be present as a cyclic ether.

Both gelsemine and dihydrogelsemine possess an amide carbonyl group, as indicated in the infrared spectra. Reduction of gelsemine with lithium aluminum hydride yields dihydrodesoxygelsemine.[232]

Degradative studies have resulted in the isolation of 2,3-dimethylindole by high-temperature treatment of gelsemine with soda lime or selenium,[233] and of skatole, 3-methyloxindole,[229] and two bases, $C_{11}H_{11}N$, and $C_{14}H_{11}N$,[225] by zinc dust distillation. These bases appear to be quinolines or isoquinolines.

[223] Moore, *ibid.*, **97**, 2223 (1910); **99**, 1231 (1911).
[224] Gerrard, *Pharm. J.*, **13**, 641 (1883).
[225] Witkop, *J. Am. Chem. Soc.*, **70**, 1424 (1948).
[226] Gibson and Robinson, *Chemistry & Industry*, **1951**, 93.
[227] Chu and Chou, *J. Am. Chem. Soc.*, **62**, 1955 (1940); **63**, 827 (1941).
[228] Forsyth, Marrian, and Stevens, *J. Chem. Soc.*, **1945**, 579.
[229] Goutarel, Janot, Prelog, Sneeden, and Taylor, *Helv. Chim. Acta*, **34**, 1139 (1951).
[230] Kates and Marion, *Can. J. Research*, **29**, 37 (1951).
[231] Chou, *Chinese J. Physiol.*, **17**, 189 (1949); *Chem. Abstr.*, **45**, 1610 (1951).
[232] Kates and Marion, *J. Am. Chem. Soc.*, **72**, 2308 (1950).
[233] Marion, *Can. J. Research*, **21**, 247 (1943).

These data indicate an oxindole nucleus, disubstituted in the 3-position (I).[229,232] This structure, the N-CH_3 and the methylene group account for

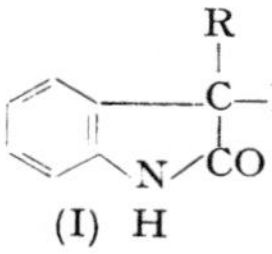

R
C—R
N CO
(I) H

ten of the twenty carbon atoms. The hydrogenated quinoline or isoquinoline moiety contains nine more, leaving but one carbon unaccounted for.

The close spatial relationship of the methylene and carbonyl groups is shown by the bromination of gelsemine.[229] The product is bromoallogelsemine hydrobromide, which, in its infrared spectrum, shows no carbonyl group, but rather a carbon to nitrogen double bond, and which is reducible catalytically to a dihydro derivative no longer containing a double bond of the imine type. The bromoallogelsemine may easily be reconverted to gelsemine by zinc and acetic acid. These transformations may be represented as follows:[229]

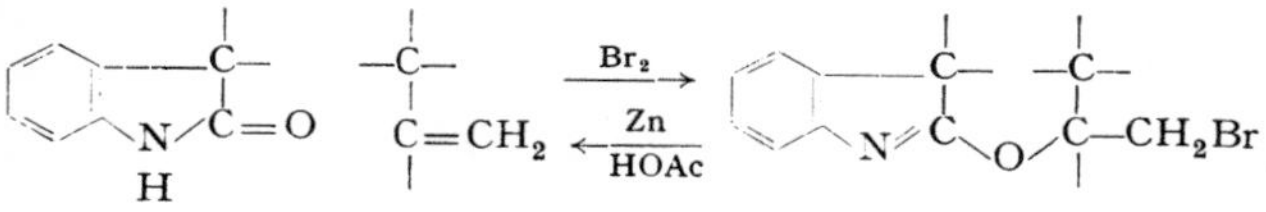

Gelsemine $\underset{Zn, HOAc}{\overset{Br_2}{\rightleftharpoons}}$ Bromoallogelsemine

Acid hydrolysis of the bromoallogelsemine causes the addition of a molecule of water, producing bromohydroxydihydrogelsemine (II). These interactions of the methylene and carbonyl groups indicate that the ring formed between them must be five or six membered, hence placing the two groups in close proximity.

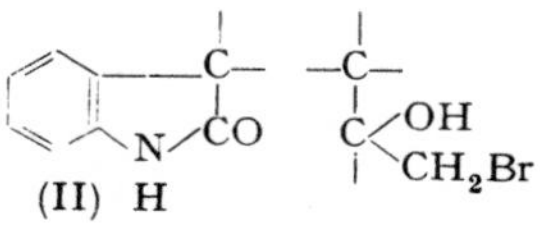

(II) H

Consideration of this evidence and the degradation products, in view of the possible biological precursors, tryptophan and dihydroxyphenylalanine, leads to a carbon skeleton such as III as a possible representation of gelsemine.[229]

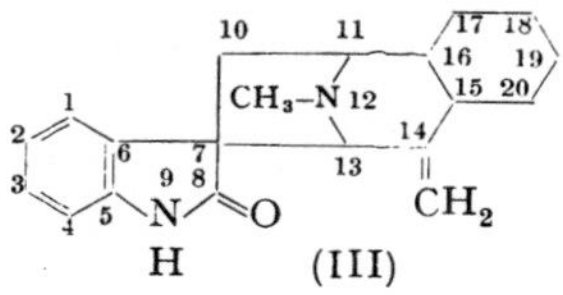

(III)

Substantiation of these ideas and evidence for the location of the ether bridge are obtained from the Hofmann degradation of gelsemine methiodide.[234] This furnishes des-*N*-methylgelsemine, which can be reduced to a dihydro derivative, also obtainable from dihydrogelsemine methodide. Since further reduction is not possible under these conditions the double bond introduced by the degradative cleavage is not reducible.

The des-base forms a methiodide, but further degradation is not possible showing the absence of a hydrogen atom β to the nitrogen atom. These results are accommodated by structure III, indicating a C_{11}–C_{16} double bond in the des-base, and providing C_{10}–C_{15} as possible attachments for the ether oxygen atom.

Gelsemine then would be IV and the des-base would be V.[234]

O CH₃N CH₂ N O H (IV)

CH—CH= O CH-C N O N CH₂ (V) H (CH₃)₂

Sempervirine

This alkaloid was first obtained from gelsemine residues,[223,235] and has since been found in *Mostea buchholzii*.[236] It has the composition $C_{19}H_{16}N_2$,[237,238] m.p. 258–260°, and is optically inactive. Sempervirine is extremely insoluble in ether and forms highly colored crystals. It yields a methiodide[228] and contains one active hydrogen atom, but no methylamino group.[239] Hydrogenation is reported to result in the absorption of three moles rapidly, followed by two moles slowly to give a decahydro derivative.[228]

When sempervirine is heated with selenium or palladium black it is isomerized to yobyrine, and on refluxing with Raney nickel in xylene it is transformed into tetrahydroyobyrine. On this basis structure VI was suggested for the alkaloid,[240] a formulation identical with that originally

[234] Goutarel, Janot, Prelog, and Sneeden, *Helv. Chim. Acta*, **34**, 1962 (1951).

[235] Stevenson and Sayre, *J. Am. Pharm. Assoc.*, **4**, 60 (1915). Sayre and Watson, *ibid.*, **8**, 708 (1919).

[236] Gellert and Schwartz, *Helv. Chim. Acta*, **34**, 779 (1951).

[237] Hasenfratz, *Compt. rend.*, **196**, 1530 (1933).

[238] Chou, *Chinese J. Physiol.*, **5**, 295 (1931); *Chem. Abstr.*, **25**, 5736 (1932).

[239] Goutarel, Janot, and Prelog, *Experientia*, **4**, 24 (1948).

[240] Prelog, *Helv. Chim. Acta*, **31**, 588 (1948).

proposed for yobyrine, and since synthesized[241] and found to be different from sempervirine. This structure does not account for the reduced ring E found in tetrahydroyobyrine, nor the color nor insolubility of sempervirine. To better satisfy these facts, VII was proposed,[242] and is substantiated by by the absence of an N-H vibration band in the infrared spectrum, and by the formation of *N*-methylyobyrine (VIII)

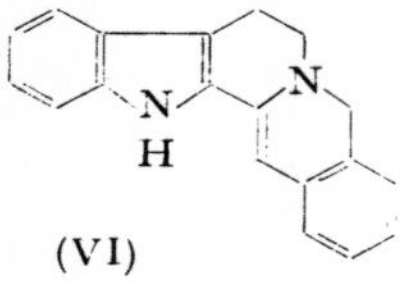

(VI) 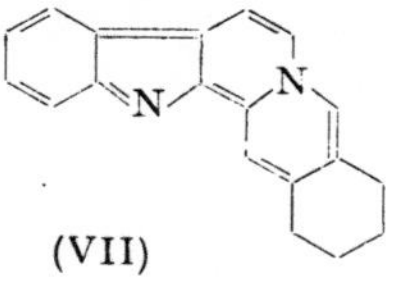 (VII) 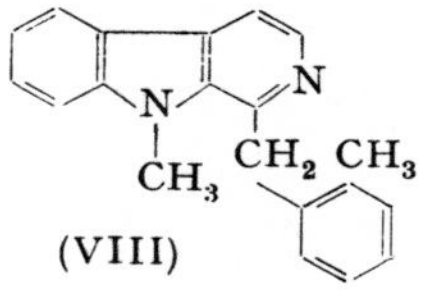 (VIII)

when sempervirine methochloride is heated with selenium, demonstrating that the metho salts have structure IX. The active H atom found in sempervirine by the Zerewitinoff determination is explained by the α-picoline structure, in which the H atoms are activated by the pyridine ring. The contribution to the resonance hydride of sempervirine of such a structure as X is observed in the intense color and insolubility of the alkaloid.

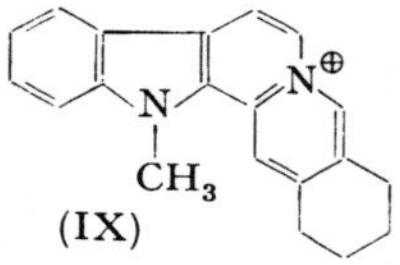

(IX) 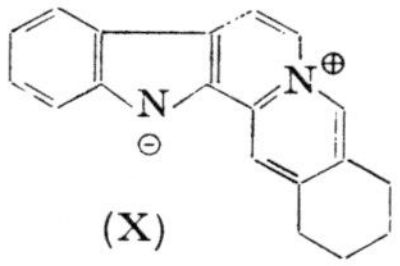 (X)

Direct confirmation of structure VII has been furnished by the synthesis of sempervirine methochloride.[243] The lithium derivative of *N*-methylharman (XI) was condensed with isopropoxymethylenecyclohexanone (XII), and on acidification of the reaction mixture, salts of the methyl sempervirine cation were isolated. By employing harman in the same reaction, the synthesis of sempervirine itself has been effected.[244]

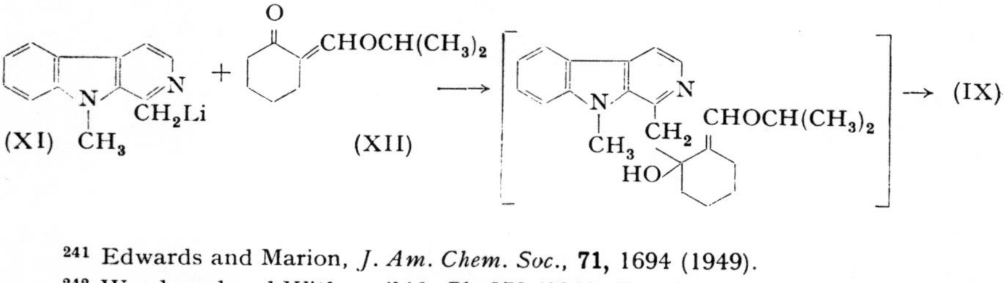

(XI) + (XII) ⟶ [ ] ⟶ (IX)

[241] Edwards and Marion, *J. Am. Chem. Soc.*, **71**, 1694 (1949).

[242] Woodward and Witkop, *ibid.*, **71**, 379 (1949). Bently and Stevens, *Nature*, **164**, 141 (1949).

[243] Woodward and Mc Lemore, *J. Am. Chem. Soc.*, **71**, 379 (1949).

[244] Woodward and Mc Lemore, Private communication.

Gelsemicine

This alkaloid, $C_{20}H_{24}O_4N_2$, m.p. 171°, $[\alpha]_D$ —140°, isolated from the root of *Gelsemium sempervirens*,[245] contains three active hydrogen atoms,[228] and yields a benzoyl derivative which is nonbasic, suggesting that the basicity is due to a secondary amino group. This is further indicated by the formation of methylgelsemicine hydriodide by the action of methyl iodide. No carbonyl derivatives are formed. On hydrogenation with Adams catalyst three moles of hydrogen are absorbed.

Other alkaloids obtained from this genus, specifically *G. elegans*, are summarized in Table IV.

TABLE IV

Alkaloid	Composition	M.p. (°C.)	$[\alpha]_D$	Ref.
Koumine	$C_{20}H_{22}ON_2$	170	—265	246, 247
Kouminidine	$C_{19}H_{25}O_4N_2$	229		246, 247
Kounidine	$C_{21}H_{24}O_5N_2$	315		248

Rauwolfia Alkaloids

Ajmaline

This alkaloid,[249] $C_{20}H_{26}O_2N_2$, m.p. 159–160°, $[\alpha]_D$ +128°, is considered identical with rauwolfine,[250] also isolated from *Rauwolfia serpentina*, but assigned the composition $C_{21}H_{26}O_2N_2$. Ajmaline contains an active hydrogen atom and an *N*-methyl group,[251] and yields a methiodide, an *O*-benzoyl, and a diacetyl derivative.[252]

A recent investigation of ajmaline[252] has shown some discrepancies in the previous work, with the result of re-establishing some of the features of the molecule. Since the alkaloid couples with diazobenzenesulfonic acid to furnish methyl orange dyes, one nitrogen atom must be attached directly

[245] Chou, *Chinese J. Physiol.*, **5**, 131 (1931); *Chem. Abstr.*, **25**, 4085 (1932).
[246] Chou, *Chinese J. Physiol.*, **5**, 345 (1931); *Chem. Abstr.*, **26**, 806 (1933).
[247] Chi, Kou, and Huang, *J. Am. Chem. Soc.*, **60**, 1723 (1938).
[248] Chou, Wang and Chen, *Chinese J. Physiol.*, **10**, 79 (1936); *Chem. Abstr.*, **30**, 4270 (1937).
[249] Siddiqui and Siddiqui, *J. Indian Chem. Soc.*, **8**, 667 (1931).
[250] van Itallie and Steenhauer, *Pharm. Weekblad*, **69**, 334 (1932); *Chem. Abstr.*, **26**, 3257 (1933).
[251] Siddiqui and Siddiqui, *J. Indian Chem. Soc.*, **9**, 539 (1932); **12**, 37 (1935).
[252] Mukherji, Robinson, and Schlittler, *Experientia*, **5**, 215 (1949).

to an aromatic ring having a free and active para position. Ajmaline bears a resemblance to strychnidine, and hence the nonbasic nitrogen atom probably bears the methyl group, is attached to a benzene ring, and is tertiary. The basic nitrogen atom is also tertiary since it yields a methiodide which forms dyes on coupling with diazobenzenesulfonic acid.

The function of one oxygen atom in ajmaline is indicated as an alcoholic hydroxyl group by the formation of *O*-acyl derivatives. The second is apparently a potential aldehyde function, since the alkaloid yields an oxime, reduces an ammoniacal silver nitrate solution, and gives a positive Angeli-Remini reaction. The infrared spectrum indicates a hydroxyl group, but no carbonyl absorption may be detected. It is possible that the diacetyl derivative is both an alcohol acetate and an enol acetate.

Soda lime fusion of ajmaline furnished *ind-N*-methylharman (I) and a zinc dust distillation yields the same methyl harman and carbazole (II).

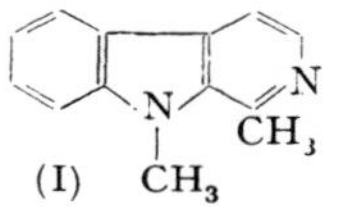

(I) 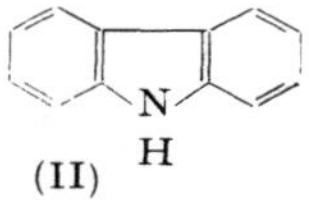 (II)

These fragments, together with the deductions above as to the nature of the functional groups, have led to two postulated structures.[252] Structure III accounts for the carboline grouping, but does not readily explain the isolation of carbazole. Furthermore, dihydroindoles which may theoretically be dehydrogenated to indoles have not been encountered in nature. Structure IV, however, more readily agrees with the degradation products. The harman may be formed by a simple migration of the bridgehead carbon from the β- to the α-position of the indole nucleus, and carbazole could be produced by dehydrogenation of the open hexane ring to give an aromatic system.

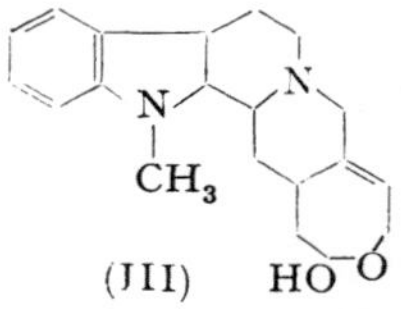

(III) 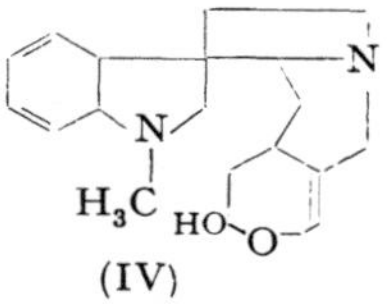 (IV)

Rauwolscine

This alkaloid, $C_{21}H_{26}O_3N_2$, m.p. 245–246°, $[\alpha]_D$ —40°, was obtained from *R. canescens*.[253] It gives color reactions similar to yohimbine,[254] and,

[253] Mookerjee, *J. Indian Chem. Soc.*, **18,** 33 (1931).
[254] Mookerjee, *ibid.*, **18,** 485 (1941).

like yohimbine and its isomerides, may be hydrolyzed to rauwolscinic acid. Re-esterification yields rauwolscine. Further parallels exist in the presence of two active H atoms and a OH group yielding an acetyl derivative.

Pyrolysis of rauwolscinic acid gives harman and 3-ethylindole, and an alkali fusion has furnished indole-2-carboxylic acid and isophthalic acid.[255] Rauwolscine, on distillation with zinc dust, gave harman, 2-methylindole, and isoquinoline.[256] These fragments further indicate a nucleus like that found in yohimbine, with only the position of the alcoholic hydroxyl and carbomethoxyl groups to be located.

Serpentine

A third alkaloid of *Rauwolfia*, serpentine,[249, 251] $C_{21}H_{22}O_3N_2$, m.p. 175°, $[\alpha]_D$ +188°, contains a carbomethoxyl group[251] but no *N*-methyl or hydroxyl group is present. The formation of colored salts bears a similarity to the behavior of alstonine, and this similarity is substantiated by the isolation of alstyrine (V) by the selenium dehydrogenation of serpentine.[257] Consideration of this nucleus, with the above data and the absence of an N-H band in the infrared spectrum suggests a structure such as VI, analogous to the structure of sempervirine.[257]

(V) CH_2CH_3 N H N CH_2CH_3 CH_2CH_3

(VI) N N CH_3OCO O

The remaining four known alkaloids of *Rauwolfia* genus are in Table V, along with such information as has been gathered as to their nature.

TABLE V

Alkaloid	Composition	M.p. (°C.)	$[\alpha]_D$	Functions	Ref.
Ajmalinine	$C_{20}H_{26}O_3N_2$	181	—97	One CH_3O, no N-CH_3, methiodide, *O*-benzoyl	249, 251
Ajmalicine	—	251	—	—	299
Serpentinine	$C_{20}H_{20}O_5N_2$	264	+167	Sec. base, one CH_3O, no N-CH_3, nitroso derivative	249, 251
Rauwolfine*	$C_{20}H_{26}O_3N_2$	237	—	Possibly phenolic OH	258

* Not the same material as that assumed identical with a ajmaline.

255 Mookerjee, *ibid.*, **20,** 11 (1943).

256 Mookerjee, *ibid.*, **23,** 6 (1946).

257 Schlittler and Schwarz, *Helv. Chim. Acta*, **33,** 1463 (1950).

Geissospermum Alkaloids

Geissospermine

From the bark of *Geissospermum vellozii* three alkaloids have been isolated. The first of these, geissospermine,[259] $C_{40}H_{50}O_3N_4$, m.p. 145–147°, $[\alpha]_D$ —102°, contains a methylamino group, a methoxyl group,[260] and two active hydrogen atoms.[261] It yields a dimethiodide, and is saturated to hydrogen.[262]

The geissospermine molecule is cleaved with cold concentrated hydrochloric acid into two fragments having the same composition, $C_{20}H_{26}O_2N_2$. Base B, m.p. 159–160° (·HCl), gives a methiodide, contains an *N*-methyl, but no methoxyl group. The methiodide yields an acetyl derivative. Base A, m.p. 205°, must then contain the methoxyl function, and its relationship to yohimbine is indicated by color reactions.

Geissospermine with dilute hydrochloric acid yields a phenolbetaine, $C_{13}H_{20}O_2N$, and methyl alcohol. The phenolbetaine, on zinc dust distillation, gives a pyridine base, probably 2-methyl-4-ethylpyridine, and an indole derivative, possibly skatole. Potassium hydroxide fusion of geissospermine yields another alkylindole, possibly 2,3-dimethylindole. A desoxy base, $C_{20}H_{26}ON_2$, m.p. 212–213°, is obtained by treating geissospermine with phosphorus and hydriodic acid.

Of three oxygen atoms in the geissospermine molecule, one is present as the methoxyl group, one probably connects the two halves as an ether linkage, but the function of the third remains undisclosed. The alkaloid will not yield acetyl or benzoyl derivatives, and it gives no reaction with carbonyl reagents.[261] The remaining oxygen is possibly present as a cyclic ether function.

The second alkaloid of the *geissospermum* group is pereirine,[259] obtained from "pereio bark". The base, $C_{20}H_{26}ON_2$, m.p. 134–135°, $[\alpha]_D$ +137°, contains no methoxyl or methylamino groups, and gives a methiodide and a methyl ether. Its purity has been questioned.[261]

The concomitant base, vellosine,[263] $C_{23}H_{28}O_4N_2$, m.p. 189°, $[\alpha]_D$ +23° contains two methoxyl groups and is a monoacidic tertiary base. Mineral acids cause the loss of water to give apovellosine, $C_{46}H_{54}O_7N_4$.

[258] Kloepfli, *J. Am. Chem. Soc.*, **54**, 2412 (1932).
[259] Hesse, *Ann.*, **202**, 141 (1880).
[260] Bertho and von Schuckmann, *Ber.*, **64**, 2278 (1931).
[261] Bertho and Sax, *Ann.*, **556**, 22 (1944).
[262] Bertho and Moog, *ibid.*, **509**, 241 (1934).
[263] Freud and Fauvet, *ibid.*, **282**, 247 (1894).

Ergot Alkaloids

The fungus *Claviceps purpurea*, occurring on certain diseased cereal grasses, especially rye, is the source of the important drug ergot. From this material 14 closely related alkaloids, among other simple bases and amino acids, have been isolated. These 14 alkaloids form seven pairs, the members of each pair being interconvertible, and only one member of each pair, the levorotary one, being appreciably physiologically active. Thus, there are two series of alkaloids, known, after the first pair discovered, as the ergotoxine series, physiologically active, and the ergotinine series dextrototary and physiologically impotent.

The great similarity of the different pairs and the difficulty encountered in the separation of the members of each pair have led to mixtures, as well as considerable confusion and misnaming among the materials. The known bases are listed in Table VI under the names holding precedence, although the constants given are in some cases those appearing to be the most accurate, rather than those first recorded.

TABLE VI

Ergotoxine series	M.p. (°C.)	$[\alpha]_D$	Ergotinine series	M.p. (°C.)	$[\alpha]_D$	Formula
Ergotoxine	183	—188	ψ-Ergotinine	228	+409	$C_{31}H_{39}O_5N_5$
Ergotamine	180	—160	Ergotaminine	242	+369	$C_{33}H_{35}O_5N_5$
Ergometrine	162	—44	Ergometrinine	196	+414	$C_{19}H_{23}O_2N_3$
Ergosine	228	—161	Ergosinine	228	+420	$C_{30}H_{37}O_5N_5$
Ergocristine	160–175	—183	Ergocristinine	226	+366	$C_{35}H_{39}O_5N_5$
Ergocryptine	212	—187	Ergocryptinine	241	+408	$C_{32}H_{41}O_5N_5$
Ergomolline*	215	—177	Ergomollinine*	241	+420	$C_{32}H_{41}O_5N_5$

* Not confirmed.

The first crystalline material isolated from ergot was ergotinine,[264] along with an amorphous material, later shown to be physiologically active[265] and investigated in detail under the names ergotoxine[266] and hydroergotinine.[267] A third base convertible into ergotoxine was discovered and named ψ-ergotinine.[268] Further examination has shown that ergotinine is

[264] Tanret, *Compt. rend.*, **81**, 896 (1875); **86**, 888 (1878).
[265] Kobert, *Arch. exptl. Pathol. Pharmakol.*, **18**, 316 (1884).
[266] Barger and Carr, *J. Chem. Soc.*, **91**, 337 (1907).
[267] Kraft, *Arch. Pharm.*, **244**, 336 (1906).
[268] Smith and Timmis, *J. Chem. Soc.*, **1931**, 1888.

not transformed into ergotoxine, and that it is identical with ergocristinine,[269] the inactive member of another pair. Hence the name ergotinine actually belongs to the base called ψ-ergotinine, but the confusion which would arise from such a change would not justify it.

The original difficulty in establishing a formula for ergotoxine[270] has been resolved by reinvestigation[269] of the pair under the name ergocornine-ergocorninine. Ergocorninine is obviously identical with the material previously named ψ-ergotinine, and, although ergocornine has constants differing from those previously assigned ergotoxine, establishment of the homogeneity of the substance does not rationalize renaming the pair, and this has been protested.[271]

The second pair of ergot alkaloids was first known as the active component, ergotamine.[272] The third active alkaloid was isolated under the names ergometrine,[273] ergotocine,[274] ergobasine,[275] and ergostetrine.[276] Comparison of samples showed the preparations to be identical.[277] The name ergometrine holds precedence, although still another name, ergonovine, has been adopted by the United States Pharmacopeia XIII.[278]

Ergosine and its ergotinine counterpart, ergosinine, were obtained[279] in 1937 and followed shortly by the fifth pair, ergocristine and ergocristinine.[280] The latest discoveries are ergocryptine and ergocryptinine,[269] and ergomolline and ergomollinine.[281]

The alkaloid ergononamine, $C_{19}H_{19}O_4N$, m.p. 132°, has been obtained from ergot,[282] but it is not related to the other ergot alkaloids in that it

[269] Stoll and Hofmann, *Helv. Chim. Acta*, **26**, 1570, 1602 (1943).

[270] Smith and Timmis, *J. Chem. Soc.*, **1930**, 1390. Soltys, *Ber.*, **65**, 553 (1932).

[271] Foster, *Analyst*, **70**, 132 (1945). Foster, Smith, and Timmis, *Pharm. J.*, **157**, 43 (1946).

[272] Spiro and Stoll, *Verhandl. schweiz. naturforsch. Ges.*, 190 (1920); *Chem. Abstr.* **16**, 2961 (1923). Stoll, *Schweiz. Apoth. Ztg.*, **60**, 341, 358, 374 (1922); *Chem. Abstr*, **16**, 3171 (1923).

[273] Dudley and Moir, *Brit. Med. J.*, **520**, 793 (1935); *Science*, **81**, 559 (1935).

[274] Kharasch and Legault, *J. Am. Chem. Soc.*, **57**, 965, 1140 (1935).

[275] Stoll and Burckhardt, *Compt. rend.*, **200**, 1670 (1935).

[276] Thompson, *J. Am. Pharm. Assoc.*, **24**, 24, 185, 748 (1935).

[277] Kharasch, King, Stoll, and Thompson, *Nature*, **137**, 403 (1936).

[278] *United States Pharmacopeia XIII*, Mack, Easton, Pa., 1947, p. 200.

[279] Smith and Timmis, *J. Chem. Soc.*, **1937**, 396.

[280] Stoll and Burckhardt, *Z. physiol. Chem.*, **250**, 1 (1937); **251**, 287 (1938).

[281] Hashimoto, *J. Pharm. Soc. Japan*, **66**, 22 (1946); *Chem. Abstr.*, **45**, 6650 (1951).

[282] Holden and Diber, *Quart. J. Pharmacol.*, **9**, 230 (1936); *Chem. Abstr.*, **30**, 6745 (1937).

gives no color reactions of the indole group and contains only one nitrogen atom. Other alkaloidal products, ergoclavine[283] and sensibamine,[284] have been found to be equimolecular mixtures of ergosine and ergosinine,[279] and ergotamine and ergotaminine,[284] respectively.

As previously mentioned, the physical characteristics, other than the optical rotation, of each member of a pair are exceedingly similar. The great difference in physiological properties within a pair, and the correlation between physiological activity and sign of rotation speak for a minor variation in structure within the pairs, the same variation being common to all the members of a series. This is empirically supported by the similarity in the absorption spectra of all the alkaloids, the fact that they all give the same indole color reactions and the extreme ease of interconversion of the members of each pair. For example, ψ-ergotinine is formed from ergotoxin by boiling with methyl alcohol, and the reverse change is effected when an ethyl alcoholic solution of the alkaloid is heated with phosphoric acid.[267] These conclusions are supported by degradative studies and other properties of the alkaloids.

The early investigations of the ergotoxine and ergotamine pairs indicated the presence of four active hydrogen atoms and one methylamino group in each of the four alkaloids, but no methoxyl group.[285] Rather drastic degradations produced only small fragments: destructive distillation of ergotinine (ergocristinine) yielded dimethylpyruvamide,[286] $(CH_3)_2CHCO$-$CONH_2$, and oxidation with permanganate or nitric acid gave benzoic or *p*-nitrobenzoic acid, respectively, and a tribasic acid, $C_{14}H_9O_8N$, containing one methylamino group.[287]

The first important clue to the structure of these alkaloids came from the basic hydrolysis of each of the four alkaloids to ergine, $C_{16}H_{17}ON_3$.[288] This product was found to contain a methylamino group, but no methoxyl group, and to give indole color reactions. It was later found[284] to be the amide of lysergic acid, $C_{16}H_{16}O_2N_2$, obtained from ergine by a similar hydrolysis.[290] It has since been observed that each of the ergot alkaloids furnishes either ergine, lysergic acid, or isolysergic acid.

283 Kussner, *Arch. Pharm.*, **272,** 503 (1934).
284 Stoll, *Schweiz. med. Wochschr.*, **65,** 1077 (1935).
285 Soltys, *Ber.*, **65,** 553 (1932).
286 Barger and Ewins, *J. Chem. Soc.*, **97,** 284 (1910).
287 Jacobs, *J. Biol. Chem.*, **97,** 739 (1932).
288 Smith and Timmis, *J. Chem. Soc.*, **1932,** 763.
289 Smith and Timmis, *Nature*, **133,** 579 (1932); *J. Chem. Soc.*, **1934,** 674.
290 Jacobs and Craig, *J. Biol. Chem.*, **104,** 547 (1934).

The lysergic acid fragment accounts for about half of the molecule of the ergot bases. Knowledge of the constitution of the remaining halves also came from hydrolytic studies. Alkaline hydrolysis[291] of ergotinine (ergocristinine)[274] was found to give, in addition to lysergic acid, ammonia and dimethylpyruvic acid, the amino acids phenylalanine and proline. Acid hydrolysis, which destroys the lysergic acid, resulted in the isolation of L-phenylalanine, a dipeptide, $C_{14}H_{18}O_3N_2$, composed of phenylalanine and proline, and D-proline. The isolation of an amino acid of the D-series is notable.

Similarly, the components of the other ergot alkaloids have been isolated by hydrolytic degradation. The fragments obtained from each pair are summarized in Table VII. The products listed are obtained in addition to lysergic acid, or isolysergic acid, and ammonia from each of the pairs except the ergometrine pair, which alkaloids are composed of only the lysergic acids and D-2-aminopropanol.

TABLE VII

Isomeric pair	Products			Ref.
Ergotoxine and ψ-ergotinine	Dimethylpyruvic acid	D-Proline	L-Valine	266
Ergotamine and ergotaminine	Pyruvic acid	D-Proline	L-Phenylalanine	292
Ergosine and ergosinine	Pyruvic acid	D-Proline	L-Leucine	279
Ergocristine and ergocristinine	Dimethylpyruvic acid	D-Proline	L-Phenylalanine	266, 291
Ergocryptine and ergocryptinine	Dimethylpyruvic acid	D-Proline	L-Leucine	266
Ergomolline and ergomollinine	Dimethylpyruvic acid	D-Proline	L-Leucine	281
Ergometrine and ergometrinine	D-2-Aminopropanol			293, 294

An indication of the structure of the peptide portion of the ergotoxine molecule is afforded by the isolation of a tripeptide from the controlled

[291] Jacobs and Craig, *ibid.*, **108,** 595 (1935); **110,** 521 (1935); *J. Am. Chem. Soc.* **57,** 383 (1935).

[292] Jacobs and Craig, *Science,* **81,** 256 (1935); *J. Org. Chem.,* **1,** 245 (1936).

[293] Smith and Timmis, *J. Chem. Soc.,* **1936,** 1166, 1440.

[294] Jacobs and Craig, *Science,* **82,** 16 (1935); *J. Biol. Chem.,* **110,** 521 (1935).

hydrolysis of the molecule.[295,296] The constitution of the peptide has been established by synthesis as dimethylpyruvyl-L-valyl-D-proline (I). Evidence

$$(CH_3)_2CHCOCONHCH(CH(CH_3)_2)CO\text{—}N\text{(pyrrolidine ring)}\text{—}COOH \quad \text{(I)}$$

that the pyruvic acid portion of the peptide does not occur as such in the molecule, but as α-hydroxyvaline and α-hydroxyalanine, is given by the fact that no nitroprusside test is obtained for pyruvic acid prior to hydrolysis.[297]

The peptide grouping may also be cleaved off by the use of anhydrous hydrazine.[296,298] This reagent reduces the α-carbonyl group of the pyruvic acids, but the tripeptide is obtained intact. The results of these degradations are summarized in Table VIII.

TABLE VIII

Alkaloid	Peptide
Ergotoxine	*N*-(*N*-Isovaleryl-L-valyl)-D-proline
Ergotamine	*N*-(*N*-Propionyl-L-phenylalanyl)-D-proline
Ergocristine	*N*-(*N*-Isovaleryl-L-phenylalanyl)-D-proline
Ergocryptine	*N*-(*N*-Isovaleryl-L-leucyl)-D-proline

Characterization of the amino acid moiety leaves the question of the constitution of lysergic acid and isolysergic acid. Lysergic acid contains one methylamino group and exhibits the same color reactions as the parent alkaloids. The high dextrorotation of ergine, $[\alpha]_D$ +514°, prompted the application of the same methods used to isomerize the dextro alkaloids to the levo forms, with the result of producing isoergine, $[\alpha]_{5461}$ +25°. Lysergic acid, $[\alpha]_D$ +40°, may similarly be converted into isolysergic acid, $[\alpha]_D$ +281°.[299] On the basis of rotations it appears that lysergic acid and isoergine correspond to the ergotoxine series, and that isolysergic acid and ergine are derived from the ergotinine series.

Lysergic acid yields a dihydro derivative,[300] also obtained by hydrolysis of dihydroergotamine.[301] Hydrogenation of isolysergic acid yields two

[295] Stoll, *Experientia*, **1**, 250 (1945).
[296] Stoll and Hofmann, *Helv. Chim. Acta*, **33**, 1705 (1950).
[297] Jacobs and Craig, *J. Biol. Chem.*, **122**, 419 (1938).
[298] Stoll, Petrzilka, and Backer, *Helv. Chim. Acta*, **33**, 57 (1950).
[299] Smith and Timmis, *J. Chem. Soc.*, **1936**, 1440.
[300] Jacobs and Craig, *J. Biol. Chem.*, **106**, 393 (1934); **111**, 455 (1935).
[301] Jacobs and Craig, *ibid.*, **115**, 227 (1936).

dihydro derivatives, dihydrolysergic acids I and II.[302,303] Both of the lysergic acids, as well as the dihydrolysergic acids, may be racemized. Resolution of both the racemic lysergic acid[304] and the dihydrolysergic acid[305] has been accomplished.

The evidence for the indole ring led to the postulation of a β-carboline structure for lysergic acid.[306] However, synthetic β-carboline carboxylic acids did not resemble lysergic acid in their color tests,[307] nor did a carboline system explain the isolation of quinoline[308] from the soda lime distillation of the tribasic acid obtained by oxidation of ergotinine.[287] The triacid is probably then an *N*-methylquinolinium betaine tricarboxylic acid.

Potassium hydroxide fusion of dihydrolysergic acid yielded 1-methyl-5-aminonaphthalene (II) and 3,4-dimethylindole (III).[308,309] Consideration

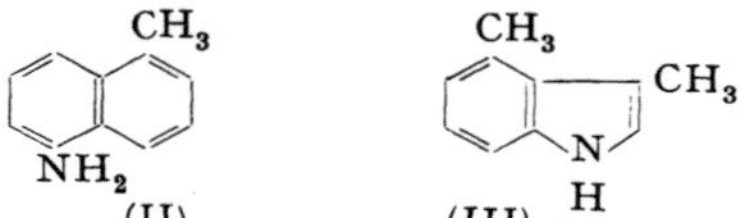

CH_3 NH_2 (II) CH_3 CH_3 N H (III)

of these fragments, together with the probable free 2-position of the indole nucleus, as indicated by the color tests, led to the conception of a tetracyclic structure such as IV,[308] for which no natural counterpart has been observed. This formulation incorporates the methylaminonaphthalene, dimethylindole, and quinoline structures.

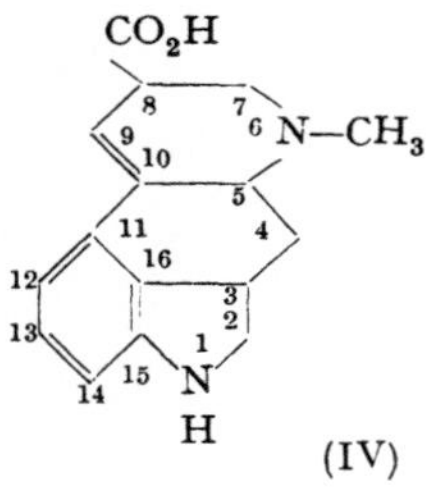

CO_2H 8 7 6 N—CH_3 9 10 5 11 4 16 12 3 2 13 1 14 15 N H (IV)

The position of the carboxyl group in lysergic acid must account for the isolation of the quinolinium betaine tricarboxylic acid, for which V is a

[302] Stoll, Hofmann, and Troxler, *Helv. Chim. Acta*, **32**, 506 (1949).
[303] Stoll, Hofmann, and Schlientz, *ibid.*, **32**, 1947 (1949).
[304] Stoll and Burckhardt, *Z. physiol. Chem.*, **250**, 7 (1937).
[305] Stoll, Rutschmann, and Schlientz, *Helv. Chim. Acta*, **33**, 67, 375 (1950).
[306] Jacobs and Craig, *J. Biol. Chem.*, **111**, 455 (1935).
[307] Jacobs and Craig, *Science*, **82**, 421 (1935); *J. Biol. Chem.*, **113**, 759 (1936).
[308] Jacobs and Craig, *J. Biol. Chem.*, **113**, 767 (1936).
[309] Jacobs and Craig, *ibid.*, **128**, 715 (1939).

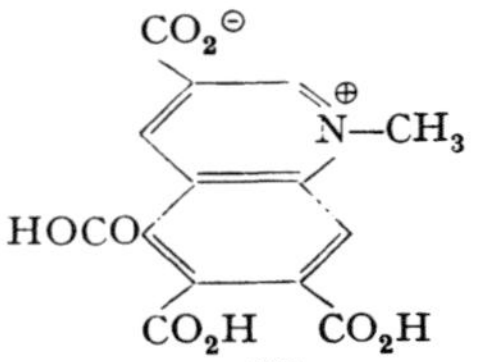

$CO_2^{\ominus}$
$\overset{\oplus}{N}—CH_3$
HOCO
CO_2H CO_2H
(V)

possible structure.[308] Thus, the possible locations of this group are C_4, C_7, C_8, and C_9. A carboxyl group at C_4 would be expected to lose carbon dioxide, as an indoleacetic acid, more readily than is observed with the lysergic acids. Of the remaining possibilities, C_7, C_8, and C_9, the second is preferred on the basis that the lysergic acids have basic dissociation constants more in agreement with those of β-amino acids than the α- or γ-isomers.[310] The characteristic loss of ammonia on heating by a β-amino acid has its counterpart in *dl*-dihydrolysergic acid, which on pyrolysis yields an unsaturated lactam (VI), the formation of which is explained[311] by elimination of the β-amino group with subsequent lactamization. This product may be transformed into *dl*-6,8-dimethylergoline (VII),[312] ergoline being the theoretical parent structure of the lysergic acids. The synthesis of ergoline,[313] of *dl*-6,8-dimethylergoline (VII),[312] and of *dl*-dihydrolysergic acid[314] confirms the proposals of the basic carbon skeleton.

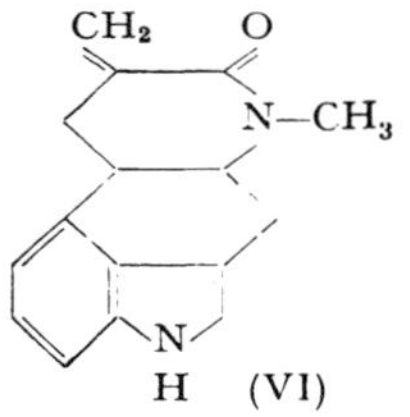

CH_2 O
$N—CH_3$
N
H (VI)

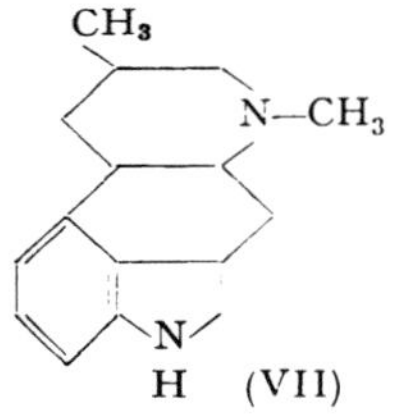

CH_3
$N—CH_3$
N
H (VII)

It was originally assumed that the isomerism of the lysergic acids was due to migration of the double bond, since isomerization of the dihydro derivatives could not be accomplished. To account for the conjugation of this linkage with the indole ring in both isomers, as evidenced in the absorption spectra,[308] and the fact that the alkaloids of the lysergic acid series are weaker bases than those of the isolysergic acid series,[301] the olefinic bond was placed at C_5-C_{10} in lysergic acid and C_{10}-C_9 in isolysergic acid.[310]

[310] Craig, Shedlovsky, Gould, and Jacobs, *ibid.*, **125**, 289 (1938).
[311] Jacobs and Craig, *J. Am. Chem. Soc.*, **60**, 1701 (1928).
[312] Jacobs and Craig, *J. Biol. Chem.*, **130**, 399 (1939); **145**, 487 (1942).
[313] Jacobs and Gould, *ibid.*, **120**, 141 (1937).
[314] Uhle and Jacobs, *J. Org. Chem.*, **10**, 76 (1945).

This formulation and explanation of the isomerism did not account, however, for the formation of two isomeric dihydroisolysergic acids. It has now been found possible to isomerize dihydroisolysergic acid into dihydrolysergic acid,[302] indicating that migration of the double bond is not the cause of isomerism in the acids. Substantiation of this is found in the fact that both of the lysergic acids are transformed into the same unsaturated lactam (VIII) on heating with acetic anhydride. Decarboxylation of both acids also produces the same product, a diolefin (IX) in which the absorption spectrum shows both double bonds to be conjugated with the indole ring.[302]

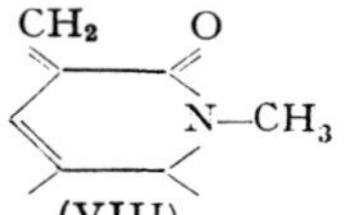

CH_2 O N—CH_3 (VIII) CH_2 $NHCH_3$ (IX)

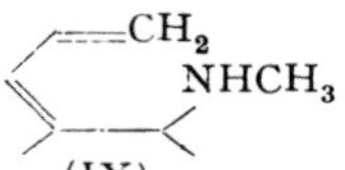

These data indicate that the only difference between lysergic acid and isolysergic acid is the configuration at C_8,[302] and that isomerization (epimerization) is caused by enolization at that center. The position of the double bond is assigned as C_9–C_{10}, and the differences in basicity of the isomeric acids is rationalized as differences in the spatial proximity of the carboxyl group and the amino nitrogen atom.

Several syntheses of the lysergic acid nucleus have been effected. Preparation of *dl*-6,8-dimethylergoline (VII) was carried out from 3-methyl-5,6-benzoquinoline-7-carboxylic acid (XI), prepared by a modified Skraup reaction on the amino naphthoic acid (X). Nitration and reduction furnished the carbostyryl (XII), which was methylated and reduced in two steps to a product identical with VII.[312]

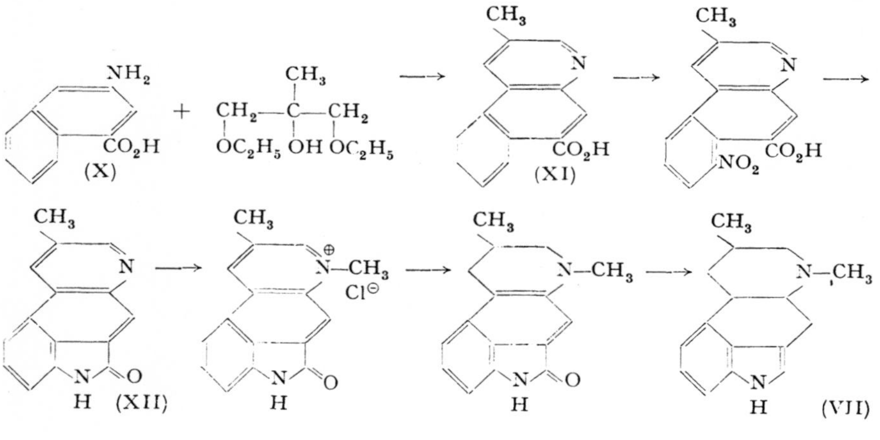

NH_2 CO_2H (X) + CH_2—C(CH_3)(OH)—CH_2 with OC_2H_5, OC_2H_5 → CH_3 N CO_2H (XI) → CH_3 N NO_2 CO_2H →

CH_3 N N H O (XII) → CH_3 $\overset{\oplus}{N}$—CH_3 $Cl^{\ominus}$ N H O → CH_3 N—CH_3 N H O → CH_3 N—CH_3 N H (VII)

The total synthesis of *dl*-dihydrolysergic acid[314] provides more complete confirmation of the structure of the lysergic acids. The sodium salt of cyanomalonaldehyde, prepared from cyanoacetal, was condensed with 3-aminonaphthostyryl (XIII). Ring closure of the cyanoformylanil (XIV) with zinc chloride, followed by hydrolysis, yielded the indoloquinoline carboxylic acid (XV). Methylation and reduction, as in the preparation of the dimethylergoline, gave XVI, identical with *dl*-dihydrolysergic acid prepared from the natural lysergic acid. Resolution of the racemic material with L-ephedrine yielded the natural isomer.[305]

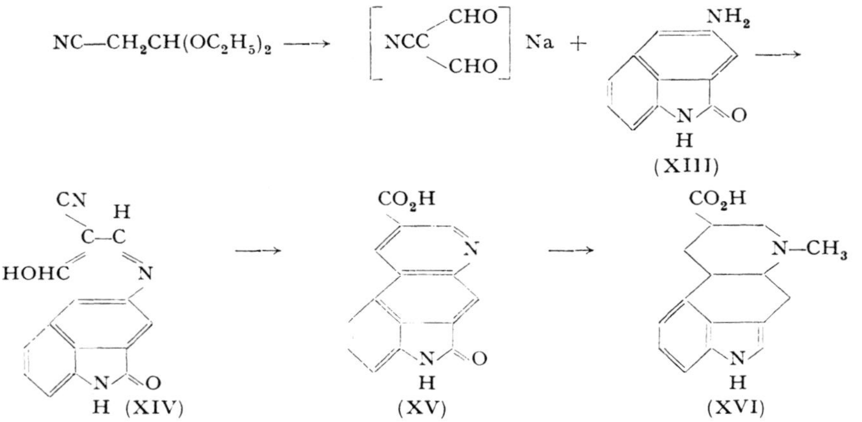

$NC—CH_2CH(OC_2H_5)_2$ ⟶ $[NCC(CHO)_2]$ Na + (XIII) ⟶

(XIV) ⟶ (XV) ⟶ (XVI)

A partial synthesis of ergometrine[315] supports the concept of the linkage between the lysergic acids and the amino acid portions of the molecules, as well as the explanation of isomerism in the alkaloids themselves. Racemic isolysergic acid hydrazide (XVII), formed by the action of hydrazine on the alkaloids,[304] was converted into the azide. Reaction with D-2-aminopropanol yielded the diasterioisomeric D-isolysergic-D-isopropanol and

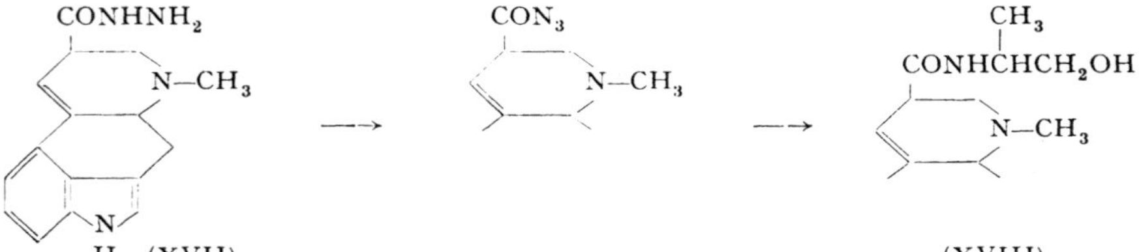

$CONHNH_2$ (XVII) ⟶ CON_3 ⟶ $CONHCH(CH_3)CH_2OH$ (XVIII)

[315] Stoll and Hofmann, *Z. physiol. Chem.*, **251,** 155 (1938).

[316] Stoll and Hofmann, *Helv. Chim. Acta*, **26,** 944 (1943). Stoll, Hofmann, Jucker, and Petrzilka, *ibid.*, **33,** 108 (1950).

L-isolysergic-D-isopropanol amides (XVIII). The D,D-compound proved identical with natural ergometrinine, and isomerization with acetic acid give ergometrine, establishing directly the relationship between the appropriate lysergic acid and its alkaloid counterpart. This general synthetic technique has been applied to the preparation of homologs and analogous peptide derivatives of the lysergic acids for pharmacological study.[316]

On the basis of the evidence for α-hydroxyvaline and α-hydroxyalanine,[297] structures have been proposed for ergocristine and ergotamine. Modified in view of the isolation of the tripeptide,[295] which places proline at the end of the peptide chain rather than in the center,[297] ergotoxine would have the structure represented by XIX.

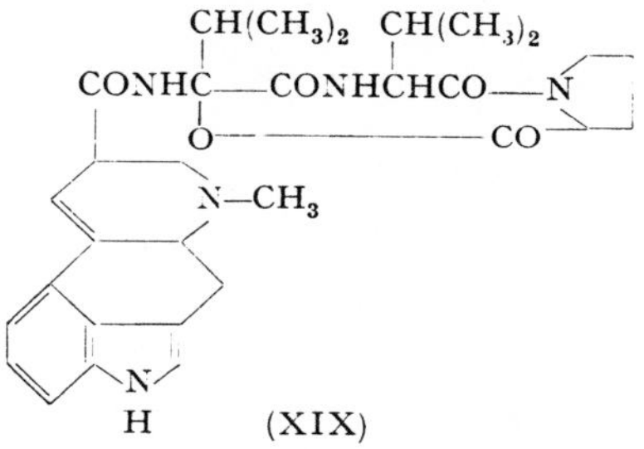

$CH(CH_3)_2$ $CH(CH_3)_2$
CONHC—CONHCHCO—N
O — CO
$N—CH_3$
N
H
(XIX)

Strychnos Alkaloids

The alkaloids of the genus *Strychnos*, principally *S. nux-vomica* and *S. Ignatii*, constitute one of the most complex groups of substances, from a structural viewpoint, of all the natural products. The problem of the structure of strychnine has occupied chemists for over sixty years, and even yet many points in the behavior of the base need further study. The investigations have resulted in the postulation of many structures for the alkaloids, only to be revised in consideration of new findings. The following account makes no attempt to include all of the more than 200 papers published during the course of the long and extensive researches, but merely to recount the salient features of the experimental results which have led to the formulation of the currently accepted structure.

Occurring with strychnine are several other bases, most notably brucine and vomicine, as well as strychnicine, struxine, α- and β-colubrine, and the recently discovered strychnospermine. The chemistry of strychnine and brucine will be presented concurrently, and followed by the known relationships of the remaining alkaloids.

Strychnine and Brucine

Strychnine, $C_{21}H_{22}O_2N_2$,[317] m.p. 268–290°, $[\alpha]_D$ —110°, was first obtained in 1817.[318] It is a monoacidic base and contains no *C*-methyl,[319] *O*-methyl, or *N*-methyl groups. The basic nitrogen atom, N(b), is tertiary since it yields a methiodide and shows no reaction with nitrous acid. The neutrality of the second nitrogen atom, N(a), is attributed to its presence in a cyclic amide structure, since strychnine may be hydrolyzed to strychnic acid,[320] $C_{21}H_{24}O_3N_2$, a dibasic amino acid, yielding a nitrosoamine,[321] and easily reconvertible to strychnine.

A single olefinic linkage is indicated by the catalytic hydrogenation of strychnine to dihydrostrychnine,[322] in which the reactions of the remaining functions are unaffected. The remaining oxygen atom is located in a cyclic ether bridge, on the basis that strychnine gives no carbonyl or hydroxyl derivatives, and treatment with hydrolytic reagents causes ether cleavage with no loss of carbon. The ether oxygen may be eliminated by reaction with phosphorus and hydriodic acid to give deoxystrychnoline.[323]

Brucine, $C_{23}H_{26}O_4N_2$,[317] m.p. 178°, $[\alpha]_D$ —80°, was isolated two years following strychnine.[324] The reactions of this alkalold closely parallel those of strychnine with the exception of those modifications caused by the presence of two methoxyl groups in the molecule. That this is the only difference between the two alkaloids was indicated by the isolation of the same acid, $C_{16}H_{20}O_4N_2$, from the chromic acid oxidation of brucine[325] and of strychnine,[326] and demonstrated conclusively by the reduction of both strychnidine and brucidine to octahydrostrychnidine.[327] The location of the methoxyl groups was shown by the production of 4,5-dimethoxy-*N*-oxalylanthranilic acid (I) from oxidation of brucine, compared to the similar prep-

(I) CH_3O / CH_3O [benzene ring] CO_2H / $NHCOCO_2H$

[317] Regnault, *Ann.*, **26**, 17 (1838).
[318] Pelletier and Caventou, *Berz. Jahresber.*, **1**, 95 (1817).
[319] Reynolds and Robinson, *J. Chem. Soc.*, **1939**, 603.
[320] Loebisch and Schoop, *Monatsh.*, **7**, 83 (1886).
[321] Perkin and Robinson, *J. Chem. Soc.*, **98**, 305 (1910).
[322] Oxford, Perkin and Robinson, *ibid.*, **1927**, 2389.
[323] Tafel, *Ann.*, **268**, 229 (1892).
[324] Pelletier and Caventou, *Berz. Jahresber.*, **3**, 171 (1819).
[325] Hanssen, *Ber.*, **17**, 2266, 2849 (1884).
[326] Hanssen, *ibid.*, **18**, 1917 (1885).
[327] Leuchs and Oberberg, *ibid.*, **66**, 951 (1933).

aration of the unsubstituted oxalylanthranilic acid from strychnine.[328] Thus an investigation of any portion of the molecule other than the methoxyl substituted ring of one alkaloid will serve an equal purpose in establishing the structure of the other.

The drastic degradations usually applied to alkaloids have been used in the case of strychnine with moderate success, but the fragments obtained have not been so helpful in the elucidation of the structure as in other cases. As will be seen, the formulation of the more probable structure was the result of the consideration of more subtle pieces of evidence.

Distillation of the alkaloid with strong bases has resulted in the isolation of indole,[329] 3-methyl-[330] and 3-ethylindole,[331] tryptamine,[331] carbazole,[329] and 3-ethyl-4-methylpyridine (II).[332] Oxidation of strychnine with nitric acid has furnished, in addition to the anthranilic acid derivatives, picric acid,[333]

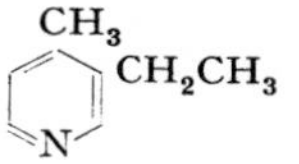

CH_3 CH_2CH_3 N (II) O_2N CO_2H CO_2H N NO_2 H (III)

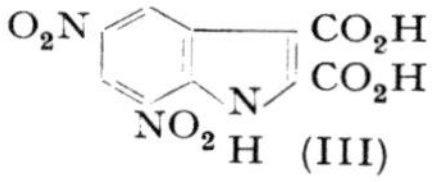

3,5-dinitrobenzoic acid,[334] and "dinitrostrychnolcarboxylic acid,"[335] $C_{10}H_5O_8N_3$, now known to be 5,7-dinitroindole-2,3-dicarboxylic acid (III).[336] While these fragments indicate an indole nucleus, the presence of an aromatic ring had previously been demonstrated by nitration,[337] halogenation,[338] and sulfonation[339] of strychnine.

That the aromatic ring in strychnine and brucine is attached to a five-membered heterocyclic ring is shown by the above degradation products. The nature of the nitrogen atom contained therein is indicated by a study of the reduction products of the alkaloids. Electrolytic reduction of strychnine furnishes, among other products, strychnidine, $C_{21}H_{24}ON_2$,[340] a diacidic

[328] Späth and Bretschneider, *ibid.*, **63**, 2997 (1930).
[329] Clemo, Perkin, and Robinson, *J. Chem. Soc.*, **1927**, 1589.
[330] Loebisch and Molfatti, *Monatsh.*, **9**, 622 (1888).
[331] Clemo, *J. Chem. Soc.*, **1936**, 1695.
[332] Oeschner de Cominck, *Bull. soc. chim.*, **42**, 100 (1884).
[333] Shenstone, *J. Chem. Soc.*, **47**, 139 (1885).
[334] Menon, Perkin, and Robinson, *ibid.*, **1930**, 842.
[335] Tafel, *Ann.*, **301**, 285 (1898).
[336] Hill and Robinson, *J. Chem. Soc.*, **1933**, 486.
[337] Claus and Glassner, *Ber.* **14**, 773 (1881).
[338] Laurent, *Ann.*, **69**, 14 (1849).
[339] Leuchs and Schneider, *Ber.*, **41**, 4393 (1908).
[340] Tafel, *Ann.*, **268**, 229 (1892).

base containing a double bond.[329] Thus it appears that the nonbasic amide function has been reduced. This is borne out by the behavior of strychnidine as a dialkylaniline, as opposed to the acylaniline characteristics exhibited by strychnine in the reactivity of the substances toward diazobenzenesulfonic acid,[322] and in the ultraviolet spectra of the two bases.[341] Therefore, it is N(a) which is attached to the aromatic ring, and which is incorporated in the amide function. Further evidence for the five-membered ring is found in the behavior of 2,3-diketonucidine, $C_{17}H_{20}O_3N_2$, produced by oxidizing strychnidine with chromic acid.[342] In this product the benzene ring no longer exists and hence the four carbons which were lost must have come from this portion of the molecule. Several other oxidation products of this type have been obtained.[343]

The diketonucidine has the properties of an α-ketoamide in that it may be reduced by zinc and hydrochloric acid to a ketohydroxynucidine, or oxidized by alkaline hydrogen peroxide with the loss of a carbon atom to an amino acid, carboxyaponucidine, $C_{16}H_{22}O_3N_2$.[344] These derivatives may be formulated as follows:

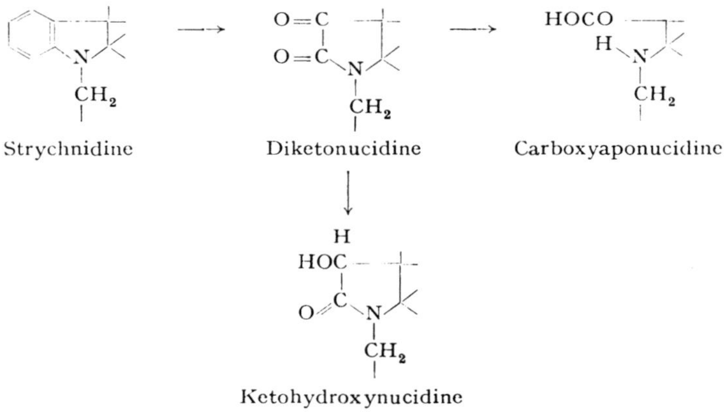

Strychnidine → Diketonucidine → Carboxyaponucidine

Diketonucidine ↓ Ketohydroxynucidine

That the carboxyaponucidine could not be recyclized to a cyclic amide shows the strong likelihood that the carboxyl and amino groups were originally present in a five-membered ring.

[341] Prelog and Szpilfogel, *Helv. Chim. Acta*, **28,** 1669 (1945).

[342] Leuchs and Krohnke, *Ber.*, **63,** 1045 (1930).

[343] Leuchs, *ibid.*, **51,** 1375 (1918); **55,** 564, 724, 2403 (1922). Wieland and Münster, *Ann.*, **469,** 216 (1929); **480,** 39 (1930).

[344] Leuchs and Wegener, *Ber.*, **63,** 2215 (1930).

Strychnine reacts with benzaldehyde to yield a benzylidene derivative[345] but no such reaction occurs with strychnidine. Thus a methylene group, activated by the amide carbonyl group must be present in the molecule. That such a system, and not a vinylog of it, is present is evident, since were the double bond conjugated with the amide carbonyl group, it would probably be reduced in the production of strychnidine, and it would be indicated in the absorption spectrum of strychnine as compared to *N*-acylhexahydrocarbazole.[341]

Oxidation of strychnine with alkaline permanganate produces strychninonic acid, $C_{21}H_{20}O_6N_2$,[346] a monobasic acid, in which both N(a) and N(b) are no longer basic, and which yields carbonyl derivatives. Accompanying this product is a small amount of strychninolic acid, $C_{21}H_{22}O_6N_2$, also produced by sodium amalgam reduction of strychninonic acid[342] and which furnishes hydroxyl derivatives. Strychninolic acid, on treatment with an excess of NaOH, is cleaved into strychninolone, $C_{19}H_{18}O_3N_2$, and glycollic acid, $HOCH_2CO_2H$.[347] Strychninolone is a neutral compound possessing a single hydroxyl group and also a double bond[348] not present in the strychninolic acid molecule. Therefore, the production of glycollic acid probably occurs by a process of elimination rather than a hydrolytic cleavage, indicating a system[399] such as:

$$-CO-\overset{H}{\underset{|}{C}}-\overset{|}{\underset{|}{C}}-O-CH_2-CO_2H \longrightarrow -CO-\underset{|}{C}=\overset{|}{\underset{|}{C}} + HOCH_2CO_2H$$

and demonstrating the presence of the ether linkage in strychnine.[350] The ultraviolet absorption spectrum of strychninolone demonstrates the conjugation of the α, β-unsaturated carbonyl system.[341]

That the carbonyl group involved in this process is the one attached to N(a) rather than that newly formed at N(b) may be shown. The simultaneous formation of strychninonic and strychninolic acids points to the proximity of the double bond attacked and N(b), since the reaction must proceed by some such route[351] as:

[345] Perkin and Robinson, *J. Chem. Soc.*, **1929**, 964.
[346] Leuchs, *Ber.*, **41,** 1711 (1908).
[347] Leuchs and Schneider, *ibid.*, **42,** 2494 (1909).
[348] Leuchs, Diels and Darnow, *ibid.*, **68,** 106 (1935).
[349] Fawcett, Perkin, and Robinson, *J. Chem. Soc.*, **1928**, 3082.
[350] Leuchs and Kanao, *Ber.*, **57,** 1799 (1924).
[351] Robinson, *Proc. Roy. Soc.*, **A 130,** 431 (1931). Compare Woodward and Brehm *J. Am. Chem. Soc.*, **70,** 2107 (1948), footnote 28.

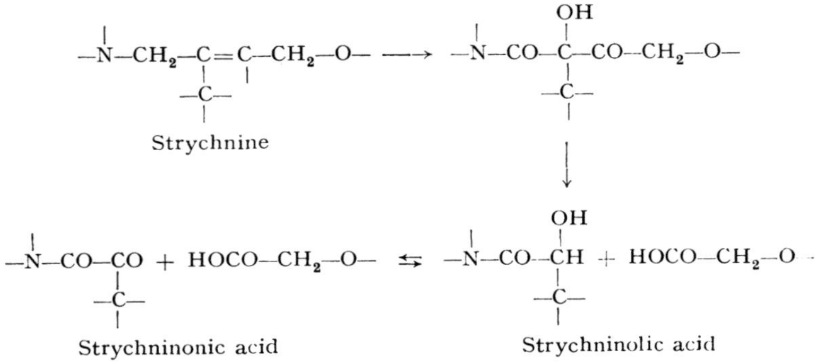

$$-\overset{|}{N}-CH_2-\underset{\underset{|}{-C-}}{C}=\overset{}{\underset{|}{C}}-CH_2-O- \longrightarrow -\overset{|}{N}-CO-\overset{OH}{\underset{\underset{|}{-C-}}{C}}-CO-CH_2-O-$$

Strychnine

$$\downarrow$$

$$-\overset{|}{N}-CO-\underset{\underset{|}{-C-}}{CO} + HOCO-CH_2-O- \rightleftarrows -\overset{|}{N}-CO-\overset{OH}{\underset{\underset{|}{-C-}}{CH}} + HOCO-CH_2-O-$$

Strychninonic acid Strychninolic acid

Were the elimination to occur α, β to the N(b) amide carbonyl, the resulting double bond would be lost, at least in part in the ketonizing of the enol, and the oxidation of dihydrostrychninolone would yield strychninolone rather than the isomeric substance, dihydrostrychninone. Therefore, the glycollic acid unit was eliminated as a result of the influence of the N(a) amide, and must be constituted thusly in strychninolic acid:[352]

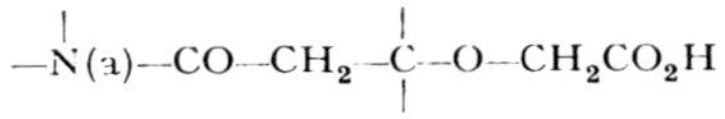

$$-\overset{|}{N}(a)-CO-CH_2-\underset{|}{\overset{|}{C}}-O-CH_2CO_2H$$

The location of the double bond from the appearance of the carbonyl group, and the characterization of strychninonic acid as an α-ketamide enables the establishment of the relationship[353] of N(a) to N(b) as:

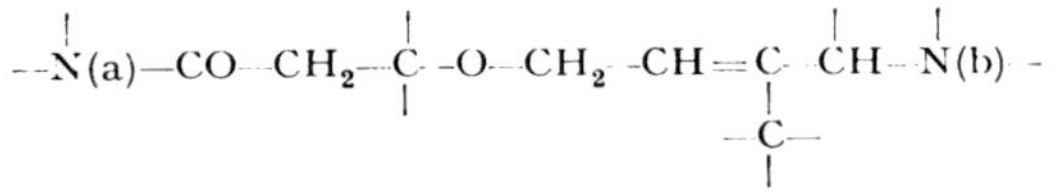

$$-\overset{|}{N}(a)-CO-CH_2-\underset{|}{\overset{|}{C}}-O-CH_2-CH=\underset{\underset{|}{-C-}}{C}-\overset{|}{C}H-\overset{|}{N}(b)-$$

and in view of the isolation of tryptamine, the part structure of strychnine may be written as IV, subject to the consideration of other experimental data.

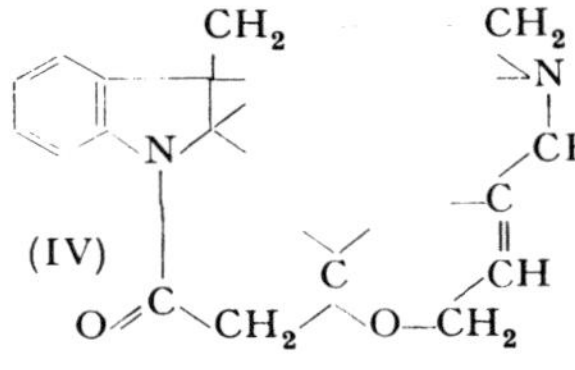

(IV)

More information regarding the substitution of the lower portion of the molecule may be gained by a study of the further reactions of strychninolone. This substance exists in three isomeric forms, differing in the posi-

[352] Briggs, Openshaw, and Robinson, *J. Chem. Soc.*, **1946**, 903.

[353] Leuchs, *Ber.*, **65**, 1230 (1932).

tion of the newly introduced double bond.[348,354] Oxidation of acetylstrychninolone-a (V), $C_{21}H_{20}O_4N_2$, with potassium permanganate yields acetylstrychninolonic acid, $C_{21}H_{20}O_8N_2$.[355] Hydrolysis of this product removes the acetyl group and furnishes an amino acid, $C_{17}H_{18}O_4N_2$, and oxalic acid, HOCOCOOH.[356] This cleavage demonstrates that in this isomeride the double bond is conjugated with the carbonyl group, and, more important, that the β-carbon atom in the N(a) lactam ring of the strychninolone bears a hydrogen atom, since the product was an acid and not a ketone.

(V) 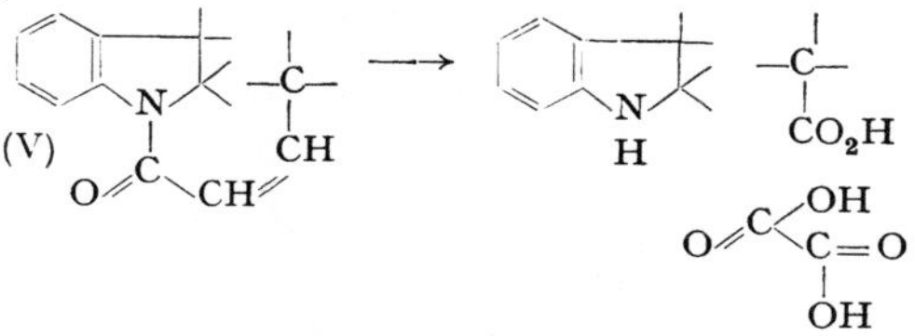 N —C— CH O=C CH= ⟶ N H —C— CO_2H O=C(OH)—C=O OH

Similarly, the analogous acetylbrucinolone-b, $C_{23}H_{24}O_6N_2$, on oxidation yields acetylbrucinolonic acid, $C_{23}H_{24}O_9N_2$ (VI),[357] which on hydrolysis gives malonic acid and the ketonic base, curbine, $C_{18}H_{20}O_5N_2$.[358] This

CH_3O CH_3O N —C— C—C— CH (VI) 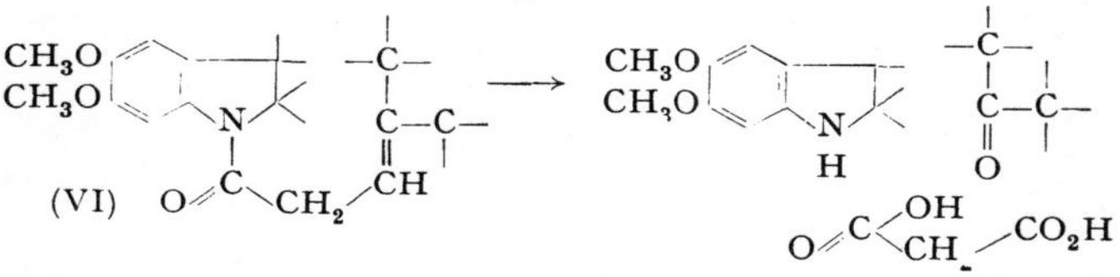 O=C CH_2 ⟶ CH_3O CH_3O N H —C— C—C— O O=C(OH) CH_2 CO_2H

demonstrates a β,γ-double bond and shows that the γ-carbon atom carries no hydrogen atom. Further, the size of the lactam ring must be five or six-membered in order to explain the facile opening and closing of the amide in strychnine and strychninic acid, respectively, and in view of the substituting just demonstrated, only the latter possibility may obtain.[359] Thus, the part structure VII may be written:

CH_2 CH_2 >C< —N N CH C CH (VII) CH C CH O CH_2 O—CH_2

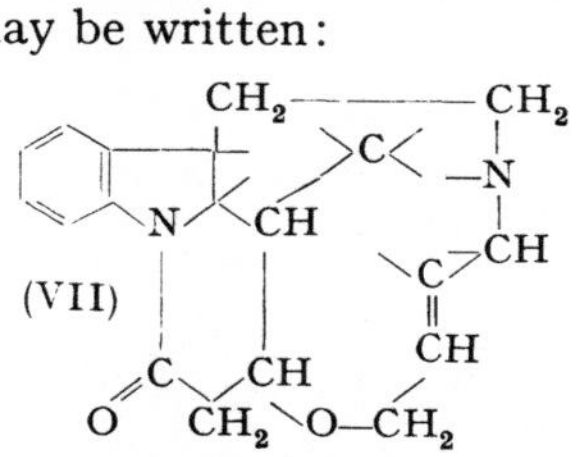

[354] Leuchs and Bendixsohn, *ibid.*, **52**, 1443 (1919).

[355] Leuchs and Schwäbel, *ibid.*, **47**, 1552 (1914).

[356] Leuchs and Schwäbel, *ibid.*, **48**, 1009 (1915).

[357] Leuchs and Brewster, *ibid.*, **45**, 201 (1912).

[358] Leuchs and Pierce, *ibid.*, **45**, 2653 (1912).

[359] See also Prelog, Kocor, and Taylor, *Helv. Chim. Acta*, **32**, 1052 (1949).

Treatment of strychnine with ammonia in methyl alcohol or with water at 160-180° causes isomerism into isostrychnine,[360] which, unlike strychnine contains a hydroxyl group.[322] Isostrychnine may be reconverted into strychnine by treatment with alcoholic potassium hydroxide.[361] An additional ethylenic linkage is present in the isomer, since reduction may be forced to give tetrahydroisostrychnine,[362] although the dihydro derivative is formed under normal conditions. This latter substance may also be formed by treating dihydrostrychnine under the same conditions, but strychnidine is not susceptible to this isomerization.[322] Therefore the amide carbonyl is essential to this isomerism, and the process bears a similarity to that encountered in the formation of strychninilone, opening the ether ring.

Isostrychnine, like the parent base, reacts with benzaldehyde to yield benzylideneisostrychnine,[363] which, unlike the yellow, easily reducible[327,364] strychnine derivative, is colorless and absorbs no hydrogen.[365] These anomalies are resolved if it is considered that the 6-membered lactam ring in isostrychnine (VIII) is capable of aromatization to an α-pyridone structure (IX).[366] For such a tautomerism to occur, a hydrogen atom must be present

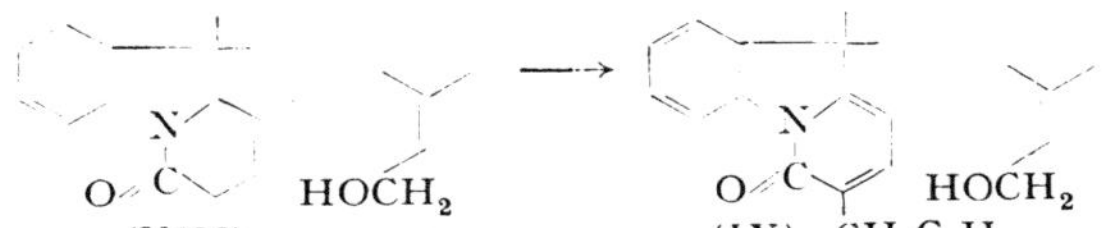

O=C HOCH$_2$ (VIII) → O=C HOCH$_2$ (IX) $CH_2C_6H_5$

on the α-position of the five-membered heterocyclic ring. Further, it may be said that the β-position of this ring is doubly substituted, since a dihydroindole nucleus having hydrogens on both the α- and β-positions is easily dehydrogenated,[367] and this is not the case with strychnine.[368] This conclusion is further substantiated by the isolation of both carbazole and tryptamine from degradative studies of strychnine.

The structure of strychnine which has been received most favorably

[360] Bacouescu and Pictet, *Ber.*, **38**, 2787 (1905).
[361] Prelog, Battegay and Taylor, *Helv. Chim. Acta*, **31**, 2244 (1948).
[362] Leuchs, Diels, and Dornow, *Ber.*, **68**, 2234 (1935).
[363] Leuchs and Dornow, *ibid.*, **69**, 1838 (1936).
[364] Leuchs and Steinhorn, *ibid.*, **71**, 1577 (1938).
[365] Leuchs and Beyer, *ibid.*, **68**, 1204 (1935).
[366] Briggs, Openshaw, and Robinson, *J. Chem. Soc.*, **1946**, 903.
[367] Menon and Robinson, *J. Chem. Soc.*, **1932**, 781.
[368] Reynolds and Robinson, *ibid.*, **1935**, 935.

until recently is X.[367, 369] A similar structure, XI,[370] has been forwarded, as well as XII,[371] as alternatives to X. It will be seen that XI and XII are

(X) (XI) (XII)

untenable in view of the isolation of tryptamine and the demonstration that the dihydroindole nucleus bears a hydrogen atom on the α-position and that the β-position is quaternary. Structure X explains the reactions of the alkaloid enumerated previously, with the exception of the formation of 4-methyl-3-ethylpyridine. This was recognized, and XIII was suggested,[372] but this structure does not account for the isolation of carbazole. Recently, a reinterpretation of the evidence has led to a new formulation which has met with general acceptance.

(XIII)

Atmospheric oxidation of strychnine in the presence of copper salts produces pseudostrychnine, $C_{21}H_{22}O_3N_2$,[373] which has also been obtained from strychnine mother liquors.[374] Pseudostrychnine is characterized by the formation of an acetyl derivative, which is neutral,[370] the easy formation of ethers,[374] and the production of an *N*-nitroso derivative. Further, pseudostrychnine may be reduced with zinc and acid to strychnine.[371] These properties are consistent with those of a carbinol amine (XIV), involving N(b) of the strychnine molecule. The single discrepancy between the behavior of pseudostrychnine and the carbinol amines is its failure to be oxidized

$$\begin{array}{ccccccc} -N(b)-CH & \longrightarrow & -N-C-OH \ (XIV) & \rightleftharpoons & -NH \ \ -C{=}O & \longrightarrow & -N-NO \ \ -C{=}O \\ & & \downarrow & & \downarrow & & \\ -\overset{\otimes}{N}-CH_3 \ \ -C-OCH_3 & \longleftarrow & -N-C-OCH_3 & & -N-COCH_3 \ \ -C{=}O & & \end{array}$$

[369] Holms and Robinson, *ibid.*, **1939**, 603.

[370] Leuchs, *Ber.*, **65**, 1230 (1932).

[371] Blount and Robinson, *J. Chem. Soc.*, **1932**, 2305.

[372] Prelog, *Experientia*, **1**, 197 (1945).

[373] Leuchs, *Ber.*, **70**, 1543 (1937).

[374] Warnat, *Helv. Chim. Acta*, **14**, 997 (1931).

to an amide.[371] This, however, is not contrary to the representation of the

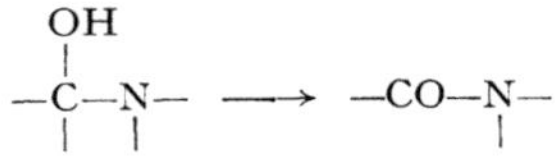

OH
—C—N— ⟶ —CO—N—

new derivative as such, but indicates that the carbon atom adjacent to N(b) and bearing the hydroxyl group must originally have been tertiary. The tertiary character of this carbon atom, as well as its probable position as a bridgehead carbon, explains the lack of formation of anhydro salts (XV)[375] but normal salts (XVI), by the carbinol base.

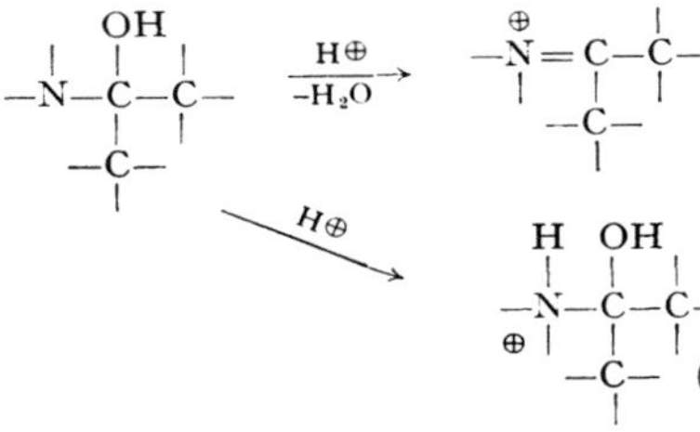

OH
—N—C—C— $\xrightarrow[-H_2O]{H^{\oplus}}$ ⊕—N=C—C— (XV)
—C—

$\xrightarrow{H^{\oplus}}$ H OH
—N—C—C—
⊕ —C— (XVI)

When pseudostrychnine is oxidized with acidic hydrogen peroxide, the product is a neutral compound, strychnone, $C_{21}H_{20}O_3N_2$,[376] which does not give an amino acid when the N(a) lactam ring is opened. The absence of an Otto reaction,[377] given by all *N*-acylhexahydrocarbazoles (XVII), and

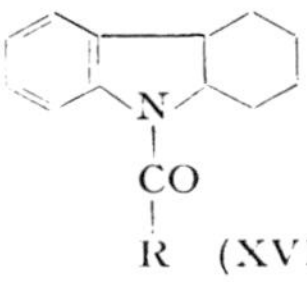

N
CO
R (XVII)

the ultraviolet absorption spectrum indicate that this compound is a true indole derivative.[378] This evidence explains the formation of strychnone (XIX)[379] from pseudostrychnine (XVIII), which must have involved a 3-hydroxyindole, subsequently dehydrated to the indole nucleus of strychnone (XIX). It is thus shown that only one carbon is situated between the

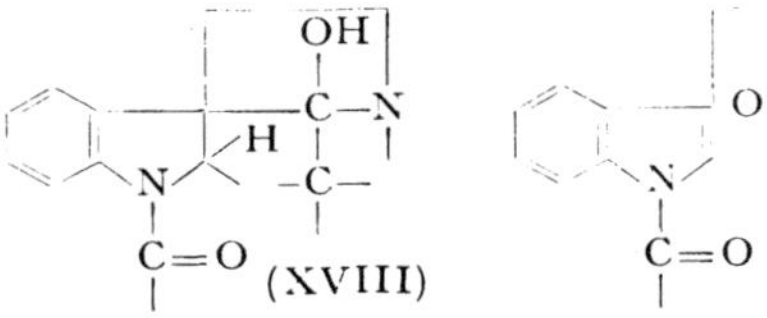

OH
—C—N
H —C—
N
C=O (XVIII)

O=C—N
—C—
N
C=O (XIX)

[375] Leuchs, Grunov, and Tessmar, *Ber.*, **70,** 1701 (1937).
[376] Leuchs, Tuschen, and Mengelberg, *ibid.*, **77,** 403 (1944).
[377] Leuchs, *ibid.*, **73,** 1392 (1940).
[378] Woodward, Brehm, and Nelson, *J. Am. Chem. Soc.*, **69,** 2250 (1947).
[379] Brehm, Dissertation, Harvard University, 1948, p. 61.

dihydroindole indole nucleus and N(b), and this fact eliminates all previously considered structures. Considering the 3-ethyl-4-methylpyridine fragment, only such a representation as XX is satisfactory for the alkaloid.[352, 378]

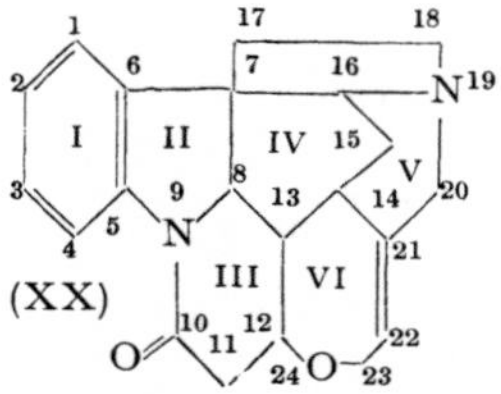

(XX)

Oxidation of *N*-acetylpseudostrychnine, $C_{23}H_{24}O_4N_2$, with permanganate gives a diketo acid, $C_{23}H_{24}O_7N_2$,[377] originally formulated as a β-diketone, on the basis of structure X, although its ethyl ester is not soluble in aqueous alkali. In terms of structure XX (and XVIII) this product would be given a structure such as XXI, a γ-diketone, which is more in line with the properties observed for the compound.

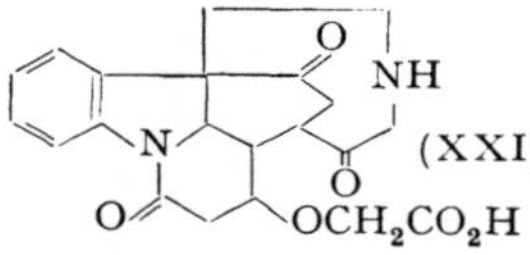

(XXI)

Basic treatment of *O*-methylpseudostrychnine methiodide, $C_{23}H_{27}O_3N_2I$ (XXII), yields a base, des-*N*, *O*-dimethylpseudostrychninium hydroxide (XXIII), and subsequent acidification furnishes *N*-methyl*chano*pseudostrychnine, $C_{22}H_{24}O_3N_2$ (XXIV).[363] The latter compound yields a dibenzylidene derivative, although *N*-methyldihydro*chano*pseudostrychnine is reported to give only a monobenzylidene derivative.[375, 380] These transformations are uniquely explained on the basis of XX for strychnine and structure XVIII for pseudostrychnine, by the following sequence of reactions, wherein the hindered carbonyl function is prevented from reaction with carbonyl

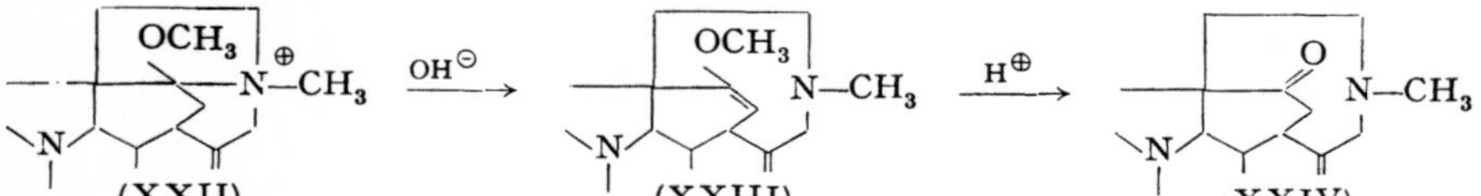

(XXII) $\xrightarrow{OH^{\ominus}}$ (XXIII) $\xrightarrow{H^{\oplus}}$ (XXIV)

reagents, and the second α-methylene group is present for the reaction with benzaldehyde.

That ring VI in strychnine is actually six- or seven-membered has recently been demonstrated unequivocally.[352] Dihydrostrychninolone-a (XXV) was oxidized to dihydrostrychninone, $C_{19}H_{20}O_3N_2$ (XXVI), and

[380] Blount and Robinson, *J. Chem. Soc.*, 1934, 595.

further oxidation with alkaline hydrogen peroxide gave carbon dioxide and cunine carboxylic acid, $C_{18}H_{20}O_3N_2$ (XXVII), having one less member in ring VI, which could lose water to give the neutral amide, $C_{18}H_{18}O_2N_2$ (XXVIII). A similar series of reactions has confirmed this result.[381]

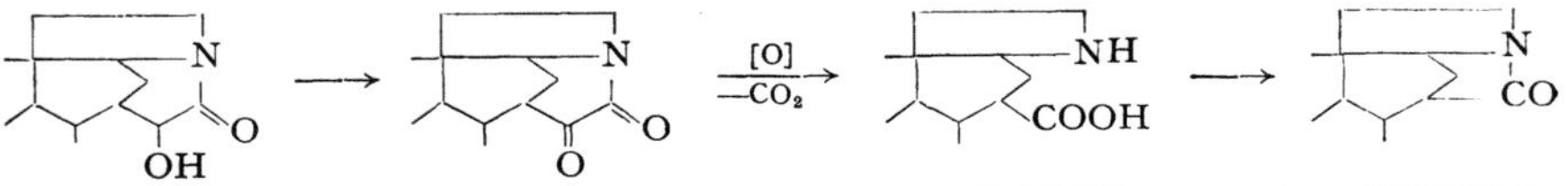

(XXV) → (XXVI) —[O], $-CO_2$→ (XXVII) → (XXVIII)

The action of bases on strychnine furnishes the isomeric neostrychnine. The corresponding compound, neostrychnidine, may be obtained from strychnidine. Since these neo-bases may be catalytically reduced to the same dihydro derivatives obtained from strychnine and strychnidine,[382, 383] it is obvious that the change has been the migration of the single double bond. The preparation of the neobases mey be accomplished in other manners. Refluxing strychnine in xylene with Raney nickel[389] produces the same base, but more insight into the nature of the changes comes from a study of strychnidine methosulfate (XXIX). Treatment of this substance

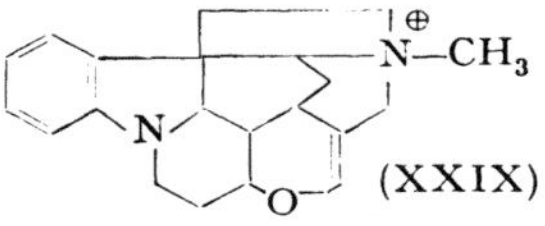

(XXIX)

with sodium methoxide in methanol does not produce a Hofmann degradation, but instead yields a base, $C_{23}H_{30}O_2N_2$, methoxymethyldihydroneostrychnidine,[329] in which a methoxyl group has been added. Treatment with acids gives methylneostrychnidinium salts $[C_{22}H_{25}ON_2]^+$, convertible to neostrychnidine on pyrolysis, and reconvertible to methoxymethyldihydroneostrychnidine by methanolic potassium hydroxide. That the isomerization is due to a vinylamine grouping in neostrychnine is shown by the formation of the *p*-nitrophenylhydrazone of diketoneostrychnine by treatment of neostrychnine with *p*-nitrobenzenediazonium chloride.[352] Diketoneostrych-

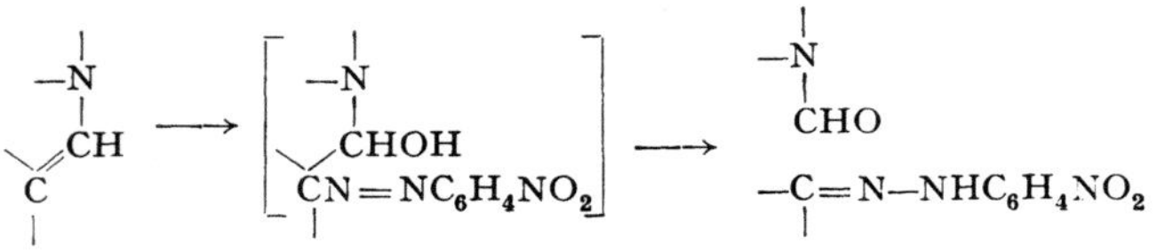

$-N-CH=C<$ → $[-N-CHOH-CN=NC_6H_4NO_2]$ → $-N-CHO$, $-C=N-NHC_6H_4NO_2$

[381] Prelog and Kocor, *Helv. Chim. Acta*, **31**, 237 (1948).

[382] Achmatowicz, Clemo, and Perkin, *J. Chem. Soc.*, **1932**, 767.

[383] Achmatowicz, Perkin, and Robinson, *ibid.*, **1932**, 486.

[384] Robinson and Chakravarti, *ibid.*, **1947**, 78.

nine, a neutral substance, and hence an N(b) amide, is also produced by permanganate oxidation of neostrychnine.[385] That the double bond should migrate further than C_{20}–C_{21} under the basic conditions seems improbable and neostrychnine may be formulated as XXX, and diketoneostrychnine as XXXI.[386]

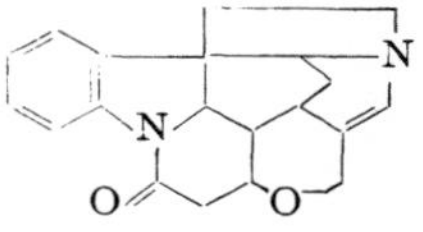

(XXX) 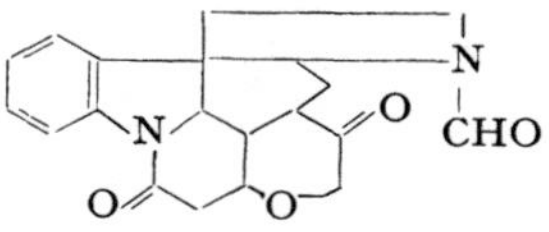 (XXXI)

Although early evidence seemed to disfavor these structures for the neo-bases,[387] proof of their correctness has been obtained. Oxidation of methoxymethyldihydroneostrychnine with perbenzoic acid resulted in the absorption of two atoms of oxygen to yield methoxymethyl*chano*-dihydrostrychnone,[388] a neutral substance containing a carbonyl function. Reduction of this compound by the Clemmensen method[389] gave methoxy-methyl*chano*dihydrostrychnane, $C_{23}H_{30}O_4N_2$, in which the carbonyl group appeared to have been replaced by hydrogen atoms, but which contained a *C*-methyl group. The latter is not consistent with XXX.

If, however, reduction is performed by a more mild procedure (conversion to the diethylmercaptal, followed by desulfurization with Raney nickel),[386] the product, desoxymethoxymethyl*chano*strychnone, isomeric with the strychnane, contains no *C*-methyl group. In view of structure XXX for neostrychnine, methoxymethyldihydroneostrychnine would be either XXXII or XXXIII[390] and the oxidation product, methoxymethyl*chano*-dihydrostrychnone would be XXXIV (or according to XXXIII).[386] It is clear that the formation of the strychnane must have been accomplished by rearrangement during the Clemmensen reduction.

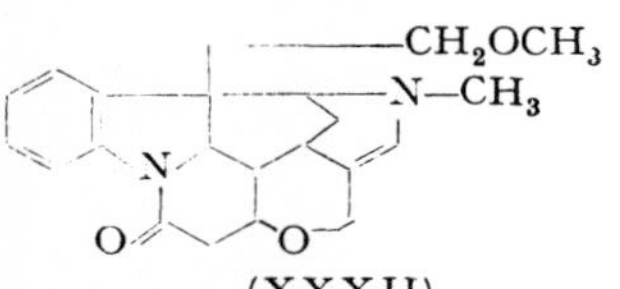

(XXXII) 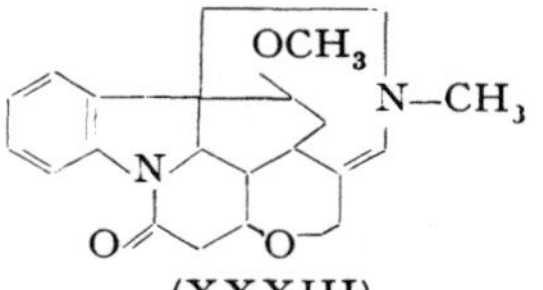 (XXXIII) 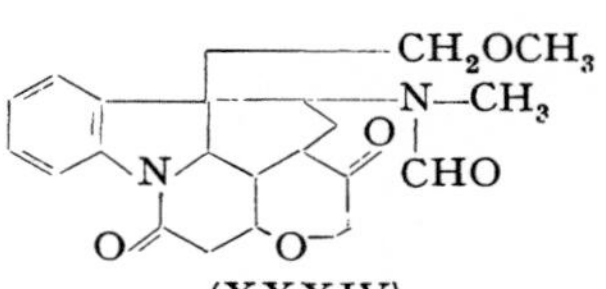 (XXXIV)

385 Kotake and Kokoyma, *Sci. Papers Inst. Phys. Chem. Res. Tokyo*, **31**, 321 (1937); *Chem. Abstr.*, **31**, 4984 (1937).

386 Woodward and Brehm, *J. Am. Chem. Soc.*, **70**, 2107 (1948).

387 Compare footnote 366, and later papers.

383 Briggs and Robinson, *J. Chem. Soc.*, **1934**, 590.

339 Leuchs and Grunov, *Ber.*, **72**, 679 (1939).

390 Woodward and Brehm,[386] footnote 21, p. 2109.

It will be noted that XXXIV predicts the presence of a formamide grouping in methoxymethyl*chano*dihydrostrychnane. Hydrolysis of this substance, as well as the desoxy derivative, has been found to give formic acid, and treatment of the desformyl compound with formic acid reproduced the strychnane,[386] thus confirming the above deductions and the proposed structure XX for strychnine.

The failure of the Hofmann degradation applied to strychnine led to the use of the Emde method on the alkaloid[391] and strychnidine,[392] and of the Hofmann treatment on dihydrostrychnidine.[393] These treatments have led to a complete mixture of products which have given no additional information toward the solution of the structural problem.

Vomicine

This *strychnos* alkaloid, $C_{22}H_{24}O_4N_2$, m.p. 282°, $[\alpha]_D$ +80°, was first described in 1929.[394] Unlike strychnine or brucine it has a replaceable hydrogen atom,[395] as a hydroxyl group, yielding an acetate.[396] It contains no methoxyl or methylenedioxy groups, and does not react with carbonyl reagents. It will yield a methosulfate, indirectly,[396,397] and a yellow benzylidene derivative, and may be reduced to a dihydro compound.[394] An aromatic ring is present,[398] and when oxidations remove this ring one of the oxygen atoms is lost.[394] No *C*-methyl group is present,[399] but the base does contain an *N*-methyl group,[400] the presence of which led to considerable confusion in the development of the vomicine structure.

The presence of one of the oxygen atoms on the aromatic ring does not lend phenolic properties to the molecule, but in the case of certain of the transformation products a ferric chloride test is obtained, and the substances are soluble in alkali, but not in carbonate.[401,402] These characteris-

[391] Perkin, Robinson, and Smith, *J. Chem. Soc.*, **1932**, 2306.

[392] Perkin, Robinson, and Smith, *ibid.*, **1934**, 574.

[393] Achmatowicz and Robinson, *ibid.*, **1934**, 681.

[394] Wieland and Oertel, *Ann.*, **469**, 193 (1929).

[395] Wieland and Holscher, *ibid.*, **500**, 70 (1932). Erlenmeyer, Apprecht, and Lobek, *Helv. Chim. Acta*, **19**, 543 (1936).

[396] Wieland and Thiel, *Ann.*, **550**, 287 (1942).

[397] Wieland and Muller, *ibid.*, **545**, 59 (1940).

[398] Wieland and Holscher, *Ann.*, **491**, 149 (1931).

[399] Wieland, Huisgen, and Bubernik, *ibid.*, **559**, 191 (1948).

[400] Wieland and Horner, *ibid.*, **528**, 73 (1937).

[401] Wieland and Moyer, *ibid.*, **491**, 129 (1931).

[402] Wieland and Kemmig, *ibid.*, **527**, 151 (1937).

tics are observed in vomicidine,[401] produced, like strychnidine, by electrolytic reduction, and having an active hydrogen atom. This has been taken to indicate a structure such as XXXV[403] or XXXVI,[402] although no conclusive evidence exists for the precise location of the hydroxyl group.

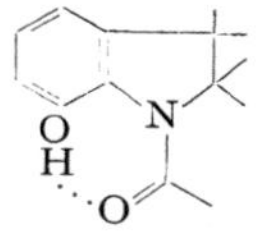

(XXXV) 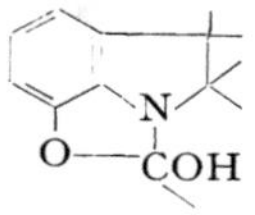 (XXXVI)

Vomicine further resembles strychnine in yielding vomicinic acid, $C_{22}H_{26}O_5N_2$,[396] which gives a nitrosoamine, and may be recyclized easily to vomicine. This demonstrates a lactam structure, and as with strychnine, the benzylidene derivative indicates an active α-methylene group.

The fact that vomicine is a congenitor of strychnine, and has, in many cases, parallel behavior, suggests a structure similar to strychnine for this alkaloid. This was recognized in the postulation of early formulas for vomicine,[404] although no direct relationship had been established. Such correlation has now been found. The oxidation of *N*-methyl*chano*pseudostrychnine (XXXVII), $C_{22}H_{24}O_3N_2$, the corresponding brucine derivative, and vomicine [399,405] by chromic acid yields in each case the same acid, $C_{17}H_{22}O_5N_2$.[403] The five carbon atoms lost were from the aromatic ring, indicating that in all three compounds the remainder of the nucleus is identical. On the basis of the structure (XXXVII) for *N*-methyl*chano*pseudostrychnine and the information previously considered regarding vomicine, the latter must be XXXIX,[403] or probably more correctly, a

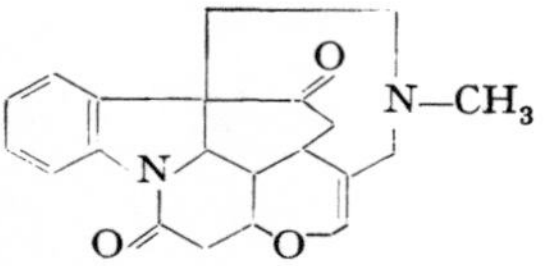

(XXXVII) 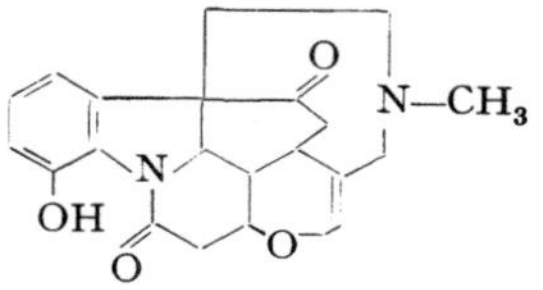 (XXXVIII) 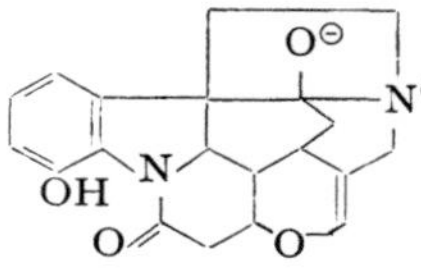 (XXXIX)

structure intermediate between XXXVIII and XXXIX.[406] This would account for the lack of ketonic properties and the difficulty encountered in producing methiodides of vomicine.[394] The acid, $C_{17}H_{22}O_5N_2$, obtained

[403] Bailey and Robinson, *Nature*, **161**, 433 (1948).
[404] Footnote 402, and subsequent papers.
[405] Wieland, Holscher, and Cortese, *Ann.*, **491**, 133 (1931).
[406] Crane, Dissertation, Harvard University, 1949.

from the three alkaloids, may be then formulated as XL. Support for the

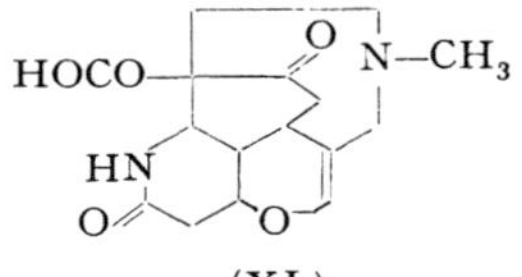

HOCO— O N—CH$_3$ HN O O

(XL)

structure (XXXVIII) for vomicine is readily found in its ability to explain the protean transformations to which the alkaloid has been subjected.

For example, when vomicine (XXXVIII) is boiled with phosphorus and hydrobromic acid in acetic acid, an isomerization occurs, producing isovomicine (XLI),[407] containing two hydroxyl groups, and retaining the lactam grouping. A similar reaction with hydriodic acid furnishes iododihydrodesoxyvomicine, $C_{22}H_{25}O_3N_2I$ (XLII).[408] Removal of the iodine with ammonia yields desoxyvomicine, $C_{22}H_{24}O_3N_2$, which exists as two interchangeable isomers, one colorless (XLIII)[402] and the other yellow (XLIV).[394] The yellow substance may be reduced to dihydrodesoxyvomicine II (XLV), $C_{22}H_{24}O_3N_2$,[409] and the colorless isomer may be transformed into dihydrodesoxyvomicine I (XLVI).[408] Formulations[406] for these reactions may be made on the following basis:

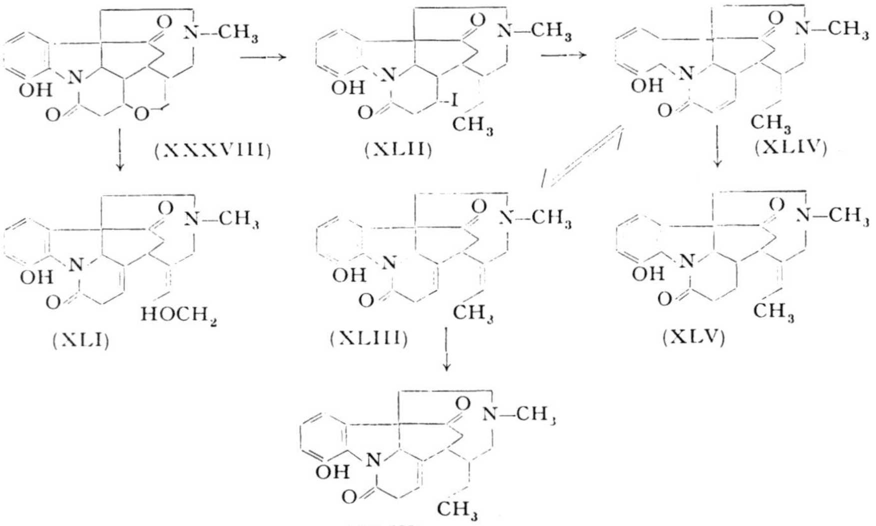

(XXXVIII) (XLII) (XLIV)

(XLI) (XLIII) (XLV)

(XLVI)

[407] Wieland and Huisgen, *Ann.*, **556,** 157 (1944).

[408] Huisgen and Wieland, *ibid.*, **555,** 9 (1943).

[409] Wieland and Varvoglis, *ibid.*, **507,** 82 (1943).

The position of the double bond in dihydrodesoxyvomicine I (XLVI) may be understood by its preparation from dihydrovomicine (XLVII) by ether cleavage with hydrogen bromide to bromodihydrodesoxyvomicine (XLVIII) having a primary bromo group, and subsequent removal of the halogen with zinc and acetic acid,[410] and by its formation of a yellow benzylidene derivative.[408]

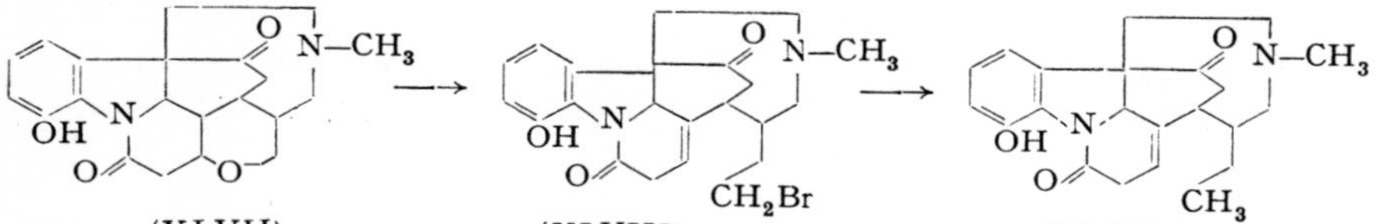

(XLVII) → (XLVIII) → (XLVI)

Vomicine methosulfate (XLIX)[408] is produced from an addition compound of vomicine and methyl sulfate, which, when heated, loses methyl alcohol to give XLIX. Application of the Emde degradation to this substance resulted, after three degradations, in the isolation of trimethylamine. The first step produces two methyl vomicines (L, LI), each containing a methoxyl and a methylamino group, and yielding a dihydro derivative. The second Emde reaction[399,411] yields dimethylvomicines (LII, LIII), each furnishing a methiodide and containing a methoxyl group, two *N*-methyl groups and one *C*-methyl group. Dimethylvomicine I has been subjected to a third degradative reaction of this type, yielding trimethylamine. This sequence of reaction may be formulated as follows:

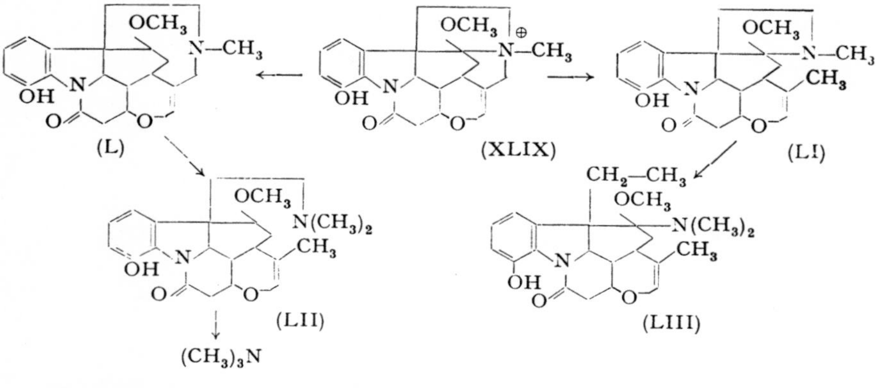

(L) ← (XLIX) → (LI)

(L) → (LII) → $(CH_3)_3N$; (LI) → (LIII)

Chromic acid oxidation of vomicine furnishes, in addition to the acid (XL), the base, $C_{16}H_{22}O_3N_2$,[405] corresponding to XL, but differing by the

[410] Wieland and Jennen, *ibid.*, **545**, 99 (1940).

[411] Wieland and Weisskopf, *ibid.*, **555**, 1 (1943). Compare also Huisgen, Wieland and Eder, *ibid.*, **561**, 193 (1949).

loss of carbon dioxide as would be expected from the β-ketoacid structure of XL. Catalytic reduction of the base saturates the single double bond present, and subsequent dehydrogenation with palladium yields oxyvomipyrine, $C_{15}H_{16}ON_2$,[407] having an α-pyridone structure. Zinc dust distillation of oxyvomipyrine gives the oxygen-free base, vomipyrine, $C_{15}H_{16}N_2$,[407] containing an *N*-methyl and a *C*-methyl group, and giving indole color reactions. Vomipyrine may also be obtained from vomicidine by oxidation and dehydrogenation.[400,412] On the basis of the vomicine structure (XXXVIII), oxyvomipyrine would be expected to have structure LIV, and vomipyrine LV.[403,413]

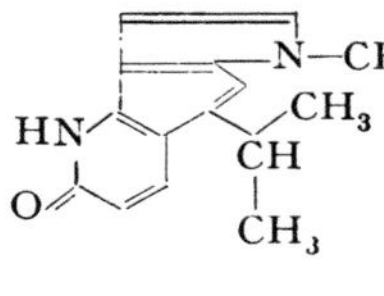

(LIV) 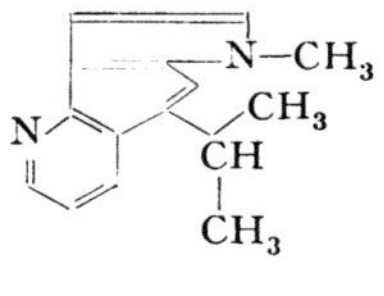 (LV) or (LV)

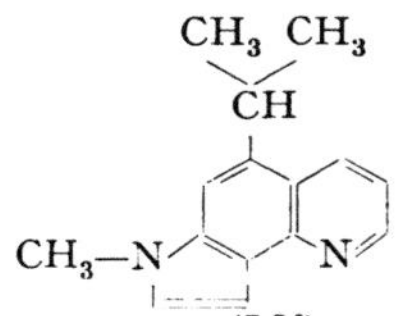

The synthesis of vomipyrine has been accomplished,[414] lending strong support to the proposed structure for vomicine. The amino-*p*-cymene (LVI), prepared by selective reduction of the dinitro compound, was converted to the methylisopropylaminoquinoline (LVII). Formylation and cyclization to the indole derivative (LVIII), followed by methylation in the presence of potassium amide, gave vomipyrine (LV).

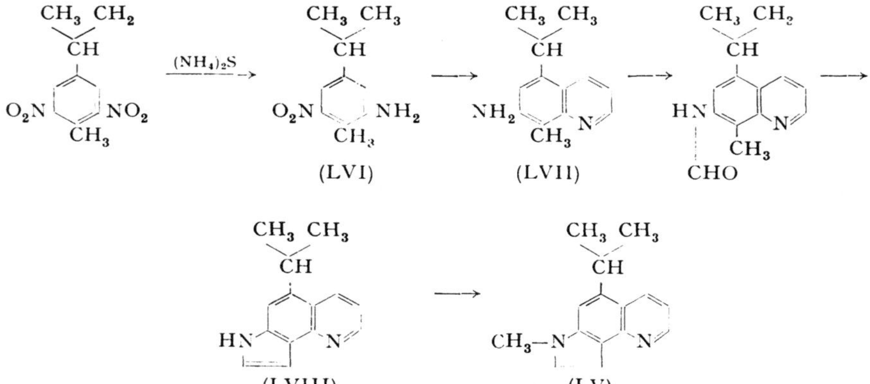

$(NH_4)_2S$

(LVI) (LVII)

(LVIII) (LV)

[412] Wieland and Horner, *ibid.*, **545,** 112 (1940).
[413] Briggs, Openshaw, and Robinson, *J. Chem. Soc.*, **1946,** 907.
[414] Robinson and Stephen, *Nature,* **162,** 177 (1948).

α-and β-Colubrine

These two isomers,[415] $C_{22}H_{24}O_3N_2$, may be closely related to strychnine and brucine, differing only in that they are monomethoxylated derivatives. α-Colubrine, m.p. 184°, $[\alpha]_D$ —76°, contains a single methoxyl group, as does the β-isomer, m.p. 222°, $[\alpha]_D$ —108°. The only existing clue to their structures is the oxidation of the two substances to *N*-oxalylanthranilic acid derivatives: α-colubrine yields the 4-methoxy compound (LIX), and the β-isomer furnishes the 5-methoxy derivative (LX).[415]

CH_3O — CO_2H, $NHCOCO_2H$ (LIX)

CH_3O — CO_2H, $NHCOCO_2H$ (LX)

Strychnospermine

This base, $C_{21}H_{28}O_3N_2$, m.p. 209°, $[\alpha]_D$ +58°, obtained from *S. psilosperma*,[416] contains an *N*-methyl and an *O*-methyl group, but no methylenedioxy grouping. That it possesses a tertiary nitrogen atom is shown by the formation of a methiodide, and the absence of a double bond is indicated by the failure to absorb hydrogen over palladium. The second nitrogen atom is present as an amide, as demonstrated by the hydrolysis to deacetylstrychnospermine, which may be reconverted to strychnospermine by the action of acetic anhydride. The third oxygen atom is inert and is assumed to be present in an ether linkage. On the basis of the above evidence, and considering known relationships of biogenesis, it has been suggested[416] that strychnospermine has a structure such as LXI.

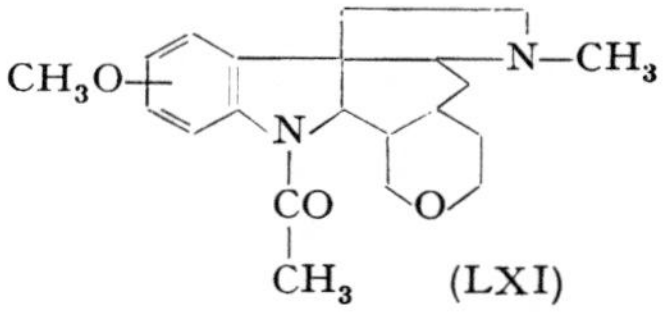

CH_3O — N—CH_3, N—CO—CH_3, O (LXI)

Holstiine

This alkaloid, $C_{22}H_{26}O_4N_2$, m.p. 247–248°, $[\alpha]_D$ +26°, was obtained from *S. holstii*.[417] It contains an *N*-methyl group, a lactam carbonyl function, and is considered[417] to be similar to vomicine, although it contains no phenolic hydroxyl group as does vomicine.

[415] Warnat, *Helv. Chim. Acta*, **14**, 997 (1931).
[416] Anet, Hughes, and Ritchie, *Nature*, **166**, 476 (1950).
[417] Janot, Goutarel, and Bosly, *Compt. rend.*, **232**, 853 (1951).

Two other *strychnos* alkaloids, struxine, $C_{21}H_{30}O_4N_2$, m.p. 250°,[418] and strychnicine,[419] have received no further attention since their isolation.

Mold Products

Gliotoxin

From *Gliocladium fimbriatum*[420] and other related molds,[421] the antibiotic gliotoxin has been isolated and related to indole and tryptophan by its ultraviolet absorption spectrum. It was found to have the composition $C_{13}H_{14}O_4N_2S_2$,[422] and m.p. 220°, $[\alpha]_D$ −290°.

Gliotoxin contains an *N*-methyl, but no *C*-methyl or methoxyl groups, and two active hydrogen atoms.[423] It yields a dibenzoate, but gives no reaction with carbonyl reagents. Basic treatment causes the loss of sulfur and some methylamine. The action of strong bases results in the isolation of indole-2-carboxylic acid (VI).

Gliotoxin may be treated with phosphorus and hydriodic acid to furnish dimethyldiketopyrazinoindole (I).[424] This degradation product was synthesized from indole-2-carboxylic acid chloride and *N*-methylalanine

COCl + $CH_3NHCHCO_2C_2H_5$ (with CH_3 on CH) → (III) C_2H_5OCO, CO, $N—CH_3$, CH, CH_3 —H⊕→ (I) C=O, $N—CH_3$, O=C, CH, CH_3

(II). The intermediate compound (III) was also obtained from the natural product by treatment with potassium hydroxide in methanol. The action of selenium on gliotoxin at 200° yields IV, degradable to the amide (V),

[418] Shaefer, *J. Am. Pharm. Assoc.*, **3**, 1677 (1914).
[419] van Boorsma, *Bull. Inst. Bot. Buitenz.*, **14**, 3 (1902).
[420] Wendling and Emerson, *Phytopathology*, **26**, 1068 (1936); *Chem. Abstr.*, **31**, 1064 (1937). Wendling, *Phytopathology*, **27**, 1175 (1937); *Chem. Abstr.*, **32**, 1737 (1938).
[421] Johnson, McCrone, and Bruce, *J. Am. Chem. Soc.*, **66**, 501 (1944).
[422] Johnson, Bruce, and Dutcher, *ibid.*, **65**, 2005 (1943).
[423] Bruce, Dutcher, Johnson, and Miller, *ibid.*, **66**, 614 (1944).
[424] Dutcher, Johnson, and Bruce, *ibid.*, **66**, 617 (1944).

from which it could be resynthesized.[425] Hydrolysis of V gave indole-2-carboxylic acid (VI).

N C=O C N–CH_3 O C O (IV) ⇄ N H $CONHCH_3$ (V) ⟶ N H CO_2H (VI)

Refluxing gliotoxin with methanolic potassium hydroxide liberates one mole of hydrogen sulfide and gives a compound, $C_{11}H_8ON_2S$, suggested to be the thiohydantoin (VII).[426] That this is correct was subsequently proved by synthesis from methyl indole-2-carboxylate and methylisocyanate.[427]

N H CO_2CH_3 ⟶ N C=O S C N–CH_3 (VII)

The action of aluminum amalgam of gliotoxin removes the two sulfur atoms to yield desthiogliotoxin, $C_{13}H_{16}O_4N_2$, having an absorption spectrum very similar to that of the parent substance.[426] Desthiogliotoxin decolorizes permanganate solution and bromine water, reduces periodic acid, and gives a positive iodoform test. It contains one *C*-methyl group, three hydroxyl groups, and is indifferent to carbonyl reagents. Hydrolysis with methanolic potassium hydroxide yields, among other products, the acid corresponding to III.[426, 428] These degradation products indicate that gliotoxin and desthiogliotoxin must be represented by structures such as VIII and IX, respectively.

H OH H N C=O HO–C N–CH_3 S C S–CH_2 OH (VIII)

H OH H N C=O HOCH N–CH_3 C CH_3 OH (IX)

[425] Dutcher, Johnson, and Bruce, *ibid.*, **66**, 619 (1944); **67**, 423 (1945).

[426] Dutcher, Johnson, and Bruce, *ibid.*, **67**, 1736 (1945).

[427] Elvidge and Spring, *J. Chem. Soc.*, **1949**, 5135.

[428] Elvidge and Spring, *ibid.*, **1949**, 2935.

Chetomin

This product, $C_{16}H_{17}O_4N_3S_2$, m.p. 215°, $[\alpha]_D$ +360°, was obtained from the *Chaetomium* genus.[429] It contains four active hydrogen atoms and one methylamino group, but no methoxyl is present.[430] The substance is nonbasic, yields an acetyl derivative, and is not affected by carbonyl reagents. No hydrogen was absorbed in the presence of platinum or palladium. Treatment of chetomin with strong acid or base liberated ammonia, assumed to arise from an amide function, since no amino nitrogen could be detected. Carbon dioxide was also liberated in the basic hydrolysis.

Chetomin, on treatment with Raney nickel or zinc and acetic acid, gave desthiochetomin, $C_{16}H_{19}O_3N_3$. This derivative gives indole color tests, and on zinc dust distillation an unidentified alkylindole was isolated. An indole nucleus is also indicated by the absorption spectrum. Note has been made of the gross similarity of chetomin to ergine.[430]

Melanine; Aminochromes

Melanine is a dark pigment, responsible for differences in skin coloration, found in the lower layer of the epidermis. The composition of the pigment has not been established since it appears not to be a well-defined material.

It has been found that melanine is formed readily by the oxidation of tyrosine (I) or 3,4-dihydroxyphenylalanine (dopa) (II) in the presence of the enzyme tyrosinase.[431] The enzyme is required for the first two steps only.[432]

HO–C₆H₄–CH₂–CH(NH₂)COOH (I) ⟶ (HO)₂C₆H₃–CH₂–CH(NH₂)CO₂H (II) ⟶ O₂C₆H₃–CH₂–CH(NH₂)CO₂H ⟶

HO, HO–indoline–2–COOH (N–H) ⟶ O, O–indoline–2–COOH (N–H) (III) ⟶ HO, HO–indole–2–CO₂H (N–H) ⟶

HO, HO–indole (N–H) (IV) ⟶ Melanine

[429] Waksman and Bugie, *J. Bacteriol.*, **48**, 527 (1944). Geiger, Conn, and Waksman *ibid.*, **48**, 531 (1944).

[430] Geiger, *Arch. Biochem.*, **21**, 126 (1949).

[431] Raper, *Biochem. J.* **20**, 735 (1926).

[432] Evans and Raper, *ibid.*, **31**, 2162 (1937).

Considerable evidence has been uncovered in support of this scheme. Thus "dopa" (II) and hallochrome (III)[433] have been isolated from natural sources; "dopa" is known to be converted into indole derivatives,[434] and into melanine[435] and 5,6-dihydroxyindole (IV) is oxidized in alkaline media to melanine-like pigments.[436]

Other substances also yield melanine-like materials on oxidation. Thus adrenaline (V) is converted into a red aminochrome, analogous to hallochrome, called adrenochrome and given structure VI.[437] This has been modified to VII to account for its solubility behavior and the lack of *o*-diketonic properties.[438] Further oxidation of adrenochrome yields melanines.

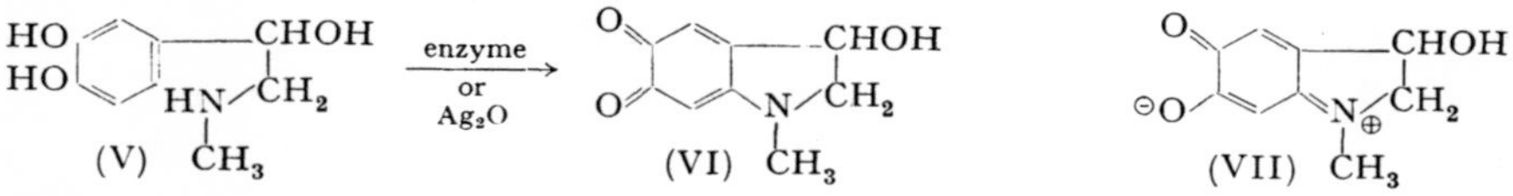

Epinochrome (desoxyadrenochrome) (IX) has also been obtained by the oxidation of epinine (VIII)[438, 439] and, like adrenochrome,[440] may be converted into true indole derivatives.[441]

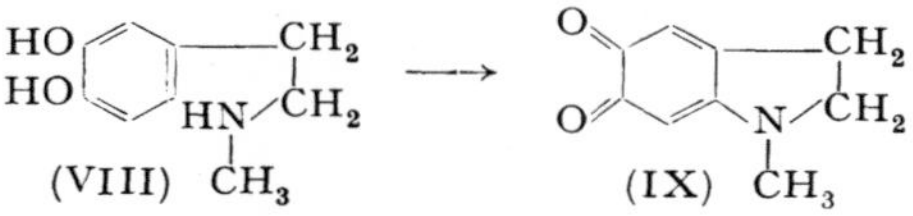

[433] Jacobs, *J. Lab. Clin. Med.*, **22**, 371, 890 (1937); *Chem. Abstr.*, **31**, 3140, 7108 (1937).

[434] Raper and Warmall, *Biochem. J.*, **17**, 454 (1923); **20**, 735 (1926); **21**, 89 (1927); **29**, 76 (1935).

[435] Veer, *Rec. trav. chim.*, **58**, 949 (1939); *Chem. Weekblad*, **37**, 214 (1940); *Chem. Abstr.*, **34**, 5102 (1940). Clemo and Weiss, *J. Chem. Soc.*, 1945, 702. Clemo, Duxbury and Swan, *ibid.*, **1952**, 3464. Clemo and Duxbury, *ibid.*, **1952**, 3844.

[436] Beer, Clarke, Khorana, and Robertson, *Nature*, **161**, 525 (1948).

[437] Green and Richter, *Biochem., J.*, **31**, 596 (1937). Braconier, Bihan, and Bedeaudet, *Arch. intern. pharmacodynamie*, **69**, 181 (1943). McCarthy, *Chimie & industrie*, **55**, 435 (1946); Veer, *Rec. trav. chim.*, **61**, 633, 646 (1942). Beer, Clark, Davenport, and Robertson, *J. Chem. Soc.*, **1951**, 2029. Macciotta, *Gazz. chim. ital.*, **81**, 485 (1951).

[438] Harley-Mason, *Experientia*, **4**, 307 (1948); *J. Chem. Soc.*, **1950**, 1276. Beaudet *Experientia*, **6**, 186 (1950). Beaudet, *Experientia*, **7**, 291 (1951). Harley-Mason, *Chemistry and Industry*, **1952**, 173.

[439] Sobotka and Austin, *J. Am. Chem. Soc.*, **73**, 3077 (1951).

[440] Harley-Mason, *J. Chem. Soc.*, **1950**, 1276. Harley-Mason and Bu'Lock, *Nature*, **166**, 1036 (1950). Bu'Lock and Harley-Mason, *J. Chem. Soc.*, **1951**, 712, 2248. Fischer, Derouaux, Lambot, and LeCompte, *Bull. soc. chim. Belges*, **59**, 72 (1950).

[441] Austin, Chanley, and Sobotka, *J. Am. Chem. Soc.*, **73**, 2395, 5299 (1951).

A number of structures for melanine as a polymer of hallochrome have been postulated[442] although no conclusive results can be drawn from the available evidence.[443]

Minor Alkaloids

Other alkaloids which are sometimes considered to possess indole nuclei are known. The *erythrina* alkaloids[444] and some derivatives of laudanosine[445] are among these. In fact, in the latter case, faint Ehrlich reactions may be observed. However, both of these groups are better considered as isoquinoline derivatives.

Uncaria Alkaloids

From *Uncaria kowakamii* have been obtained two isomeric bases, $C_{21}H_{24}O_4N_2$, designated uncarine-A, m.p. 120–130°, $[\alpha]_D$ +106°, and uncarine-B, m.p. 216–217°, $[\alpha]_D$ +91°.[446] Both contain a tertiary nitrogen atom, an active hydrogen atom, and a carbomethoxy grouping. Either may be converted into a mixture of the two by heating with acetic anhydride.

Palladium dehydrogenation has resulted in the isolation of 3-ethyl-4-methylpyridine, isoquinoline, and 3-methyl- and 3-ethyloxindole.[447] Other alkyl pyridines were obtained by zinc dust and soda lime distillations.[448] Nitric acid oxidation produced oxalic and *p*-nitrobenzoic acids.

Since the absorption spectrum of uncarine-B is similar to that of 3-ethyloxindole and unrelated to that of yohimbine, structures have been proposed containing an indole nucleus oxygenated in the 2-position.[449]

Kamassine

This alkaloid, $C_{19}H_{26}N_2$, m.p. 143–144°, $[\alpha]_D$ —99°, occurs in *Gonioma kamassi*.[450] It contains one *N*-methyl and one *C*-methyl group, and one

[442] Cohen, *Compt. rend.*, **220,** 927 (1945); *Bull. soc. chim. biol.*, **28,** 104, 107, 354 (1946). Clemo and Weiss, *J. Chem. Soc.*, **1945,** 702; *Nature*, **159,** 339 (1947). Harley-Mason, *ibid.*, **159,** 338 (1947); *J. Chem. Soc.*, **1948,** 1244. Burton, *Chemistry & Industry*, **1948,** 313. Bu'Lock and Harley-Mason, *J. Chem. Soc.*, **1951,** 703; Morton and Slaunwhite, *J. Biol. Chem.*, **179,** 259 (1949).

[443] Beer, Brown, and Robertson, *J. Chem. Soc.*, **1951,** 2426.

[444] Folkers and Major, *J. Am. Chem. Soc.*, **59,** 1580 (1937) and later papers.

[445] Henry, *The Plant Alkaloids*. 4th ed., Blakiston, New York, p. 187.

[446] Kondo and Ikeda, *J. Pharm. Soc. Japan*, **61,** 416, 453 (1941); *Chem. Abstr.*, **45,** 2960 (1951).

[447] Ikeda, *J. Pharm. Soc. Japan*, **61,** 460 (1941); *Chem. Abstr.*, **45,** 2960 (1951).

[448] Ikeda, *J. Pharm. Soc. Japan*, **63,** 393 (1943); *Chem. Abstr.*, **44,** 7332 (1950)

active hydrogen atom. Kamassine has recently been shown to be identical with quebrachamine[451] (see p. 235).

Akuammine and its relatives from *Picralima klaineana*[452] have shown indications[453] of being related to some of the yohimbine group. Perloline and perlolidine from *Lobium perenna* have afforded evidence of having an indole nucleus,[454] as has ibogaine from *Tabernanthe iboga*.[455] The alkaloids of *Mitragyna* are usually assumed to contain an indole moiety.[456]

Biogenesis

The concept that the amino acids or related substances are the starting materials in the process of the biogenesis of alkaloids[457,458] has had outstanding success as the foundation of extended theoretical projections and in the facilitation of structure determinations of these bases. The flexibility of this general theory omits binding restrictions on further predictions or possible biological synthetic routes.

With respect to the indole alkaloids, the most probable precursor is, of course, tryptophan (I), since, as explained at the first of this chapter,

$CH_2\text{–}CH(NH_2)CO_2H$ (I)

essentially all of this group of alkaloids contain the β-aminoethylindole nucleus. Although the details of the biological processes are obscure and extremely difficult to elucidate, the most plausible possibility is that tryptophan is enzymatically decarboxylated to tryptamine, which then condenses with an aldehyde fragment, or its biogenetic equivalents (possibly α-keto acids or α-imino acids) to form the *N*-alkyl derivatives; further condensations then produce the products as they are known.

[449] Ikeda, *J. Pharm. Soc. Japan,* **62,** 15, 38 (1942); *Chem. Abstr.,* **45,** 2961 (1951)

[450] Schlittler and Gellert, *Helv. Chim. Acta,* **34,** 920 (1951).

[451] Gellert and Witkop, *ibid.,* **35,** 114 (1952).

[452] Henry and Sharp, *J. Chem. Soc.,* **1927,** 1950. Henry, *ibid.,* **1932,** 2759.

[453] Raymond-Hamet, *Compt. rend.,* **221,** 699 (1945).

[454] White and Reiker, *New Zealand J. Sci. Tech.,* **27,** 38 (1945); *Chem. Abstr.,* **40,** 1826 (1946); *New Zealand J. Sci. Tech.,* **27,** 242 (1945); *Chem. Abstr.,* **40,** 3760 (1946).

[455] Raymond-Hamet, *Bull. soc. chim. biol.,* **25,** 205 (1943); *Bull. soc. chim.,* **9,** 620 (1942).

[456] Field, *J. Chem. Soc.,* **119,** 887 (1921). Ing and Raison, *ibid.,* **1939,** 986. Barger, Dyer, and Sargent, *J. Org. Chem.,* **4,** 418 (1939).

[457] Robinson, *J. Chem. Soc.,* **111,** 876 (1917).

Specifically this basic condensation is related to the Mannich reaction, and support for the fundamental thesis is furnished by the fact that most alkaloids contain the grouping $-\overset{|}{N}-\underset{H}{\overset{|}{C}}-CH_2-$, the production of which may be pictured[459] as occurring by the condensation between an amino group, an aldehyde function, and an anionoid center:

$$-\overset{|}{N}-H + \underset{\underset{|}{CH_2}}{\overset{H}{C}}=O + {}^{\ominus}\overset{|}{\underset{|}{C}}- \xrightarrow{H^{\oplus}} -\overset{|}{N}-\underset{\underset{|}{CH_2}}{\overset{H}{C}}-\overset{|}{\underset{|}{C}}-$$

Application of the concept to the indole alkaloids has been made in the case of the reaction of tryptamine with aldehydes under "physiological conditions" – dilute buffered solutions at acidities of *p*H 5–7 – to produce tetrahydroharman derivatives (II).[460] A further reaction of the same type with formaldehyde, or again, its biological equivalent, has been shown to yield the yohimbine nucleus (III).[461] Reaction to give II then employs the anionoid 2-position of the indole nucleus; the second process utilizes the similarly reactive 2-position of the aromatic nucleus of II.

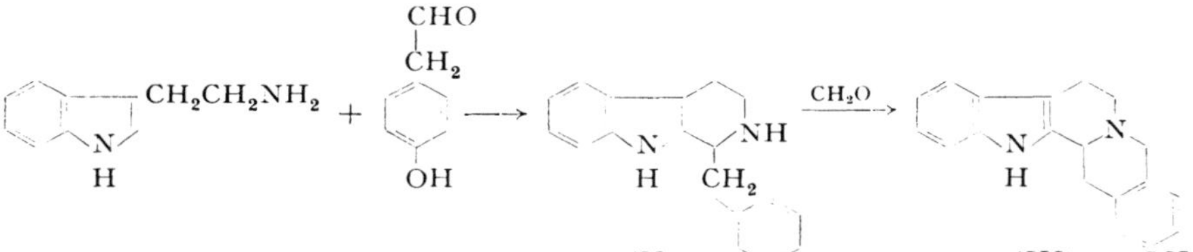

$CH_2CH_2NH_2$ N H + CHO CH_2 OH ⟶ N H NH CH_2 OH $\xrightarrow{CH_2O}$ N H N OH

(II) (III)

It is interesting to note that the methylation of amino and hydroxyl groups probably occurs through the function of formaldehyde, in the absence of the anionoid center, followed by a reductive process to yield the *N*- and *O*-methyl groups. This view finds support in the great prevalence

$$-\overset{|}{N}-H + CH_2O \longrightarrow -\overset{|}{N}-CH_2OH \longrightarrow -\overset{|}{N}-CH_3$$

of such methyl groups in natural materials, with only one case of an *N*-ethyl group having been reported.[462] The formation of the higher alkyl

[458] Schöpf, *Z. angew. Chem.*, **50**, 787 (1937).

[459] Woodward, *Nature*, **162**, 155 (1948).

[460] Hahn and Ludewig, *Ber.*, **67**, 2031 (1934). Hahn and Werner, *Ann.*, **520**, 123 (1935).

[461] Hahn and Hansel, *Ber.*, **71**, 2192 (1938).

[462] Freudenberg, *ibid.*, **69**, 1962 (1936).

derivatives, however, is not beyond the basic concepts, since it could conceivably be easily effected through the analogous agency of acetaldehyde, or more probably, pyruvic acid or glycine.

A further striking application of the theory in the indole series has been made in the case of strychnine.[459] Since the indole nucleus possesses anionoid activity at the β- as well as the α-position, tryptamine would furnish both the anionoid center and the amino function for reaction with 3,4-dihydroxyphenylalanine or its transformation products, to yield IV.

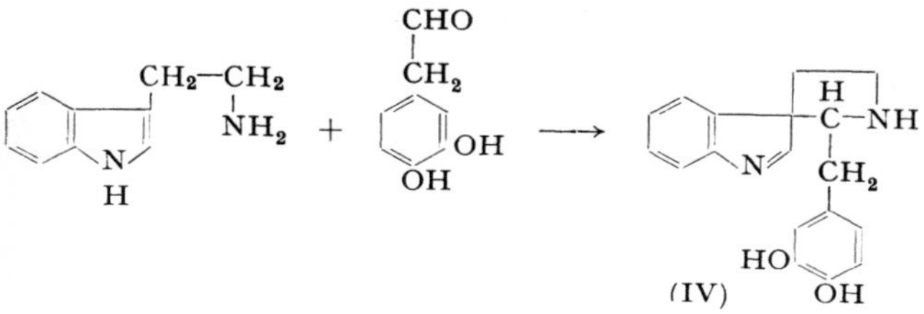

(IV)

This molecule now possesses the required functions for a further reaction with formaldehyde to give V. Closure to VI (or the tautomeric VII) might occur at this, or at a later stage, but the process is of the same type as the previous reactions of the biogenetic process. Conversion of VI to strychnine

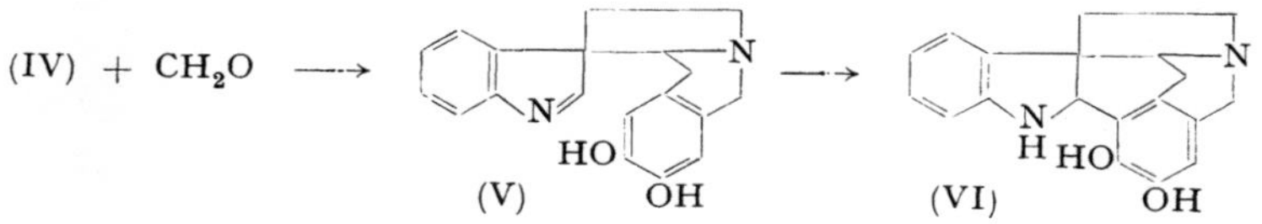

(IV) + CH_2O ⟶ (V) ⟶ (VI)

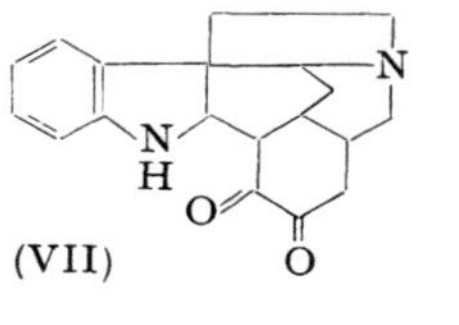

(VII) 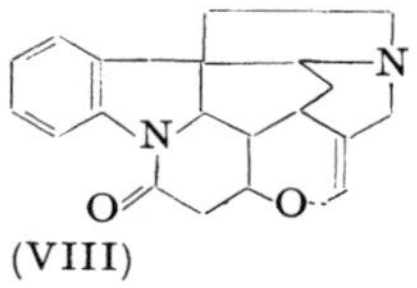 (VIII)

(VIII) may be pictured as occurring by several routes.[459] Cleavage of the diketonic ring (VII) provides the necessary requirements for formation of the ether function, and the addition of the remaining two carbon atoms of the lactam ring could come about by acetylation, with subsequent condensation on the active α-methyl group of the amide to produce the lactam ring. The possible variations of the scheme and of the order of occurrence lead to several plausible routes.

It is of importance that biogenetic considerations such as these were instrumental in the development of the currently accepted structure of the strychnine molecule,[459,463] and have been applied with great success to alkaloids possessing the structural features of other heterocyclic nuclei.[457,458]

Absorption Spectra of Indole Bases

The use of the absorption spectra, both ultraviolet and infrared, has been of invaluable assistance in the structure determinations of the indole group of alkaloids, and in the comparison of natural and synthetic materials. The spectra of these bases present, on the whole, general characteristics, illustrated by the consideration of typical examples of the ultraviolet curves.

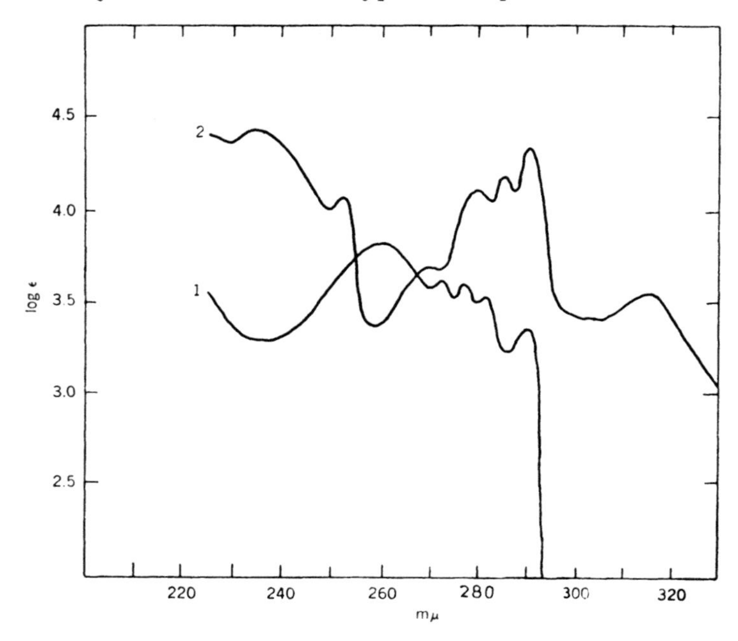

Fig. 1. (1) Indole.[464] (2) Carbazole.[465]

The absorption curves for indole (I) and carbazole (II) (Fig. 1) show similarity in their general shapes. The carbazole spectrum is shifted to

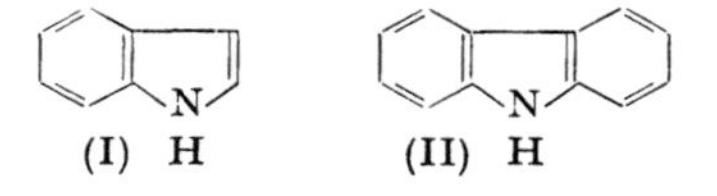

[463] Woodward, Brehm, and Nelson, *J. Am. Chem. Soc.*, **69,** 2250 (1947).

[464] Janot and Berton, *Compt. rend.*, **216,** 564 (1943).

[465] Pruckner and Witkop, *Ann.*, **554,** 127 (1943).

longer wave lengths, corresponding to the increased resonance energy of this compound (91 kcal./mole) as compared to indole (54 kcal./mole).[466] It will be observed that when the resonating system present in the indole nucleus is extended, a corresponding shift in the absorption spectrum of the molecule may be expected.

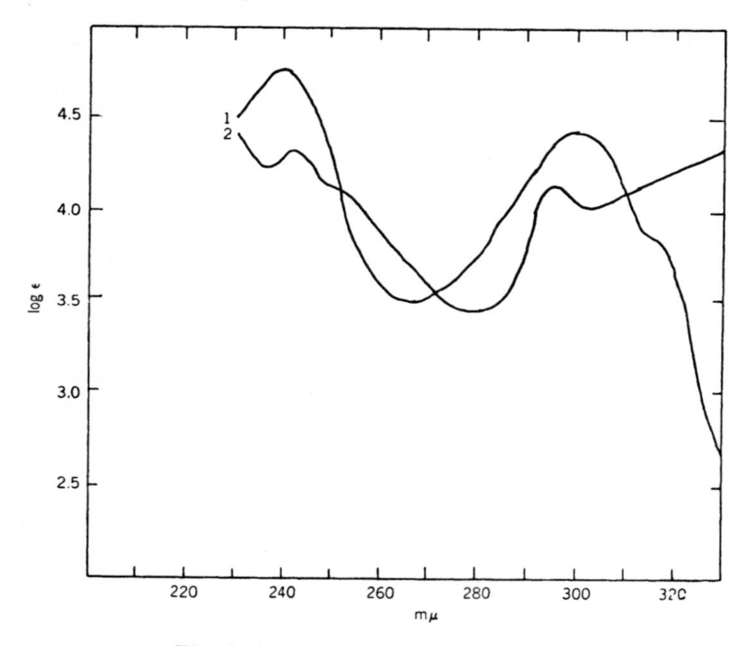

Fig. 2. (1) Harmine.[465] (2) Harmaline.[465]

This is demonstrated in the spectra of the two harmala bases, harmaline (III) and harmine (IV) (Fig. 2), both of which exhibit the typical indole curve with maxima at slighty longer wave lengths.

CH_3O … N … N … H … CH_3 (III) $\qquad$ CH_3O … N … N … H … CH_3 (IV)

With respect to general structural similarity, the curves for yohimbine (V) and corynanthine (Fig. 3) illustrate the inferences which may often be found in the study of spectra, since these two isomers have been demonstrated to have the same basic skeleton, in that they both may be con-

[466] Pauling and Shermen, *J. Chem. Phys.*, **1**, 606 (1933).

[467] Goutarel and Berton, *Compt. rend.*, **217**, 71 (1943).

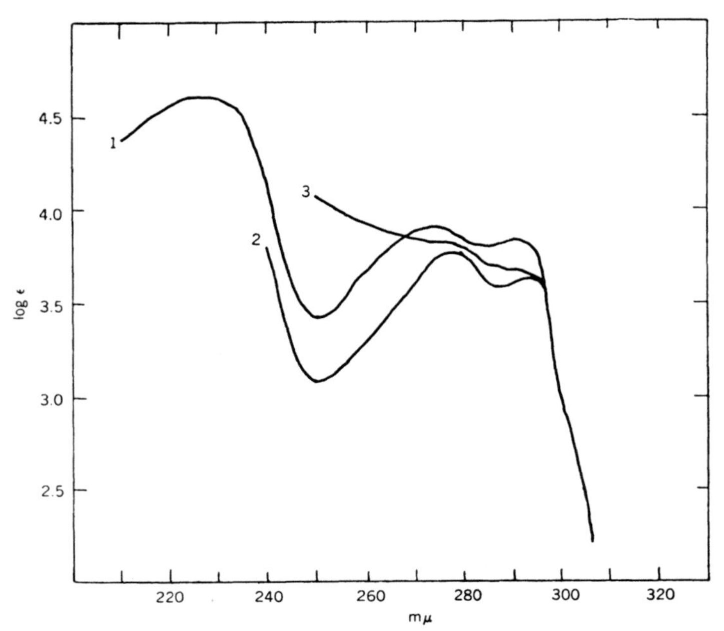

Fig. 3. (1) Yohimbine.[467] (2) Corynanthine.[467] (3) Corynantheine.[467]

verted into yohimbone.[468] The corynantheine curve indicates the basic difference of this substance as compared to the above two alkaloids.

N H N CH_3OCO OH (V)

CH_2CH_2OH N H N $CH{=}CH_2$ (VI)

The relationship between the cinchona indole alkaloids and the other members of the indole group is demonstrated by the absorption spectrum of cinchonamine (VI) (Fig. 4) in the close resemblance of this curve and those of indole and yohimbine. Alstonine, on the other hand, has a spectrum more like those of the harmala group, and the resemblance is confirmed by the isolation of harman and norharman by degradative procedures.[469]

The spectra of the *Gelsemium* alkaloids, sempervirine and gelsemine (Fig. 5), indicate a relationship to each other, with that of the former

[468] Janot and Goutarel, *ibid.*, **220,** 617 (1945); *Bull. soc. chim.*, **13,** 535 (1946).
[469] Leonard and Elderfield, *J. Org. Chem.*, **7,** 562 (1942).
[470] Prelog, *Helv. Chim. Acta,* **28,** 1671 (1945).
[471] Jacobs, Craig, and Rothen, *Science*, **83,** 166 (1936).

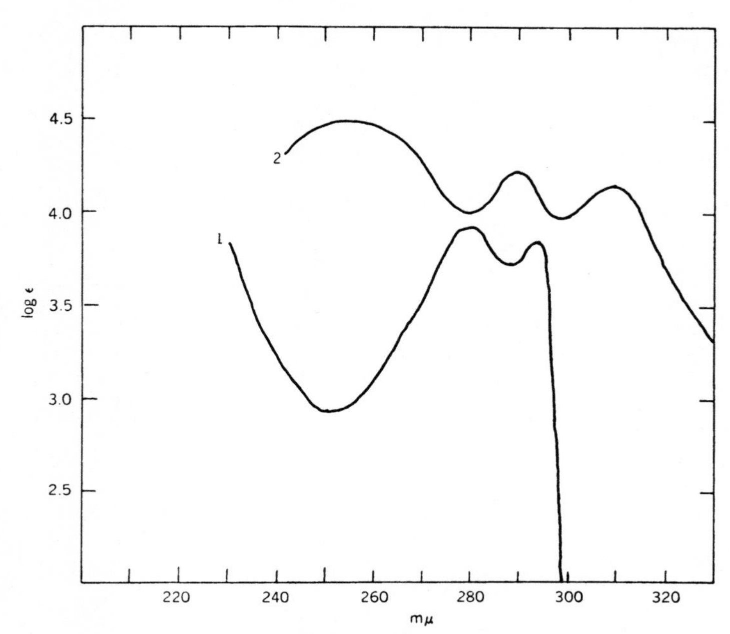

Fig. 4. (1) Cinchonamine.[464] (2) Alstonine.[439]

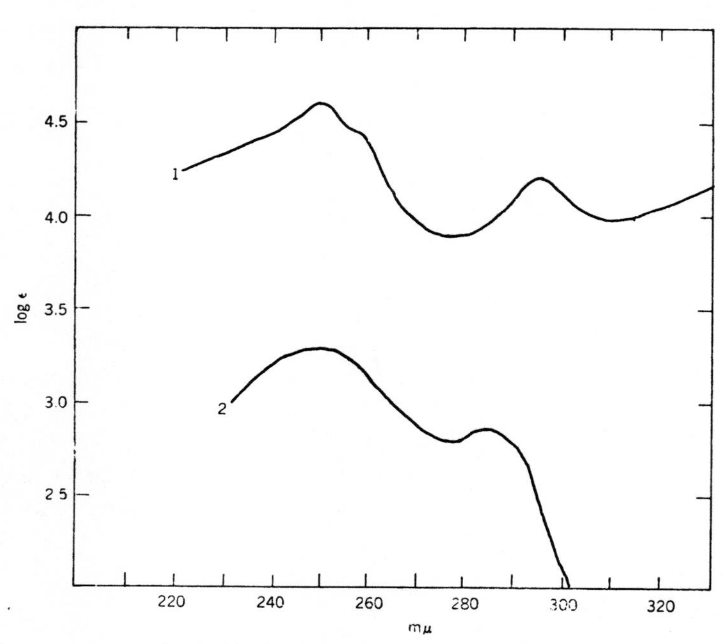

Fig. 5. (1) Sempervirine.[464] (2) Gelsemine.[464]

being much more intense, and extended toward the visible. This is in agreement with the known high degree of aromaticity of the heterocyclic system of sempervirine (VII).

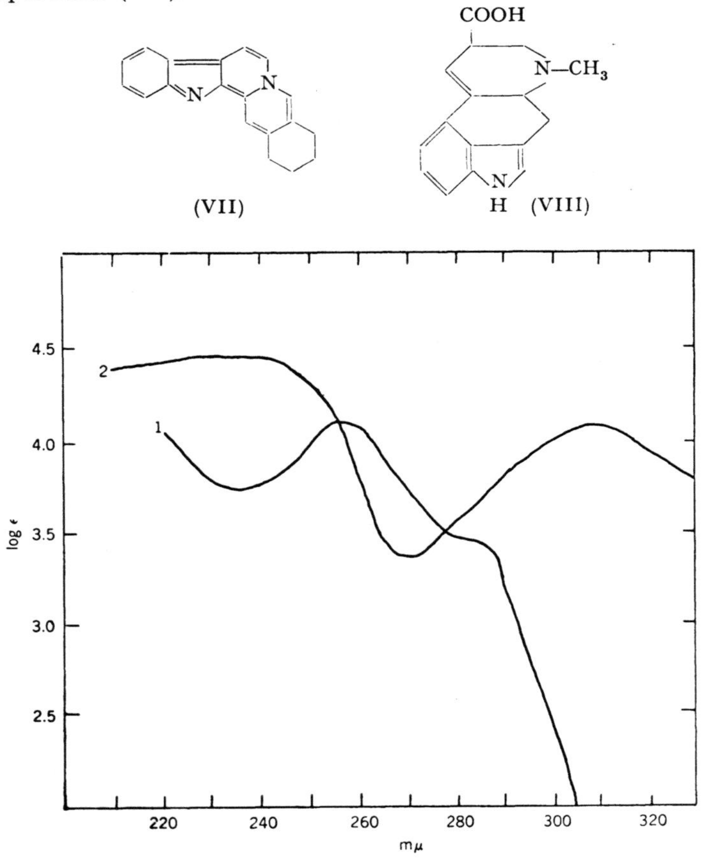

Fig. 6. (1) Strychnine.[470] (2) Lysergic acid.[471]

Lysergic acid (VIII) possesses an absorption curve (Fig. 6) of interest in view of the known nature of the molecule, since the spectrum is of the typical indole type, but shifted somewhat, due to the conjugation of the olefinic double bond with the aromatic nucleus. The curve for strychnine (IX) is typical of those alkaloids possessing a dihydroindole nucleus:

(IX)

SUBJECT INDEX

B

C

F

G

H

I

J

K

L

M

N

O

P

Q

R

U

V

W

X

Y

Breinigsville, PA USA
05 April 2011
R4082200001B/R40822PG258843BVX1B/1/A